Microbiome in Human Health and Disease

Pallaval Veera Bramhachari
Editor

Microbiome in Human Health and Disease

Springer

Editor
Pallaval Veera Bramhachari
Department of Biotechnology
Krishna University
Machilipatnam, Andhra Pradesh, India

ISBN 978-981-16-3155-9 ISBN 978-981-16-3156-6 (eBook)
https://doi.org/10.1007/978-981-16-3156-6

This Springer imprint is published by the registered company Springer Nature Singapore Pte Ltd.
The registered company address is: 152 Beach Road, #21-01/04 Gateway East, Singapore 189721, Singapore

Foreword

The microbial world is largely invisible to the human eye, but it is almost beyond imagination. There are hundreds, thousands of different kinds of bacteria (leaving aside other kinds of microbes: archaea, viruses, fungi, and protists), living in every possible environment including deep seabed, high in the clouds, and in the boiling hot springs. Multicellular organisms created an entirely new set of habitats, in and on all those animals and plants.

Research data suggested that during the last two decades, extensive research has been carried out on endophytic fungi and several biologically active compounds have been isolated from endophytic fungi. This book makes all the readers generally conversant in the language of microbiomes and metagenomics. It also provides excellent examples of how microbial communities affect health and cure diseases and dole out typical practical examples of how medical interventions interact with the microbiome and change outcomes.

Understanding the Host–Microbiome Interactions in Human Health. The volume published by the Springer Nature Biomedicine is an important volume, and I strongly believe that it will attract readers working in the field. The present volume has 14 chapters contributed by diversified academicians and scientists working on microbiome research throughout the world. I must congratulate the editor of this book for bringing out this volume with excellent contributions from the most talented scientists working on microbiome research and their applications in understanding the host microbiome interactions in human health.

Krishna University
Machilipatnam, India

K. B. Chandra Sekhar

Preface

The human body is a beautifully complex system, hosting trillions of microbial cells that colonize epithelial surfaces such as those present in the mouth and gut are included. There is growing evidence that these microbiomes, which play important roles in human physiology and organ development, do more than just lie on tissues. Indeed, relative to our human genome, there are 100 times more genes in our microbiome and these microbial genes code for proteins that affect different processes such as digestion, immunity, and development. The purpose of this book is to provide a summary of human microbial diversity and to explain attempts to connect microbial communities to human beings.

The Human Microbiome Project (HMP) projects in the USA, Canada, Europe, and Asia have provided a boost to attempt to understand human-associated microbial populations. More than $200 million has been invested by the NIH in the Western World to fund these studies, which are rising now with the introduction of evidence and a flurry of publications. This is a time of extraordinary breakthroughs, and the area is sufficiently mature, while still new to merit a book highlighting advancement. Responses to several issues are appearing now. How do various microbial populations vary across locations in the body? What is the heterogeneity of microbial composition at the same specific location in stable and diseased humans?

To maximize well-being and mitigate disease risk, how can microbial populations be manipulated? Over the course of human evolution, how do microbial populations change? What are the internal factors (genetic, anatomical, hormonal, and physiological) and external environmental factors that form human-associated microbial communities (diet, sexual behavior, and hygiene)? Many additional concerns regarding the personal interactions between human and microbial cells in the body come with partial answers to these questions. The researchers continue to make noteworthy and exhilarating contributions to our understanding of the basic biology of human health in the area of microbiome.

Yet the practical translational applications of this fascinating and enthralling area of science are outstanding. The book also discusses that research on microbiomes provides a more comprehensive view of the genetics of humans and other species, and how it can include innovative human health treatments and new techniques. For the beginner and microbiome enthusiasts, this book may be an essential reading of its importance with existing applications in medicine, agriculture, and climate. With these aims in mind, the material of this textbook has been structured from basic to more advanced topics in a sequential progression. Finally, this book also reviews advancement from fundamental research to relationships between immune–microbiomes and human health microbiomes: clinical applications.

We hope that your creativity is inspired by this book and wish you luck in your experiments. This book illustrates astonishingly the urgency with which the numerous scientific brains are committed to the welfare of the scientific world. I am immensely grateful to the contributors for consistently paying attention to my request and expressing confidence in my skills. I will still be forever highly obliged to all the contributors forever. The worthlessness of their efforts cannot be explained by these terms.

Because of the heartfelt interest and painstaking effort of many other well-wishers whose names are not listed, but they are already in our hearts, we have effectively compiled our innovative and reflective research work. So, the reward for their sacrifices is definitely worth it. I want this book to be devoted to my mum, S. Jayaprada (late). From the bottom of our souls, I and the contributing authors hope this book will be a good guide and guidance for scientific studies to understand the host microbiome relationships in human health.

Machilipatnam, India Pallaval Veera Bramhachari

Acknowledgements

My sincere thanks are extended to all the academicians and scientists who have contributed chapters and happily agreed to share their work on Understanding the Host–Microbiome Interactions in Human Health in this volume.

This book is a stunning reflection of the seriousness with which several scientific minds are dedicated from the scientific community. I am extremely thankful to the contributors for paying continuous attention to my request and showing faith in my competencies and capabilities. I shall always remain highly obliged to all of them forever. These words cannot justify the worthiness of their efforts. We appreciate the excellent work of the authors and co-authors who were invited to contribute chapters to this book. The credit for making this book a reality goes to them. As an editor and the review team for the chapters especially appreciate sharing expertise with the contributors. Each chapter is informative and written as a stand-alone, so the reader can begin anywhere in the book depending upon his/her interests and needs.

At the same time, I also express my deepest gratitude to my family members especially my wife (Ramadevi Ramaswamy) and my kids (Ruthvik and Jayati) for their kind support which has prompted me to complete the assignment on time. I am also thankful to the Department of Biotechnology, Krishna University, for the support. I am equally thankful to the Springer Nature Publishing group for their full cooperation during the peer review and production of the volume.

I am thankful to my beloved teachers and mentors, for their constant support and motivations at all stages of the progress.

About the Book

The book provides an overview on how the microbiome contributes to human health and disease. The microbiome has also become a burgeoning field of research in medicine, agriculture and environment. The readers will obtain profound knowledge on the connection between intestinal microbiota and immune defense systems, medicine, agriculture and environment. The book is addressed to several researchers, clinicians, and scholars working in biomedicine, microbiology, and immunology. The application of new technologies has no doubt revolutionized the research initiatives providing new insights into the dynamics of these complex microbial communities and their role in medicine, agriculture and environment shall be more emphasized. Drawing on broad range concepts of disciplines and model systems, this book primarily provides a conceptual framework for understanding these human–microbe, animal–microbe, and plant–microbe interactions while shedding critical light on the scientific challenges that lie ahead. Furthermore, this book explains why microbiome research demands a creative and interdisciplinary thinking—the capacity to combine microbiology with human, animal, and plant physiology, ecological theory with immunology, and evolutionary perspectives with metabolic science.

This book provides an accessible and authoritative guide to the fundamental principles of microbiome science, an exciting and fast-emerging new discipline that is reshaping many aspects of the life sciences. These microbial partners can also drive ecologically important traits, from thermal tolerance to diet in a typical immune system, and have contributed to animal and plant diversification over long evolutionary timescales. Also, this book explains why microbiome research presents a more complete picture of the biology of humans and other animals, and how it can deliver novel therapies for human health and new strategies.

Contents

Editor and Contributors

About the Editor

Pallaval Veera Bramhachari is currently a Head and Senior Faculty member of the Department of Biotechnology at Krishna University, Machilipatnam, India. He has earlier served as Postdoctoral Research Scientist at the Bacterial Pathogenesis Laboratory, Queensland Institute of Medical Research (QIMR-Berghofer Research Center), Brisbane, Australia and DBT Postdoctoral Fellow at the Department of Microbiology and Cell Biology, Indian Institute of Science (IISc), Bangalore. He has completed Ph.D. in Microbiology from Goa University, India. His major research interests are applied researches in microbiology and cell biology including biochemical mechanism of bacterial–biofilm formations, structure–function relationship of bacterial rhamnolipids, basic and translational cancer research. He has published more than 140 research articles in Scopus-indexed and web of science journals, authored and co-authored 6 international books, and scores of invited book chapters in prestigious peer-reviewed international edited volumes and presented 47 abstracts at various national and international conferences. He has served as editorial board and reviewer for a number of national and international journals. He is a member of numerous international scientific societies and organizations, most importantly, Indian Science Congress and Society of Biological Chemists, India. He has successfully completed two major research projects.

He was awarded with Travel scholarship from QIMR-2007, Australia for attending Fourth Indo-Australian Biotechnology Conference at Queensland, Brisbane, Australia, and young scientist travel fellowship (2007) from the DST, Govt. of India for attending XVII Lancefield International Symposium at Porto Heli–Greece-2008. He was conferred with various prestigious awards, notably, Science Education Research Board (SERB), Government of India-Young Scientist award (2011) with a research project and Nominated as Associate Fellow of Andhra Pradesh Academy of Sciences (APAS)-2016, MASTER TRAINER: Andhra Pradesh English Communications Skills Project. British Council and APSCHE 2017, Andhra Pradesh State Best Scientist (APCOST) award-2017 by State Govt of Andhra Pradesh, Dr.V. Ramalingaswamy Memorial award for Biomedical Sciences-2019 by Science city of Andhra Pradesh. He also obtained two Indian patents in 2017. He has more than 13 years of teaching and research experience at the university level.

List of Contributors

S. Anju Bhavan's Vivekananda College of Science, Humanities, and Commerce, Secunderabad, Telangana, India

K. Anuradha Bhavan's Vivekananda College of Science, Humanities, and Commerce, Secunderabad, Telangana, India

Y. Aparna Bhavan's Vivekananda College of Science, Humanities, and Commerce, Secunderabad, Telangana, India

Chanda Vikrant Berde Marine Microbiology, School of Earth, Ocean and Atmospheric Sciences (SEOAS), Goa University, Taleigao Plateau, Goa, India

B. Berde Vikrant Department of Zoology, Arts, Commerce and Science College, Lanja, Maharashtra, India

Pallaval Veera Bramhachari Department of Biotechnology, Krishna University, Machilipatnam, Andhra Pradesh, India

Gautam Das miBiome Therapeutics LLP, Flexcel Park, C Wing, Mumbai, Maharashtra, India

Nandini Dasgupta miBiome Therapeutics LLP, Flexcel Park, C Wing, Mumbai, Maharashtra, India

Kishore Kumar Godisela Department of Biotechnology, SRR Govt. Arts & Science PG College, Karimnagar, Telangana, India

Kuraganti Gunaswetha Department of Microbiology, Kakatiya University, Warangal, Telangana, India

A. Joshi Suyoj Gogate Jogalekar College, Ratnagiri, Maharashtra, India

V. Kajarekar Kunal Gogate Jogalekar College, Ratnagiri, Maharashtra, India

Badithala Siva Sai Kiran College of Pharmaceutical Sciences and Research, Krishna University, Machilipatnam, Andhra Pradesh, India

Harish Babu Kolla Department of Biotechnology, Vignan's Foundation for Science Technology and Research (Deemed to be University), Guntur, Andhra Pradesh, India

Anusha Konatala University of Windsor, Windsor, ON, Canada

Chandrajit Lahiri Department of Biological Sciences, Sunway University, Petaling Jaya, Selangor, Malaysia

Vrishali Murugkar miBiome Therapeutics LLP, Flexcel Park, C Wing, Mumbai, Maharashtra, India

Dhana Lakshmi Padi Department of Biotechnology, Andhra University, Visakhapatnam, Andhra Pradesh, India

Fain Parackel University of Windsor, Windsor, ON, Canada

Prerna Pathak Jawaharlal Nehru University, New Delhi, India

Shrikant Pawar Yale Center for Genome Analysis (YCGA), Yale University, New Haven, CT, USA

Sudhakar Pola Department of Biotechnology, Andhra University, Visakhapatnam, Andhra Pradesh, India

Amrita Rao miBiome Therapeutics LLP, Flexcel Park, C Wing, Mumbai, Maharashtra, India

Prakash Narayana Reddy Department of Biotechnology, Vignan's Foundation for Science Technology and Research (Deemed to be University), Guntur, Andhra Pradesh, India

P. Salvi Sagar Gogate Jogalekar College, Ratnagiri, Maharashtra, India

Pavani Sanapala Department of Biotechnology, Andhra University, Visakhapatnam, Andhra Pradesh, India

J. Sarada Bhavan's Vivekananda College of Science, Humanities, and Commerce, Secunderabad, Telangana, India

Ravi Shroff miBiome Therapeutics LLP, Flexcel Park, C Wing, Mumbai, Maharashtra, India

Alka Srivastava miBiome Therapeutics LLP, Flexcel Park, C Wing, Mumbai, Maharashtra, India

Pola Sudhakar Department of Biotechnology, AUCST, Andhra University, Visakhapatnam, Andhra Pradesh, India

Edla Sujatha Department of Microbiology, Kakatiya University, Warangal, Telangana, India

Yong Chiang Tan Department of Biological Sciences, Sunway University, Petaling Jaya, Selangor, Malaysia

Abbreviations

ACT	Adoptive cell treatments
AI diseases	Autoimmune diseases
AID	Activation-induced cytidine deaminase
ALT	Alanine aminotransferase
AMPK	Adenosine monophosphate-activated protein kinase
AMPs	Antimicrobial peptides
Ang4	Angiogenin-4
AvrA	Avirulence protein A
BAL	Bronchioscopic alveolar lavages
BB	Bio-breeding
BB-DP	Bio-breeding diabetes-prone
BCG	*Bacillus Calmette-Guerin*
BD	Bayesian redistribution
BIOM	Biological Observation Matrix
BLAST	Basic local alignment search tool
BOS	Bronchiolitis obliterans syndrome
BURST	Based upon related sequence types
c.f.u.	Colony-forming units
CagA	Cytotoxin associated gene A
CAPD	Continuous ambulatory peritoneal dialysis
CAR	Chimeric antigen receptor
CD	Celiac disease
CD	Crohn's disease
CDI	Clostridium difficile *infection*
CF	Cystic fibrosis
CFTR gene	Cystic fibrosis transmembrane conductance regulator gene
CIA	Co-inertia analysis
CIA	Collagen induction arthritis
CKD	Chronic kidney disease
CLA	Conjugated linoleic acid
CNS	Central nervous system
COPD	Chronic obstructive pulmonary disease

CRC patients	Colorectal cancer patients
CRDs	Chronic respiratory disorders
CRP	C-reactive protein
CTLA-4	Cytotoxic T lymphocyte-association protein 4
CVDs	Cardiovascular diseases
CYP	Cytochrome P450
DAG	Directed acyclic graph
DC	Dendritic cells
DIAMOND	Double Index Alignment of NGS Data
DM	Diabetes mellitus
DMARDs	Disease-modifying anti-rheumatic drugs
EC	Enzyme Commission
ED	Enteric dysbiosis
eHOMD	Human Oral Microbiome Database
EPS	Exopolysaccharides
ESRD	End-stage renal disease
EVs	Extracellular vesicles
FadA	*Fusobacterium nucleatum* effector adhesin A
FDA	Food and Drug Administration
FGID	Functional gastrointestinal disease
FIT	Fecal immunochemical test
FMT	Fecal microbiota transplantation
FOBT	Fecal occult blood test
FODMAPs	Fermentable oligo-, di-, and monosaccharide and polyol
FOS	Fructo-oligosaccharides
FUT2	Galactoside 2-alpha-L-fucosyltransferase2
FXR	Farnesoid X receptor
GALT	Gut-associated lymphoid tissue
GF animals	Germ-free animals
GF mice	Germ-free mice
GF	Green fluorescent
GIT	Gastrointestinal tract
GLP-1	Glucagon-like peptide-1
GMO	Genetically modified organisms
GO	Gene ontology
GOS	Galactooligosaccharides
GPCRs	G-protein-coupled receptors
GUI	Graphical user interface
HDAC inhibitors	Histone deacetylase inhibitors
HFGP	Human Functional Genomics Project
HM	Hematological malignancies
HMOS	Human milk oligosaccharides
HMP	Health Human Microbiome Project
HRMS	High-resolution mass spectrometry
IARC	International Agency for Cancer Research

IBD	Inflammatory bowel disease
IBS-C	IBS with constipation
IBS-D	IBS with diarrhea
IBS-M	IBS with mixed bowel habits
ICIs	Immune checkpoint inhibition
ICIs	Immune checkpoint inhibitions
IECs	Intestinal epithelial cells
IELs	Intraepithelial lymphocytes
IgA	Immunoglobulin A
IGC	Integrated human gut microbial Gene Catalog
iHMP Research Network	Integrative HMP Research Network
ILCs	Innate lymphoid cells
ILFs	Isolated lymphoid follicles
ILFs	Lymphoid follicles
IPF	Idiopathic pulmonary fibrosis
IS	Indoxyl sulfate
ISCs	Intestinal stem cells
ITFs	Inulin-type fructans
JNK pathway	Jun N-terminal kinase (JNK) pathway
KO	KEGG ortholog
LCA algorithm	Last-common ancestor (LCA) algorithm
LCA	Lowest common ancestor
LCMS	Liquid chromatography-mass spectrometry
LGG	*Lactobacillus rhamnosus GG*
LHMP	Lung HIV microbiome project
LoD	Limit od detection
LP	Lamina propria
LPS	Bacteria lipopolysaccharide
LPS	Lipopolysaccharide
LTi	Lymphoid tissue-inducer
MALTs	Mucosa-associated lymphoid tissues
MAMPs	Microbe-associated molecular patterns
MCP-1	Mmonocyte chemoattractant protein-1
MDP	Marker Data Profiling
MDP	Muramyl dipeptide
MLH	Mesenteric lymph hubs
mLNs	Mesenteric lymph nodes
MNS	Microbiome novelty score
MOS	Mannan oligosaccharide
MP toxin	Metalloproteinase toxin
MPA	MetaProteomeAnalyzer
MRSA	Methicillin-resistant *Staphylococcus aureus*
MS	Mass spectrometry
MS	Multiple sclerosis
mTOR	Mechanistic target of rapamycin

MyD88 protein	Myeloid differentiation primary response 88 protein
NADPH	Micotinamide adenine dinucleotide phosphate
NAFLD	Non-alcoholic fatty liver disease
NASH	Non-alcoholic steatohepatitis
NCDs	Non-communicable diseases
NGS	Next-generation sequencing
NIH	Human Microbiome Project
NK cell	Natural killer cell
NLRs	Nod-like receptors
NMDS	Non-metric multidimensional scaling
NO	Nitric oxide
NOD 1/2	Nucleotide-binding oligomerization domain 1/2
NOD mouse	Non-obese diabetic mouse
NSAF	Normalized spectral abundance factors
$ONOO^-$	Peroxynitrite
ONT	Oxford Nanopore Technology
OTU	Operational taxonomic units
PA approaches	Procrustes analysis (PA) approaches
PacBio	Pacific biosciences
PCA	Principal component analysis
PCA	Protocatechuic acid
PCD-1	Programmed-cell-death protein 1
PCDB	ProteoClade Database
PCoA	Principal-coordinate analysis
PPD	Projection with Public Data
PRRs	Pattern recognition receptors
PSM	Peptide-spectrum match
QoL	Quality of life
R value	Correlation coefficient
RA	Rheumatoid arthritis
RCT	Receptive cell treatment
RF	Random forest
RNS	Reactive nitrogen species
ROS	Reactive oxygen species
RSAI	Respiratory system associated infections
SCFA-GPCR	G-protein-coupled receptors
SCFAs	Short-chain fatty acids
SDP	Shotgun Data Profiling
SFB	Segmented filamentous bacteria
SG	Sleeve gastrectomy
SIBO	Small intestine bacterial overgrowth
SLE	Systemic Lupus Erythematosus
SODs	Superoxide dismutase
SPF	Specific-pathogen free
SR-A1	Scavenger receptor A1

SRM	Selected reaction monitoring
SVM	Support vector machine
T1DM	Type 1 diabetes
Th17	T helper 17
TLR2	Toll-like receptors 2
TLRs	Toll-like receptors
TMA	Trimethylamine
TMAO	Trimethylamine oxide
TME	Tumor microenvironment
TNF-α	Tumor necrosis factor-α
TPL	The monophosphoryl lipid A
TPP	Trans-proteomic pipeline
Tregs	T regulatory cells
TSEA	Taxon Set Enrichment Analysis
TSLP	Thymic stromal lymphopoietin
UC	Ulcerative colitis
WNT	Wingless-related integration site

Part I

Introduction to Microbiomes

Microbiomes and Its Significance with the Current Applications in Human Health and Disease: Goals and Challenges of Microbiome Research Today

1

Pallaval Veera Bramhachari

Abstract

Human beings are home to a massive invisible microbial community, which powers nearly all processes in the body. Bacteria, archaea, viruses, protozoa, and fungi are the most common organisms in or on our bodies. These fascinating microbial species are called our microbiota collectively. The human body includes a wide number of bacteria, both inside and outside. In particular, it is the microbial genome set that leads to an overall human genetic picture. We know very little about how the pendulum between health and illness swings in our microbiome. The diversity and balance of our microbiome and vulnerability to diseases are inextricably linked. Here, we try to consolidate our existing knowledge about microbiome evolution and ecology in future studies, as well as to consider the relationships between host microbiomes and human health and disease. Finally, we highlight modern methods and technology to human health advancement. The present book on microbiomes is therefore intended to give readers a broad understanding and encouragement for future research and multidisciplinary cooperation on the goals and challenges of microbiome research today.

Keywords

Microbial communities · Microbiome's diversity · Microbiomes · Human health

P. V. Bramhachari (✉)
Department of Biotechnology, Krishna University, Machilipatnam, India

P. V. Bramhachari (ed.), *Microbiome in Human Health and Disease*,
https://doi.org/10.1007/978-981-16-3156-6_1

1.1 Introduction

The set of microbiomes inside and within the human body can be classified as microbiome items. Scientists begin to recognize that microbes play a significant role in our human health. A standard microbiome has ten times as many cells as humans. Indeed, our body is a complex microbial ecosystem; a broad variety of bacteria, fungi, viruses, and microeucaryotes is contained here. The unique microbial combination that exists at each level and we sound like we have created a very significant and precise task along with our microbial partners in every niche. The human microbiome makes up more than 50% and it can influence our mood, appetites, and immune responses to a range of biological functions. Human beings host a huge invisible microbial environment, which affects almost every system of the body. Bacteria, archaea, viruses, protozoa, and fungi are the most common microbes that live within or on our bodies. This fascinating microbial group is known as our microbiota collectively.

1.2 The Complexity of the Microbiomes

The abundance and complexity of the microbiome are staggering with more than one million genes versus 23,000 in the human genome; microbiome communities have unique profiles in different body sites, as do each organism, affected by diet, medications, and other environmental factors (Coyte et al. 2021). The overwhelming proliferation of microbial organisms means that a supra-organism is the human body. Microbiomes are not microorganisms living alongside each other, but instead, form highly regulated complex structurally and functionally organized communities attached to the surfaces as biofilms that contribute to their ecological stability through interspecies and interspecies collaboration. Disease prevention is not the future, but health is the future, free of essential diseases (Shanahan et al. 2021). Multifaceted illness like autoimmune disorders, the human body is an extremely complex system in humans. It is getting more and more difficult to find medications today. It is time to alter the paradigm shift so that scientists across the globe begin focusing on the common theme that microbiomes in combinations of virus bacteria fungi and bacteriophage are related to any disease. In the microbes that we bring, we are each unique, yet they reflect another molecular fingerprint present in each individual. We deal with good and poor microbes in such a way that we know about dental caries caused by bacteria in the mouth, aches, or body odor in the skin and now we have a chance to interact with good bacteria. What these microbial species are doing within the human body is a fascinating issue, and there is a simpler way to cope with tooth cavities, aches, and digestive disorders to treat more serious health problems such as obesity, type 2 diabetes, irritable bowel syndrome, and depression. Given the pros and cons, we need to naturally achieve better health by using microbes that have co-evolved with us.

Otherwise, the human microbiome could be referred to as an undiscovered world that gets benefited in many respects viz. (a) Synthesizes and excretes vitamins

Vitamin B12 and Vitamin K. (b) Prevents contaminants that are difficult for attachment sites or essential nutrients. (c) From colonizing. Could probably alienate in the production of substances that inhibit or destroy nonindigenous species with other bacteria (nonspecific fatty acids, peroxides, bacteriocins). (d) Stimulate the growth of certain tissues, such as intestines, lymphatic tissues, capillary density. (e) The production of cross-reactive antibodies is enkindled. In addition to the ability to extract nutrients, microbiomes generate extra energy that is otherwise unavailable to the host, produce vitamins, metabolize trivial xenobiotics, and provide resistance to cancer and tumor-causing neoplasms, and help grow a mature immune system (Parida and Sharma 2021). Instead of researching the relationship between the microbiota, health, and disease, several studies have shown a connection between various microbial consortia and certain disease states; however, there is still little evidence that a specific disease is triggered by some sophisticated microbial group. The quest was subjugated by the early history of microbiology to discover the microbes responsible for disease and uncover ways to impede them. The first step in setting up ways to prevent and cure infectious diseases was the discovery and analysis of causative agents (Honda and Littman 2012).

1.3 Co-Evolution of Microbiomes with Humans

It looks like we have co-evolved with our microbial partners with very critical and complex tasks that exist in any given section. We are each special in the microbes we bear, but in each individual, they constitute another molecular fingerprint (Hooks and O'Malley 2020). Without any exception, we may map the microbes on the fingerprints and microbes on the machine keyboards to map the keyboard owner and verify how special these microbial signatories are. Microbiome habitats with high diversity are more robust and more resistant and able to return to their safe state of perturbation. We know that over the past few decades, there has been a rising prevalence of allergies, hypersensitivity disorders, and asthma in children. Changes in eating patterns, access to packaged food, decreased interaction with the natural world, the usage of antibacterial soaps, and the possibility of cleansers living in a sterile environment should be taught.

Unfortunately, we have pushed the development of our microbial species to a less complex and inherently less secure condition. Strikingly, microbiota modifications are known to be associated with different diseases, i.e., diabetes, obesity, cardiovascular diseases, carcinogenesis, host anatomy, metabolism, irritable bowel syndrome, and immune dysfunction (Tang et al. 2017). An increasing body of evidence indicates a strong correlation with changes in various disease states in microbiomes. To decipher the deeper understanding of the interactions of microbe disease, multiple studies are ongoing. This poses the exhilarating possibilities that we might think of different therapeutic goals if illnesses are correlated with changes in microbiota and begin to consider ways to transition back to a healthier state.

In multiple environments, microbiomes live and display remarkable differences within and between individuals. With different health states and phenotypes, variability is associated. In several roles, the microbiome is involved viz. production of vitamins, metabolic rate, digestion, odor, behavior, parasite or pathogen defense and Immune Regulation (Costello et al. 2009). Nevertheless, within the human body, hundreds of beneficial microbes reside and are critically important for sustaining human health. They are very functional in educating our immune system in the early years of life to identify them like themselves, but not themselves. They harvest energy from all of the food sources we consume in the (Gastrointestinal) GI Tract. Strikingly, we do not break down all the polysaccharides that we consume, for example. In the gut, microbes develop synthetic vitamins, metabolites, nourish the cells that line the GI tract, and most likely in all environments in the human body.

Researchers are now exploring circumspectly what our microbial inhabitants are doing and how they are contributing to or defending against disease. Research on microbiomes is evolving pretty rapidly to test how we see ourselves as humans. The body harbors at least as many microbial cells as human cells, and there are more than a million genes in our microbial gene database. Nevertheless, we still know very little about the function of most of the second genome and how it affects human health. Besides, microbiome types preserve innate immunity and adaptive immunity by building up the host's metabolic capacity to digest plant carbohydrates, milk products (glycans), vitamin endowments (e.g., B2, B12, K, and folic acid), protect against pathogenic bacteria colonization, and create resistance to colonization (Thaiss et al. 2016). Microbiota metagenome may also function with rapid modification of various strains, exchange of genetic elements, and occurrence of mutations in response to multiple environmental stimuli. The following variables have a vital effect on the microbiota (Nayfach et al. 2019). Host biology, individual lifestyle, diseases, antibiotic exposure, at-time colonization, and birth delivery type, respectively.

1.4 Microbiome Dysbiosis

The proliferation of our beneficial thriving microbes holds in check the pathogenic microbes and maintains a harmonious equilibrium. However, this equilibrium is disrupted when pathogenic microbes govern and we reach a state of dysbiosis. Dysbiosis of the microbiome has been associated with a large number of health problems and is causally concerned with metabolic, immunological, and developmental disorders, as well as susceptibility to the development of infectious diseases. Several illnesses, including cancer, inflammatory bowel disease, obesity, and asthma, are associated with dysbiosis. The composition of the microbiome can be affected by our lifestyle choices, our diet, our use of antibiotics and drugs, and the climate in which we live.

Microbiome dysbiosis has been associated with surplus health problems and has been associated with metabolic, immunological, and developmental disorders as well as susceptibility to infectious disease growth. Genetic and environmental factors causing impaired barrier function, overgrowth of pathogenic bacteria, and subsequent inhibition of defensive bacteria are proposed mechanisms that lead to dysbiosis. Translocation of bacteria and bacterial products into cells, immune activation and development of proinflammatory cytokines, chronic inflammation, the leaky gut term, leads to tissue destruction and complications. In the human body, the gut comprises the largest, densest, and most diverse microbial group. As a crucial determinant of nutrient uptake, energy regulation, and eventually, weight gain and metabolic disorders, recent research has also implicated the gut microbiota. In the future, gut flora modulation could be an important part of weight loss services and various therapies for illnesses. Some metabolites processed by gut microbes that drive the progression of many cardiovascular pathologies, such as atherosclerosis, hypertension, heart failure, and type 2 diabetes, have recently been identified by researchers (Lee and Hase 2014). "These findings suggest that by generating bioactive metabolites that can directly or indirectly affect host physiology, the gut microbiome functions like an endocrine organ" (Tang et al. 2019). Besides, in cancer development affecting predisposing conditions, particularly initiation, progression, response to therapy, microbiota also plays an imperative role (Matson et al. 2021).

The complexity of the fecal microbiota is actively being established and recent studies have shown that microbiota-related dysregulation results in the pathogenesis of several diseases. In the future, gut flora modulation could be an important part of weight loss services and various therapies for illnesses. A healthy intestinal microbiota is restored by fecal microbiota transplantation (FMT) and results in remarkable cure rates for many diseases and is likely to achieve widespread therapeutic advantage for several diseases in the future (Kelly et al. 2021). To work at the petri dish, genomics, and clinical results stage, such studies need translational science.

1.5 Significance of Microbiome Research.

The considerable quantity of research on the microbiome has led to a better understanding of the microbiome and its function in the human, urban, and natural environments in recent years (Cullen et al. 2020). Host-microbe studies such as interactions between the gut and diet offer a major insight into how the microbiome reacts over time to the introduction of new microbes and changes and may potentially serve as the roadmap for microbiome-based intervention and diagnostic technology.

The future microbiome diagnostics and therapeutics armamentarium provide broad and deep possibilities for the monitoring and treatment of a range of diseases-personalized diets, prebiotics, postbiotics, microbiota transplantation, engineered bacteriophages, microbial metabolites, precision editing of microbiota, and modulation of the intestinal barrier. Great data from such microbiome populations have led to the advent of computational genomics that helps to explore the enduring uncultured microbes attributable to current developments in metagenomic sequencing technologies and other omics platforms. Microbiome research has been catapulted into an exciting new frontier in medicine and human health by enormous developments in next-generation sequencing and Omics technologies. Promising knowledge indicates that the microbiome, the rich ecosystem of more than 100 trillion bacteria, fungi, and viruses in and on the human body, is important to all aspects of human health and disease, and that 1 day microbiome analysis may play an imperative role in clinical practice.

In recent years, microbiome research has improved noticeably, powered by developments in technology and a substantial reduction in study costs. Such research has unlocked a wealth of data that has provided significant insight into the existence of microbial communities, including their interactions and consequences, both as part of an ecological community within a host and in an external environment. Understanding the function of microbiota, including its complex interactions with its hosts and other microbes, will make it easier for new diagnostic techniques and interventional methods to be established that can be used in a variety of fields, from ecology and agriculture to medicine and forensics to exobiology.

It would also be very amazing to understand what microbes harbor each tissue and organ, what they encode, what they create as communication signals and how they transform over time, and how human health could be advanced by manipulating these signals. In order to facilitate tailored preventive and therapeutic approaches, systematic understanding and application of microbiome heterogeneity hold great promise. Increasing our experience in this area will afford proof of concept and implement new therapeutic pathways. The ultimate objective should facilitate to reinstate the status quo, using probiotics to replace vital missing and/or extinct microbiome species and strains that respond to essential developmental pathways, likely with accompanying prebiotics.

1.6 The Way Forward

To decide whether there are sets of microbes common to each person, several challenges are still underway. To learn whether modifications to our microbiota result in various health or disease states. New technologies for the study of complex microbial systems and the study of complex microbial systems within their natural environments should be developed (Carr et al. 2020; Galloway-Peña and Hanson 2020). Therefore, therapeutically targeting the makeup of the microbiota will put forward new methods for disease prevention and treatment. The unidentified taxa of bacteria and the analogous genetic levels at which they function are still the greatest impediment. For the next scientific frontier of human health that needs a lot of study, this may be resolutely indispensable (Liu et al. 2020). In addition, the convergence of multiple scientific disciplines and the use of innovative technical methodologies in microbiome research are expected to pave the way for the conception of evidence-based clinical treatments for the health problems of modern life (Methé et al. 2012; New and Brito 2020).

An overview of how the microbiome contributes to human health and illness is given in the book. The microbiome has also become a flourishing field of medical, agricultural, and environmental study. Readers can gain in-depth knowledge of the connection between microbiota, medicine, agriculture, and the environment, and immune defense systems. A number of researchers, physicians, and scholars working in biomedicine, microbiology, and immunology will be discussed in the book. There is no question that the execution of emerging technologies has revolutionized research initiatives by bringing new insights into the complexities of these complex microbial communities and their role in the fields of medicine, agriculture, and the environment. Centered on a wide variety of discipline principles and model systems, this book mainly offers a conceptual outline for understanding these interactions between human microbes, animal microbes, and plant microbes, while flaking fundamental light on the scientific challenges ahead. In addition, this book discusses why microbiome study warrants the opportunity to combine microbiology with human, animal, and plant physiology, evolutionary perspectives with metabolic science, ecological theory with immunology, and innovative and interdisciplinary thought (Berg et al. 2020).

This book offers an accessible and definitive guide to microbiome science's fundamental concepts, an exciting and rapidly evolving new field that transforms many aspects of life sciences. From thermal tolerance to diet in a typical immune system, these microbial partners can also drive ecologically essential traits and have led to animal diversification over long evolutionary timescales. This book also discusses why microbiome research offers a more comprehensive image of human and other animal biology, and how it can produce innovative human health treatments and new strategies.

Acknowledgments DR. PVBC is grateful to Krishna University, Machilipatnam for the support extended.

Conflict of Interest The author declares that they have no competing interests.

References

Berg G, Rybakova D, Fischer D, Cernava T, Vergès MCC, Charles T et al (2020) Microbiome definition re-visited: old concepts and new challenges. Microbiome 8(1):1–22

Carr V, Shkoporov A, Hill C, Mullany P, Moyes D (2020) Probing the mobilome: discoveries in the dynamic microbiome. Trends Microbiol 29(2):158–170

Costello EK, Lauber CL, Hamady M, Fierer N, Gordon JI, Knight R (2009) Bacterial community variation in human body habitats across space and time. Science 326:1694–1697

Coyte KZ, Rao C, Rakoff-Nahoum S, Foster KR (2021) Ecological rules for the assembly of microbiome communities. PLoS Biol 19(2):e3001116

Cullen CM, Aneja KK, Beyhan S, Cho CE, Woloszynek S, Convertino M et al (2020) Emerging priorities for microbiome research. Front Microbiol 11:136

Galloway-Peña J, Hanson B (2020) Tools for analysis of the microbiome. Dig Dis Sci 65 (3):674–685

Honda K, Littman DR (2012) The microbiome in infectious disease and inflammation. Annu Rev Immunol 30:759–795

Hooks KB, O'Malley MA (2020) Contrasting strategies: human eukaryotic versus bacterial microbiome research. J Eukaryot Microbiol 67(2):279–295

Kelly CR, Yen EF, Grinspan AM, Kahn SA, Atreja A, Lewis JD et al (2021) Fecal microbiota transplantation is highly effective in real-world practice: initial results from the FMT National Registry. Gastroenterology 160(1):183–192

Lee WJ, Hase K (2014) Gut microbiota-generated metabolites in animal health and disease. Nat Chem Biol 10(6):416–424

Liu Z, Ma A, Mathé E, Merling M, Ma Q, Liu B (2020) Network analyses in microbiome based on high-throughput multi-omics data. Brief Bioinform 22(2):1639–1655

Matson V, Chervin CS, Gajewski TF (2021) Cancer and the microbiome—influence of the commensal microbiota on cancer, immune responses, and immunotherapy. Gastroenterology 160(2):600–613

Methé BA, Nelson KE, Pop M, Creasy HH, Giglio MG, Huttenhower C et al (2012) A framework for human microbiome research. Nature 486(7402):215

Nayfach S, Shi ZJ, Seshadri R, Pollard KS, Kyrpides NC (2019) New insights from uncultivated genomes of the global human gut microbiome. Nature 568(7753):505–510

New FN, Brito IL (2020) What is metagenomics teaching us, and what is missed? Annu Rev Microbiol 74:117–135

Parida S, Sharma D (2021) The microbiome and cancer: creating friendly neighborhoods and removing the foes within. Cancer Res 81(4):790–800

Shanahan F, Ghosh TS, O'Toole PW (2021) The healthy microbiome—what is the definition of a healthy gut microbiome? Gastroenterology 160(2):483–494

Tang WW, Kitai T, Hazen SL (2017) Gut microbiota in cardiovascular health and disease. Circ Res 120(7):1183–1196

Tang WW, Li DY, Hazen SL (2019) Dietary metabolism, the gut microbiome, and heart failure. Nat Rev Cardiol 16(3):137–154

Thaiss CA, Zmora N, Levy M, Elinav E (2016) The microbiome and innate immunity. Nature 535 (7610):65–74

Part II

Microbiome for Human Health: From Basic Science to Immune-Microbiome Interactions

Modulation of Systemic Immune Responses Through Genital, Skin, and oral Microbiota: Unveiling the Fundamentals of Human Microbiomes

2

Pavani Sanapala and Sudhakar Pola

Abstract

The human system is encountered by several microbes within and on the external surface of the cells. The prevalence of microbes is more compared to the body cells. The microbial organisms have both beneficial and harmful effects on the system. Studying the microbiome of the system is an accessible and escalating field of research. These organisms eventually involve in modulation. Many studies reported the presence of microbes in the gut, vaginal, oral, skin, urinary, and on the lining of the respiratory tract. Factors such as external environment, internal modulation, and epidemiological factors signify the distinct function and composition of microbes. It is necessary to understand the topography, mechanism, and action of microbes so that it would be beneficial to known the pathophysiology, alteration of metabolic events, and progression of many diseases which may pave a pathway for biomarker discovery or drug delivery that can cure or treat or manage the disease with clinical practices.

Keywords

Microbiota · Genital microbiome · Skin · Topography · Oral cavity · Modulation

2.1 Introduction

Human beings are super organisms consisting both microbial cells and their cells. The resident of microbes on the body is reported to be ten folds higher than the body cells (Turnbaugh et al. 2007). These organisms colonizing modulate the immune

P. Sanapala · S. Pola (✉)
Department of Biotechnology, Andhra University, Visakhapatnam, Andhra Pradesh, India
e-mail: sudhakar@andhrauniversity.edu.in

P. V. Bramhachari (ed.), *Microbiome in Human Health and Disease*,
https://doi.org/10.1007/978-981-16-3156-6_2

system and maintain homeostasis in the human system (He et al. 2015). Microbiota of the human body is classified into core and variable microbiome. The core is the usual and universal familiar for all the individuals, while the variable is the distinct microorganism unique to each person depending on their physiological and lifestyle character variations. According to the literature, the body habitat, the skin, the oral, the gastrointestinal, the urinary and the genital tracts, and the lining of the respiratory tract (mucosal membrane) often have discrepancy bacterial communities. The microbial floras often present are usually referred to as healthy flora; however, indigenous flora is expressed alternatively.

Over thrice a 100 million symbiotic microbes exist such as archaea, bacteria, viruses, and fungi inhabit within and on human beings. Over a decade, the role of microbiomes in the health and disease of an individual has been strongly evident (Ley et al. 2006). The human microbiota or microbiomes affect the host bodily processes to an enormous level basing on various mechanisms. Foremost, the microbiota can boost energy extraction from food sources (Turnbaugh et al. 2006) after that increases the nutrient harvest (Gill et al. 2006; Roberfroid et al. 1995) and later altering the appetite (Perry et al. 2016; Cani et al. 2004), besides it benefits the host in metabolic processes in either xenobiotic processing or in gaining nutrients. Next, the human microbiomes act as a barrier in protecting the host against foreign pathogens in the course of competitive eliminations and the production of antimicrobial material (Cash et al. 2006; Hooper et al. 2003; Schauber et al. 2003). Moreover, lastly, the microbiota is crucial in the progress of intestinal mucosa and boosts the immune system (Cash et al. 2006; Bouskra et al. 2008). The human system consists of no less than a thousand different species of already identified bacteria and known to cart 150 times added microbial genes compared to the whole genome of humans (Ursell et al. 2014).

The function and composition of microbiota vary as per age, gender, race, and diet. The composition of the microbiome is distinct in each individual, with the difference being more substantial in individuals than the biochemical differences that occur within a human being eventually (Lax et al. 2014). Interactions of humans with the surrounding environment or even other individuals build up possible microbes on the body sites either indulged in influencing the immune system or as a source to other organisms that can take over the human system (Gilbert et al. 2018). The human microbiome track specific body sites adapting pathway during the body growth, maturity, and development so that each site develops defined biogeography, for example, skin, colon, vaginal, esophagus, urogenital tract, and the oral cavity where the uniqueness of an individual is characterized and confounded by the differing microbiota (Fig. 2.1).

Here we present the up-to-date status of comprehension relating the microbiomes in modulation to the human oral, vaginal, and skin. A detailed note of the function, diversity of species colonized with the human, factors for growth and development of microbes, and the different modulation process is helpful in improving the health of the human population.

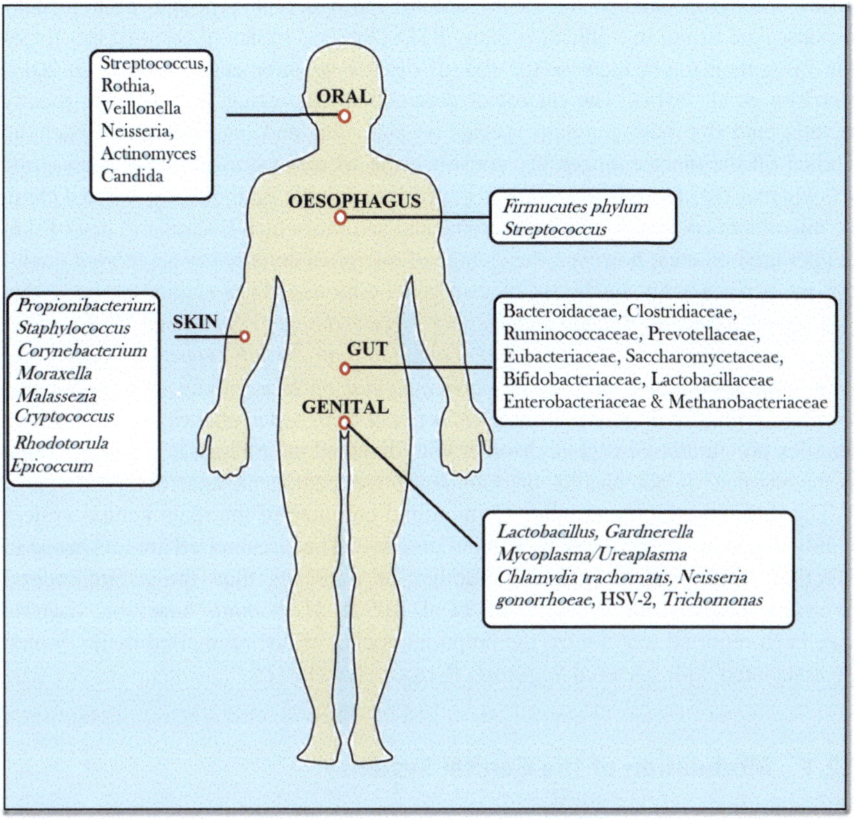

Fig. 2.1 Represents distribution of various microorganisms; that defines uniqueness of human's microbiome

2.2 Genital Microbiome

The ecosystem of the genital refers to the vagina that includes the uterine cervix, the periurethral area, and the preputial pocket of uncircumcised men. Being the significant parts of the reproductive system in both men and women, they are more accessible for infectious microbes during sexual intercourse.

The female genital system (FGT) is divided into upper and lower tracts. The lower FGT constitutes vaginal and ectocervix. The FGT plays a significant role as a protective physical and immunological barrier and as structural support with multiple leukocytes present beneath the epithelial-stromal fibroblasts. The upper FGT comprises the endocervix, uterus, fallopian tube, and ovaries (Wira et al. 2005). Vagina has a unique microbiota, the region is heavily loaded with microbes that as a complex, varied, and eventually active ecosystem that decides the vaginal health. The feminine genital system keeps up a well-maintained immune response that poises tolerance against infections during reproduction. Though inflammatory responses are useful in eradicating many STDs, the occurrence of elevated inflammation in women might increase the risk of disease acquirement (Lajoie et al. 2012; Morrison et al. 2014). The microbial community colonizing the FGT are mainly bacteria, and the most common species be anaerobic and microaerophilic bacteria. Studied till the nineteenth century constitute the species *Lactobacilli* for the vaginal pH, whereas *Gardnerella vaginalis* as imprecise vaginitis. In the pre-menarchal child, the microbiota colonized is both anaerobic and aerobic, which is similar to that of skin and periurethral area; however, the change of microbial community is reported gradually in increase with the levels of estrogen hormone. Many studies reported that pathogens *Chlamydia trachomatis, Neisseria gonorrhoeae*, HSV-2, and *Trichomonas vaginalis are more related to the onset of inflammation. A study report by Delaney and Onderdonk exemplified the role of cervicovaginal bacteria modulating an immune response in the female genital tract (Delaney et al.* 2001*)*. Bacterial vaginosis (BV) is a complex polymicrobial vaginal disorder with diverged microbiota. BV is exemplified by the alternate of lactobacillus predominance over pathogenic anaerobes.

The difference in the number of microbial colonize of the male genital system mainly depends on circumcised or uncircumcised. The circumcised are less prone to infections with preferably a low number of microbes than the uncircumcised (Wiswell and Roscelli 1986; Spach et al. 1992). *M. hominis* anaerobic bacteria have been reported to colonize the preputial pocket of uncircumcised males, which are associated with bacterial vaginosis (Serour et al. 1997).

2.2.1 Modulation of the Genital System

Vaginal microbiota (VMB) and endometrial microbiota can modulate inflammation. Cultures of endometrial cells in combination with *Neisseria gonorrhoeae* (pathogenic bacteria) provoke proinflammatory mediators, whereas *L. crispatus and G. vaginalis* colonizing reproduction did not show proinflammatory activity signifying endometrial microbes modulating inflammation in the host (Łaniewski et al. 2017). Additionally, the epithelial cells of lower FGT do modulate leukocyte function by fabricating cytokine and chemokines, which in turn induce inflammation (Fahey et al. 2005).

Studies showed endogenous sex steroid hormones (Estradiol) playing a role in the modulation of FGT immunity. Estradiol modifies the down streaming gesture of receptors as well as the NκF-β function. However, modulation by human contraceptives of VMB has also been explored. These modifications are vulnerable to HIV-1 in women (Ghisletti et al. 2005). Facultative anaerobes, namely *Gardnerella vaginalis, Prevotella, Mobiluncus,* and *Atopobium* are linked with reproductive health outcomes (Atashili et al. 2008; Zevin et al. 2016). The process of vaginal drying correlating through an increased bacterial variety and also improved inflammatory marks shows significance for HIV risk. Modulation in vaginal microbiota through vaginal drying might cost complications on the mucosal barrier and promotion of cellular inflammatory practices (Anahtar et al. 2015).

The vaginal microbiota plays an essential role as a mutualistic relationship with the host system and involves modulating the threat of acquiring and transmitting sexually transmitted infections (STI). STIs are usually caused by *Chlamydia trachomatis.* Nowadays, these infections are prevalent worldwide, especially in the USA (Newman et al. 2015).

Studies reported the role of lactic acid in modulating the infection caused by *C. trachomatis* (Gong et al. 2014). Nevertheless, the species *Lactobacillus* inhibits *C. trachomatis* and intracellular development (Mastromarino et al. 2014). Both the forms D and L isomers institute the microbiome of the vagina. The ratio of the two forms depends on the *Lactobacillus* that predominates vaginal microbiota (Boskey et al. 2001). But still, the mechanism is unclear; however, the vaginal microbiome imparts confrontation to infection through modulating host cellular function.

2.3 Skin Microbiome

Skin, the largest organ of the human body, acts as a surface barrier protecting against invading pathogens or foreign bodies. The human skin a trillion of microbiota that constitutes mainly of bacteria, fungi, and viruses in various proportions. The skin microbes play significant roles such as protection (defense mechanism), enhancing immune response once in a while, and helps in the breakdown of innate products (Scharschmidt and Fischbach 2013; Belkaid and Segre 2014; Grice 2015). The skin is the first line of defense in opposition to pathogens, while concurrently wharf a diverge surroundings of commensalism, including bacteria, fungi, and viruses. These organisms take part in vital roles in lipid metabolism, colonization resistance to transient organisms, and edification of the immune system (Scharschmidt and Fischbach 2013; Belkaid and Segre 2014; Grice 2015). Genetical strategies show more considerable variation in the organism. The phyla *Actinobacteria*, *Firmicutes*, *Bacteroidetes*, and *Proteobacteria* are the most common inhabitants on the skins. The species Propionibacterium is dominant in sebaceous areas, while species *Corynebacterium* and *Staphylococcus* reside in moist sites. However, most of the diversification of microbiomes is observed in dry sites with gram-negative organisms more likely to colonize the skin (Gao et al. 2007). In contrast to the colonization of bacterial communities on the skin, non-bacterial microorganisms fungi and arthropods have been isolated from the skin. The most predominant fungal genus is *Malassezia* at sebaceous sites, whereas Aspergillus colonized other sites, *Cryptococcus*, *Rhodotorula*, *Epicoccum,* and many more (Findley et al. 2013). The presence of *Demodex folliculorum* and *Demodex brevis* microscopic arthropod mites on the skin illustrates that these arthropods are also among the healthy skin flora. These mites colonize sebaceous areas of the face and also nourish on epithelial cell lining of the pilosebaceous unit. Viruses also occupy a place in the flora of skin microbiota, but studies have not yet reported its promising role (Costello et al. 2009). The microbiota of the skin falls under two groups: transient or resident. The residents are the healthy commensal fixed flora that is homeostatic with the host and are not harmful but are beneficial to the host where transient are microbes from the environment that lives for the time being (Chiller et al. 2001). The assortment of transient and resident microbes on the skin depends on topographical sites of the human body, specific characteristics such as pH, temperature, humidity, sebum content, intrinsic factors (age, sex, genome), and many extrinsic factors such as an epidemiological parameter that is lifestyle, domicile and occupation and antidrug, and many more (Fierer et al. 2010). Both the groups are non-pathogenic in normal conditions such as proper hygiene, healthy immune response, and functioning properly yet, these groups proliferate, colonize, and cause disease after perturbation.

2.3.1 Skin Architecture

Skin is self-possessed into two distinct layers: the epidermis and dermis. The layer outmost is known epidermis, which constitutes distinct keratinocyte layers. The stratum corneum, the top layer consists of mortally differentiated enucleated keratinocytes that act as a bridge to strengthen the skin barrier (Segre 2006). Besides these structural layers, the locations of the body make available to diverge microenvironments that differ in UV light exposure, temperature, moisture, pH, sebum content, and topography (Grice and Segre 2011). Specific skin sites have specific habitats of the microbiome; knowing the composition of the microbiota is useful to define the etiology of the skin disorders, which will be a boon to the dermatologist in analyzing the infections. For instance, hairy, moisture underarms that are distant from the smooth, dry forearms have a different ecological niche that is habitant for several devoid microbial communities (Costello et al. 2009). Factors persuading the composition of the skin microbiome are divided into intrinsic and extrinsic. The intrinsic factors are age, immunity, and genetic manipulation, while extrinsic factors climate and hygienic conditions within the surroundings do affect the microbial community. Basing on these characteristics, human skin is typified into three types; that is, it may be oily (sebaceous), moist or dry. The site's face, chest, and back are composed of oily or sebaceous while bending of the elbow, knee back and groin make up the moist, and forearm and palm fall under dry. All these sites are predisposed by appendages, such as sweat glands, hair follicles, and sebaceous glands. The role of sweat glands in moist areas is thermoregulation by water evaporation that acidifies the skin that makes it unsuitable for the growth and colonization of microorganisms (Grice and Segre 2011).

2.3.2 Association of Microorganisms and Skin Diseases

Skin diseases occur either if the barrier of the skin is damaged or if any stability flanked by commensalism and pathogen is disturbed. The skin is populated by a vast number of assorted microbes, which have either beneficial or harmless (Grice et al. 2009). Over time the composition of species becomes moderately stable (Kong et al. 2012). Nevertheless, skin disorders, namely acne vulgaris (Lomholt and Kilian 2010), eczema (Kong et al. 2012; Kobayashi et al. 2015; Chng et al. 2016; Myles et al. 2016), psoriasis (Alekseyenko et al. 2013), or dandruff (Wang et al. 2015; Clavaud et al. 2012) are linked with alteration in the microbiome.

Skin disease with modified microbiota is termed dysbiosis that is determined by commensal species (Iebba et al. 2016). Acne vulgaris is the common inflammatory skin disorder seen in teenagers and healthy adults, caused mainly by the presence of species *P. acnes* (Leyden et al. 1975; Fitz-Gibbon et al. 2013). Other disorders such as atopic dermatitis (AD) and chronic eczema alter several factors of skin most likely worsening the inflammatory disease, impairment of epidermal barrier, immune cell activation, and alteration in inhabitant of skin microbes. The disease AD has nearly 30 mutations in the host genome along with filaggrin, a barrier protein, and the genes associated with immune response (Palmer et al. 2006). The species *S. aureus* is colonized with AD (Paternoster et al. 2015). Studies reported microbes to instruct the immune system, one such example is primary immunodeficiency (PID). PID, when investigated for colonization, showed opportunistic fungi *Candida* and *Aspergillus* species and also bacteria *Serratia marcescens* that have not appeared in controls. Besides the classical skin diseases, microbes were shown to affect the curing of chronic wounds in the adult population, diabetic and obese people. The study of a diabetic foot ulcer (DFU) is an example, DFU occurs in every 15–20% of individuals having diabetic nephropathy. Sequencing of 16S rRNA reported bacterial species, namely *Staphylococcus sps* and *Proteobacteria sps* showed shorter and longer duration for shallow and deeper ulcers, respectively (Valensi et al. 2005; Ramsey et al. 1999; Gardner et al. 2013).

2.3.3 Topography of Skin

Studies based on cultures illustrated variation of skin microbiota due to different topographical regions, and the variation shows distinct sets of microbes on the skin site. Sites such as the groin, axillary vault, and toe web are partly occluded. The microbes that can withstand temperate and humid conditions such as gram-negative *Bacilli*, *Coryneforms,* and *S. aureus* grow in these areas of moist conditions (Roth and James 1988). Highly sebaceous sites such as the face, chest, and back facilitate the growth of lipophilic microbes, especially *Propionibacterium* spp. and *Malassezia* spp. (Roth and James 1988), whereas other regions, arms, and legs show fluctuation in temperature and hence experience lower habitats for microbial growth (Marples 1965).

Nevertheless, the topography of skin differs equally at microscopic and macroscopic points. Microscopically uneven skin surface, thickness, folds, and mass of hair follicles present and density of glands exemplify the skin microbiota. Macroscopical levels comprise plateaus of the forearm and back, elbow, shoulder, and cervices. Besides, skin appendages that are ununiform are also staged under the macroscopic level (Kong 2011). The mass of the epidermis and dermis lining also varies which affects skin texture. Other factors such as ambient humidity, seasonal variations, clothing, usage of cosmetics, antidrug, and other environmental surfaces affect the topography of microbes (Fierer et al. 2010).

The distinct changes in the skin at the time of birth in an infant is that it switches to gaseous or environmental microbes interface from a sterile condition marking the habitant for several microorganisms. As soon as few days from birth, the infant undergoes quick changes with surface colonization that plays a vital role as a skin barrier. The changes usually notified are water loss and gain and change in pH and sebaceous activity, which characterize the colonization of microbiota (Chiou and Blume-Peytavi 2004). At the time of development, the structure of skin shows the difference from that of adult function and biochemical composition (Stamatas et al. 2010). Recent studies reported that over time the point of stability poses topographical and temporal differences (Peterson et al. 2009).

2.3.4 Skin Modulation

Additionally, acting as a physiological barrier, the skin plays a vital role as an immunological barrier (Borkowski and Gallo 2011). The immune response of the skin modulates the commensal microbiota that inhabits the skin. Specific receptors, namely PRRs (pattern recognition receptors) colonize the skin incessantly by keratinocytes. PRRS, especially toll-like receptors (TLRs), mannose, and NOD-like receptors, recognize pathogen-associated molecular models with flagellin and nucleic acids, LPS (lipopolysaccharides) of gram-negative bacteria, mannan and zymosin of fungal cell walls, peptidoglycan and teichoic acid of gram-positive species. Cytokines, chemokines, and antimicrobial peptides aid in the activation of keratinocyte PPRs. Afar affecting the adaptive immune response, AMPS slaughter the growth of microbes (Braff et al. 2005). For that reason, the continuous interactions of keratinocytes, immune cells, and microbes modulated by AMPs, cytokines, chemokines, and peptides. Differentiating between transient and resident microbes is still unclear; however, TLRs interact on protracted exposure to commensal microbes. A commensal bacterium *S. epidermidis* has reported modulating the host immune response by inhibiting the skin pathogens such as *S. aureus* and *Streptococcus* and also assist the host AMPs with alcoholic soluble modulins (Cogen et al. 2010a; Cogen et al. 2010b). Another modulin lipoteichoic acid from *S. epidermidis* modulates restrain inflammation by TLRs 2 and 3 (Lai et al. 2009).

The complement system modulates the cutaneous microbiome and inflammatory response. The complement system is triggered by either of the three pathways, classical, lectin, or alternative. Upon activation, the active proteins trigger the defense mechanisms, for instance, opsonization, lysis of microbial cells through membrane attack complex, production of effector molecules that arbitrate activation of inflammatory cells, and phagocytosis (Ricklin et al. 2010). The activation of inflammatory cells by the complement components C3a and C5a fragments modulates the innate and acquired immune responses through C5a receptor and pattern recognition signaling (Hajishengallis and Lambris 2010). A study by Chehoud et al. through a culture-based sequencing method suggested C5aR signaling directs essential changes in the skin microbiota eventually, as well as reducing variety and distorted taxonomic composition (Chehoud et al. 2013).

Studies recently reported that the application of healthy flora into the skin lowers the pH and enhances the moisture withholding activity (Nodake et al. 2015). Mixing different microbial components of the skin microbiota and altering the composition of the beneficiary unlock the probability of expanding probiotic solutions that aid or modulate the skin reverting the microbiome in a healthy person (Paetzold et al. 2019).

Vitamin B12 biosynthesis in *Propionibacterium acnes* showed significant downregulation in acne patients. Vitamin B12 is known to modulate the activity of the skin microbiome and put in acne pathogenesis. Analysis suggested that supplementation reticent the expression of the vitamin B12 pathway in *P. acnes* that modified the transcriptome of the skin microbiome (Kang et al. 2015).

2.4 Oral Microbiome

The oral microbiome is one of the critical components among other parts of the human system, and the second most common complex microbiota is followed to colon (Wade 2013). The oral microbiome refers to the population of microbes residing in the oral cavity (Dewhirst et al. 2010). Over 700 varieties of microbes are present in the human mouth. Studies manifest the co-relation of oral microbiota with systemic diseases, besides rheumatoid arthritis (RA), cardiovascular disease, and poor pregnancy effects (Graves et al. 2019; Chen et al. 2018; Cobb et al. 2017; Ray 2017). Several microbes enter the downstream digestive tract from the mouth through saliva, causing diseases related to the digestive system (Dewhirst et al. 2010). Nevertheless, microbiota aids in curing oral and systemic diseases. The oral cavity is dispersed under two types of facades: the hard and the soft tissues. Hard tissue implies the teeth, while oral mucosa falls under soft tissue (Zaura et al. 2014). The diversity of microorganisms at different sites of the oral cavity is reported. Bacteria usually plethora are coated onto the surface, forming a biofilm (Zhao et al. 2017). Temperature range around 37 °C without significant changes is apt for bacterial growth; also, pH of 6.5 to 7 serves as the growth factor, which is equivalent to that of saliva pH where the medium aids in carrying of nutrients to microbes (Lim et al. 2017). Influencing factors of the oral microbiome are time (Costello and Relman 2014), age (Anukam and Agbakoba 2017), diet (Lassalle et al. 2018), exterior environment (Brown et al. 1976), and other factors such as gender, education, community, breastfeeding in infants (Galvão-Moreira et al. 2018).

2.4.1 Ecology of the Oral Microbiome

The cavity of the mouth is a composite location that covers different and minutes microbial locales with diversifying heterogeneous ecological systems on parts of the oral cavity that are teeth, buccal mucosa, hard and soft palate, and tongue (Kilian 2018). All varieties of phylum reside or inhabit the mouth, of which the major are bacteria, fungi, and viruses. Table 2.1 lists the various microbiota in association with

Table 2.1 Microbiota in association with oral diseases

Organisms	Site of colonization	Type of disorder	Related microbial species	References
Bacteria	Dental plaque, teeth	Caries	*Streptococcus mutans*, *Staphylococcus*, *Prevotella spp*, *Dialister spp*, *Filifactor spp*, *Veillonella*, *Lactobacillus*, *Bifidobacterium*, *Propionibacterium*, *Actinomyces spp.*, and *Atopobium spp.*	(Dzidic et al., 2018; Gomez et al., 2017; Hujoel et al., 2018)
	Periodontal	Fall of teeth losing gums	*Porphyromonas gingivalis*	(Gomez et al., 2017)
	Saliva	Hyposalivation	*Firmicutes (genus Streptococcus and Veillonella) and Bacteroidetes (genus Prevotella)*	(Keijser et al., 2008)
	Tongue dorsal	Halitosis	*Streptococcus salivarius*, *Rothia mucilaginosa*	Scully and Greenman (2008)
	Supragingival plaque	Dental decay on occlusal	*Corynebacterium and Actinomyces*	Keijser et al. (2008)
	Subgingival plaque	Prostatic inflammation and periodontitis	*Obsidian Pool OP11*, *TM7*, *Deferribacteres*, *Spirochaetes*, *Fusobacteria*, *Actinobacteria*, *Firmicutes*, *Proteobacteria*, and *Bacteroidetes*	Marsh (2006)
Fungi	Periodontal and gingival plaque	Oral candidiasis, periodontitis, gingivitis	*Candida*, *Cladosporium*, *Aureobasidium*, *Saccharomyces*, *Aspergillus*, *Fusarium*, and *Cryptococcus*	Ghannoum et al. (2010)
Archaea	Periodontal	Periodontitis	*Thermoplasmatales*, *Methanobrevibacter*, *Methanobacterium*, *Methanosarcina*, and *Methanosphaera*	Dridi et al. (2011), Lepp et al. (2004), Nguyen-Hieu et al. (2013)

(continued)

Table 2.1 (continued)

Organisms	Site of colonization	Type of disorder	Related microbial species	References
Viruses	Gingival plaque	Benign-like oral papillomas, oral condylomas, and focal epithelial hyperplasia	*Herpes simplex virus*	Woo and Challacombe (2007)
	Gingival plaque, periodontal	Oral hairy leukoplakia, linear gingival erythema, necrotizing ulcerative periodontitis, and Kaposi's sarcoma	*Human immunodeficiency viruses (Retrovirus)*	Reznik (2005)

oral diseases. Of all the bacteria, species of Bacillus, Firmicutes, Proteobacteria, and Actinomycetes are more likely to colonize the oral cavity (Welch et al. 2016). Distinct to the gut microbiota, the bacteria of the oral cavity act no change considerably. Factors such as diet and environment have a negligible effect on the composition of bacteria. Nearly 85 species of fungi reside in the human mouth of which fungi *Candida* the most common and significant (Baker et al. 2017). The activity of fungi is standard at regular times, but it may cause infections or harass oral tissues in case of imbalance in the composition. Studies reported nearly 101 fungal species of which each individual ranged between 9 and 23 in number (Ghannoum et al. 2010). The pathogenicity of *Candida* is shown up in combination with Streptococcal species (Wang et al. 2012). Viruses' especially phages, act on oral microbiota, which is stable all through the stages of life (Wang et al. 2016; Dudek et al. 2017). Other viruses, namely HIV and mumps, often show in the composition of the oral microbiome (Sällberg 2009; Presti et al. 2018).

2.4.2 Association of Microbiota and Oral Diseases

Microbiota of the oral cavity generates metabolites that influence the development of a series of oral diseases. Oral diseases such as dental caries, periodontal diseases, oral tumors, and recurrent Aphthous Stomatitis are the most common diseases caused by microbiota and are discussed elaborately in the sections below.

2.4.2.1 Caries

Dental caries is the most prevalent infection in the mouth, causing oral pain and tooth loss (Bowen et al. 2018; Selwitz et al. 2007). The rate of incidence is high, occurring from children to adult ages. However, caries are reported more in children than adults. In children with caries, the root cause was recommended to be eating sweets at bedtime (Cao et al. 2017). Species, namely *Prevotella spp.*, *Lactobacillus spp.*, *Dialister spp.*, and *Filifactor spp* are likely to colonize and show progression and pathogenicity for dental caries. Comparing the healthier and caries individuals, the results had shown increased complexity and lower diversity in caries due to acidic nature (Lu et al. 2019). This feature is noticeable in the microbiota of saliva, which shows a considerable number of *S. acidophilus* in caries. Nutrients in the form of glycoproteins are supplied to microbiota through saliva and gingival crevicular fluid when people are under fast or starvation conditions (Cao et al. 2017). Glycoproteins are broken down to saliva and peptide which are under the neutral condition when the host is starving, later after ingesting sugars or starch the acid-generating bacteria reign. Weak acid decay the teeth. The rate of corroding depends on the rate of teeth regeneration and the repair activity they trigger by themselves, which finally results in caries (Lu et al. 2019).

A study by Aas et al. stated the species of *Propionibacterium*, *Lactobacillus*, *Veillonella*, *Bifidobacterium*, and low-pH *non-S. mutans streptococci, and Atopobium spp* and *Actinomyces spp.* genera take part in the progression of caries (Aas et al. 2008). In contrast, a study investigation revealed *Selenomonas*, *Streptococcus mitis,* and *Neisseria,* and high frequency of *Propionibacterium* FMA5 is seen in younger individuals with severe caries progression (Coventry et al. 2000). Apart from these species, *S. pneumoniae, S. infants group*, *Eubacterium* IR009 strain, *Corynebacterium matruchotii*, *Streptococcus cristatus*, *Capnocytophaga gingivalis*, *Streptococcus gordonii, and Lachnospiraceae sps.* C1 strain and *Campylobacter rectus* showed to lower in number as the disease progressed (Gross et al. 2010).

2.4.2.2 Periodontal Diseases

Periodontitis and gingivitis are the two periodontal diseases of which, gingivitis is the most common, affecting 90% of adults (Coventry et al. 2000). Gingivitis is a reversible inflammatory disease rooted by a local bacterial plaque that occurs at the gingival lining. Gingivitis plaque is caused by *Veillonella*, *Leptotrichia*, *Prevotella*, *Streptococcus*, *Haemophilus*, *Selenomonas*, TM7 strain, and *Lautropia*. If gingivitis is unrestrained at the earliest, it may develop into periodontitis.

Periodontitis is a chronic irreparable inflammatory disease for the period in which infiltrates of immune cells penetrate persuading obliteration of connective tissue, vascular propagation, and alveolar bone devastation (Pihlstrom et al. 2005). A study by Costalonga et al. reported that the development and progression of periodontitis are related to *Treponema denticola*, *Porphyromonas gingivalis,* and *Tannerella forsythia* (Socransky et al. 1998). The development of periodontal pockets happens in predisposed individuals at times when the connection between the gingivae and the teeth is lost where anaerobic bacteria colonize (Darveau 2010). Recent studies with sequencing technologies revealed few more organisms associating with periodontal diseases. They are *Peptostreptococcus stomatis*, *Filifactor*, *Prevotella*, *Desulfobulbus*, *Megasphaera,* and *Synergisters* (Kumar et al. 2003; Kumar et al. 2005). *Candida albicans* was practically declared to be highly allied with chronic periodontitis (Canabarro et al. 2013). However, viruses such as *Herpes simplex*, *Human cytomegalovirus,* and HIV are described to biofilm periodontitis.

2.4.2.3 Oral Tumor and Recurrent Aphthous Stomatitis

Recurrent aphthous stomatitis is the oral mucosal disease affecting 20% population and is a common disease (Akintoye and Greenberg 2014). The disease features a painful ulcer in the mouth and is also associated with dysbiosis of mucosa and saliva (Marchini et al. 2007; Seoudi et al. 2015; Hijazi et al. 2015). Microbiota associated with RAS was found to be *Prevotella* (Akintoye and Greenberg 2014), *Actinobacteria sps*., (Seoudi et al. 2015), *Streptococceae* (Kim et al. 2016).

Oral tumors termed oral squamous cell carcinoma sited from the epidermis of the oral cavity (Markopoulos 2012). Aerobes and anaerobes are the general species infecting the cavity (Nagy et al. 1998). Association of *Streptococcus salivarius, Peptostreptococcus stomatis, Gemella haemolysans, Streptococcus gordonii, Streptococcus parasanguinis, Gemella morbillorum,* and *Johnsonella ignava* was more prevalent in tumor sites where *Granulicatella adiacens* was seen in non-tumor sites (Hooper et al. 2007; Pushalkar et al. 2012).

2.4.2.4 Association with Systemic Diseases

Diabetes and periodontitis are interrelated; both are reversible mechanisms. Periodontitis is one of the complication factors for uncontrolled diabetes; likewise, diabetes is one of the root causes of periodontitis (Preshaw et al. 2012; Hintao et al. 2007; Casarin et al. 2013). Diabetes coupled with supragingival plaque is reported to be colonized by *Streptococcus Intermedius, Prevotella nigrescens, Streptococcus sanguinis, T. denticola,* and *Streptococcus oralis* (Hintao et al. 2007). Few other studies revealed *Veillonella, Eikenella, Fusobacterium, Gemella, Actinomyces, Neisseria, Capnocytophaga, Aggregatibacter and Streptococcus, Selenomonas* and *TM7* genera and lower levels of *Filifactor, Synergistetes, Porphyromonas, Tannerella, Eubacterium* and *Treponema genera, Eikenella corrodens, Fusobacterium nucleatum, Veillonella parvula, and V. dispar* (Casarin et al. 2013).

The microbiota plays an influential role in cardiovascular diseases; these species are closely related to that of periodontal pathogens (Ramirez et al. 2014). Genera *Veillonella* and *Streptococcus* were associated with atherosclerotic plaques that are similar to the oral cavity (Koren et al. 2011). Besides these genera, *Aggregatibacter actinomycetemcomitans, F. nucleatum, Campylobacter rectus, P. gingivalis, T. forsythia,* and *E. corrodens* were also detected in plaque (Figuero et al. 2011; Pucar et al. 2007).

Rheumatoid arthritis (RA) is a systemic autoimmune malady. Both RA and periodontitis share a common pathogenic mechanism that is bone loss and inflammation. Patients with periodontitis and RA are enriched with anaerobic bacteria species such as *Atopobium, Cryptobacteriumcurtum, Lactobacillus salivarius, Prevotella,* and *Leptotrichia* with high frequency, where Corynebacterium is present in reduced levels. Individuals with only RA are associated with *Prevotella* periodontitis bacteria (Corrêa et al. 2016).

2.4.3 Modulation of the Oral Microbiome

The function of oral microbes in the causation and pathogenesis of oral and systemic diseases is crucial to improve the protection of oral pathogens so that it sustains homeostasis. Understanding the mechanism of oral microbiome is necessary to reduce oral pathogen development and progression.

Disproportion in the oral microbiota is said to be associated with condensed CVD and metabolic health. The relation between oral microbiota and health is probably the nitrate, nitrite, and nitric oxide pathways that lower the three compounds regulating vascular endothelial function and after that increased blood pressure. Inorganic nitrate, the natural source to humans through dietary vegetables is itself purely inert. Human cells lack nitrate reductase ability yet, commensal bacteria in the oral cavity use the inorganic nitrate as an energy source for the synthesis of ATP where the nitrate is reduced to nitrite with an electron transfer. The nitrite obtained is further reduced to nitric oxide, a potent vasodilator (Larsen et al. 2006). This pathway highlights the findings that dietary nitrate supplementation through nitrate salts or vegetables lowers blood pressure in old and adolescents (Webb et al. 2008). Sequencing 16S rRNA genes of the bacterial species detailed the correlation of oral microbiome and nitrate mechanism. *Neisseria* and *Rothia, Veillonella,* and *Prevotella are* present in high and low abundance, respectively, were correlated with a more substantial raise in plasma in response to nitrate supplementation. The finding of Vanhatalo et al. demonstrated these findings in his study (Vanhatalo et al. 2018).

Pro and prebiotics are served to modulate the oral microbiome. The probiotic methods are reported to treat caries that intrusive with the oral microbiota of cariogenic pathogens. Species of *Lactobacillus* (*L. rhamnosus)* (Näse et al. 2001), *L. paracasei* (Holz et al. 2013), *Lactobacillus reuteri* (Caglar et al. 2006), *B. animalis* (Cildir et al. 2009), *Bifidobacterium* (Busscher et al. 1999) have the inhibiting ability in vivo in reducing carcinogenic bacteria preventing dental caries. *L. reuteri* is reported to reduce gum bleeding and gingivitis (Cildir et al. 2009). Probiotics, namely inulin, lactose, galactose, xylo, and fructooligosaccharides, are common probiotics used as modulates. Prebiotics majorly xylitol, xylose, and arabinose are the most potent that suppress the activity of *S. mutans* (Busscher et al. 1999).

Antibiotics are the targeted drugs designed especially for pathogenic bacteria in mammals (Krasse et al. 2006). Studies described that administration of amoxicillin and metronidazole drugs before scaling and rooting significantly enhanced periodontal factors and lowered the levels of *P. gingivalis*, *P. intermedia,* and *T. forsythensis* (Kojima et al. 2016). A study by Haffajee and his colleagues detailed the reduction of periodontal microbiota at subgingival through antibiotic administration (Haffajee et al. 2006). Knowing the mechanism of antibiotics and genome analysis may identify the response of the treatment which is necessary to reduce the development and progression of oral diseases (Zarco et al. 2012).

Certain mechanical debridement such as self-cleaning that is brushing teeth may improve or control the level of plaque (Petersilka et al. 2002). Professional practices such as scaling, rooting, and surgeries probably reduce the number of pathogenic bacteria and maintain ecological balance. Pathogens causing periodontal disorders, namely *Tannerella forsythia*, *Treponema socranskii*, *P. gingivalis,* and *Tannerella denticola* are marked to vanish after mechanical debridement (Sakamoto et al. 2004).

2.5 Conclusion

In recent years, the expansion of molecular methods in identifying the composition and function of the microbiome in health and disease is of huge interest. Exploring and understanding the interaction between the species are a highly complex chore. However, several studies have drawn many conclusions on the microbiome interaction with the human system, it is still in an emerging stage. Further research with vast sample sizes and novel techniques is essential to get reliable models to create concrete data. This will further engage in the discovery of novel drugs, biomarkers and assist in targeted therapies and customized medicines for healthier management in clinical applications.

Acknowledgments The authors **Pavanai Sanapala** and **Sudhakar Pola** are grateful to the authorities of Andhra University Visakhapatnam, India for providing the necessary facilities to carry out the research work and for extending constant support.

Conflict of Interest The authors declare that they have no competing interests.

References

Aas JA, Griffen AL, Dardis SR, Lee AM, Olsen I, Dewhirst FE et al (2008) Bacteria of dental caries in primary and permanent teeth in children and young adults. J Clin Microbiol 46(4):1407–1417
Akintoye SO, Greenberg MS (2014) Recurrent aphthous stomatitis. Dental Clin 58(2):281–297
Alekseyenko AV, Perez-Perez GI, De Souza A, Strober B, Gao Z, Bihan M et al (2013) Community differentiation of the cutaneous microbiota in psoriasis. Microbiome 1(1):31
Anahtar MN, Byrne EH, Doherty KE, Bowman BA, Yamamoto HS, Soumillon M et al (2015) Cervicovaginal bacteria are a major modulator of host inflammatory responses in the female genital tract. Immunity 42(5):965–976
Anukam KC, Agbakoba NR (2017) A comparative study of the oral microbiome compositions of healthy postmenopausal, premenopausal, and prepubertal Nigerian females, using 16s rrna metagenomics methods. Niger J Clin Pract 20(10):1250–1258
Atashili J, Poole C, Ndumbe PM, Adimora AA, Smith JS (2008) Bacterial vaginosis and HIV acquisition: a meta-analysis of published studies. AIDS 22(12):1493
Baker JL, Bor B, Agnello M, Shi W, He X (2017) Ecology of the oral microbiome: beyond bacteria. Trends Microbiol 25(5):362–374
Belkaid Y, Segre JA (2014) Dialogue between skin microbiota and immunity. Science 346 (6212):954–959
Borkowski AW, Gallo RL (2011) The coordinated response of the physical and antimicrobial peptide barriers of the skin. J Investig Dermatol 131(2):285–287
Boskey ER, Cone RA, Whaley KJ, Moench TR (2001) Origins of vaginal acidity: high D/L lactate ratio is consistent with bacteria being the primary source. Hum Reprod 16(9):1809–1813
Bouskra D, Brézillon C, Bérard M, Werts C, Varona R, Boneca IG, Eberl G (2008) Lymphoid tissue genesis induced by commensals through NOD1 regulates intestinal homeostasis. Nature 456(7221):507–510
Bowen WH, Burne RA, Wu H, Koo H (2018) Oral biofilms: pathogens, matrix, and polymicrobial interactions in microenvironments. Trends Microbiol 26(3):229–242
Braff MH, Bardan A, Nizet V, Gallo RL (2005) Cutaneous defense mechanisms by antimicrobial peptides. J Investig Dermatol 125(1):9–13

Brown LR, Fromme WJ, Handler SF, Wheatcroft MG, Johnston DA (1976) Effect of Skylab missions on clinical and microbiologic aspects of oral health. J Am Dent Assoc 93(2):357–363
Busscher HJ, Mulder AFJM, Van der Mei HC (1999) In vitro adhesion to enamel and in vivo colonization of tooth surfaces by lactobacilli from a bio-yoghurt. Caries Res 33(5):403
Caglar E, Kavaloglu Cildir S, Ergeneli S, Sandalli N, Twetman S (2006) Salivary mutans streptococci and lactobacilli levels after ingestion of the probiotic bacterium Lactobacillus reuteri ATCC 55730 by straws or tablets. Acta Odontol Scand 64(5):314–318
Canabarro A, Valle C, Farias MR, Santos FB, Lazera M, Wanke B (2013) Association of subgingival colonization of C andida albicans and other yeasts with severity of chronic periodontitis. J Periodontal Res 48(4):428–432
Cani PD, Dewever C, Delzenne NM (2004) Inulin-type fructans modulate gastrointestinal peptides involved in appetite regulation (glucagon-like peptide-1 and ghrelin) in rats. Br J Nutr 92 (3):521–526
Cao X, Wang D, Zhou J, Yuan H, Chen Z (2017) Relationship between dental caries and metabolic syndrome among 13 998 middle-aged urban Chinese. J Diabetes 9(4):378–385
Casarin RCV, Barbagallo A, Meulman T, Santos VR, Sallum EA, Nociti FH et al (2013) Subgingival biodiversity in subjects with uncontrolled type-2 diabetes and chronic periodontitis. J Periodontal Res 48(1):30–36
Cash HL, Whitham CV, Behrendt CL, Hooper LV (2006) Symbiotic bacteria direct expression of an intestinal bactericidal lectin. Science 313(5790):1126–1130
Chehoud C, Rafail S, Tyldsley AS, Seykora JT, Lambris JD, Grice EA (2013) Complement modulates the cutaneous microbiome and inflammatory milieu. Proc Natl Acad Sci 110 (37):15061–15066
Chen B, Zhao Y, Li S, Yang L, Wang H, Wang T et al (2018) Variations in oral microbiome profiles in rheumatoid arthritis and osteoarthritis with potential biomarkers for arthritis screening. Sci Rep 8(1):1–8
Chiller K, Selkin BA, Murakawa GJ (2001) Skin microflora and bacterial infections of the skin. J Invest Dermatol Symp Proc 6(3):170–174
Chiou YB, Blume-Peytavi U (2004) Stratum corneum maturation. Skin Pharmacol Physiol 17 (2):57–66
Chng KR, Tay ASL, Li C, Ng AHQ, Wang J, Suri BK et al (2016) Whole metagenome profiling reveals skin microbiome-dependent susceptibility to atopic dermatitis flare. Nat Microbiol 1 (9):16106
Cildir SK, Germec D, Sandalli N, Ozdemir FI, Arun T, Twetman S, Caglar E (2009) Reduction of salivary mutans streptococci in orthodontic patients during daily consumption of yoghurt containing probiotic bacteria. Eur J Orthodon 31(4):407–411
Clavaud C, Jourdain R, Bar-Hen A, Tichit M, Bouchier C, Pouradier F et al (2012) Correction: dandruff is associated with disequilibrium in the proportion of the major bacterial and fungal populations colonizing the scalp. PLoS One 8(10):e58203
Cobb CM, Kelly PJ, Williams KB, Babbar S, Angolkar M, Derman RJ (2017) The oral microbiome and adverse pregnancy outcomes. Int J Women's Health 9:551
Cogen AL, Yamasaki K, Muto J, Sanchez KM, Alexander LC, Tanios J et al (2010b) *Staphylococcus epidermidis* antimicrobial δ-toxin (phenol-soluble modulin-γ) cooperates with host antimicrobial peptides to kill group A Streptococcus. PLoS One 5(1):e8557
Cogen AL, Yamasaki K, Sanchez KM, Dorschner RA, Lai Y, Mac Leod DT et al (2010a) Selective antimicrobial action is provided by phenol-soluble modulins derived from Staphylococcus epidermidis, a normal resident of the skin. J Investig Dermatol 130(1):192–200
Corrêa JD, Saraiva AM, Queiroz-Junior CM, Madeira MFM, Duarte PM, Teixeira MM et al (2016) Arthritis-induced alveolar bone loss is associated with changes in the composition of oral microbiota. Anaerobe 39:91–96
Costello EK, Lauber CL, Hamady M, Fierer N, Gordon JI, Knight R (2009) Bacterial community variation in human body habitats across space and time. Science 326(5960):1694–1697

Costello EK, Relman DA (2014) Population health: immaturity in the gut microbial community. Nature 510(7505):344–345

Coventry J, Griffiths G, Scully C, Tonetti M (2000) Periodontal disease. BMJ 321(7252):36–39

Darveau RP (2010) Periodontitis: a polymicrobial disruption of host homeostasis. Nat Rev Microbiol 8(7):481–490

Delaney ML, Onderdonk AB, Microbiology and Prematurity Study Group (2001) Nugent score related to vaginal culture in pregnant women. Obstet Gynecol 98(1):79–84

Dewhirst FE, Chen T, Izard J, Paster BJ, Tanner AC, Yu WH et al (2010) The human oral microbiome. J Bacteriol 192(19):5002–5017

Dridi B, Raoult D, Drancourt M (2011) Archaea as emerging organisms in complex human microbiomes. Anaerobe 17(2):56–63

Dudek NK, Sun CL, Burstein D, Kantor RS, Goltsman DSA, Bik EM et al (2017) Novel microbial diversity and functional potential in the marine mammal oral microbiome. Curr Biol 27 (24):3752–3762

Dzidic M, Collado MC, Abrahamsson T, Artacho A, Stensson M, Jenmalm MC, Mira A (2018) Oral microbiome development during childhood: an ecological succession influenced by postnatal factors and associated with tooth decay. ISME J 12(9):2292–2306

Fahey JV, Schaefer TM, Channon JY, Wira CR (2005) Secretion of cytokines and chemokines by polarized human epithelial cells from the female reproductive tract. Hum Reprod 20 (6):1439–1446

Fierer N, Lauber CL, Zhou N, McDonald D, Costello EK, Knight R (2010) Forensic identification using skin bacterial communities. Proc Natl Acad Sci 107(14):6477–6481

Figuero E, Sánchez-Beltrán M, Cuesta-Frechoso S, Tejerina JM, del Castro JA, Gutiérrez JM et al (2011) Detection of periodontal bacteria in atheromatous plaque by nested polymerase chain reaction. J Periodontol 82(10):1469–1477

Findley K, Oh J, Yang J, Conlan S, Deming C, Meyer JA et al (2013) Topographic diversity of fungal and bacterial communities in human skin. Nature 498(7454):367–370

Fitz-Gibbon S, Tomida S, Chiu BH, Nguyen L, Du C, Liu M et al (2013) Propionibacterium acnes strain populations in the human skin microbiome associated with acne. J Investig Dermatol 133 (9):2152–2160

Galvão-Moreira LV, de Andrade CM, de Oliveira JFF, Bomfim MRQ, Figueiredo PMS, Branco-de-Almeida LS (2018) Sex differences in salivary parameters of caries susceptibility in healthy individuals. Oral Health Prev Dent 16(1):71–77

Gao Z, Tseng CH, Pei Z, Blaser MJ (2007) Molecular analysis of human forearm superficial skin bacterial biota. Proc Natl Acad Sci 104(8):2927–2932

Gardner SE, Hillis SL, Heilmann K, Segre JA, Grice EA (2013) The neuropathic diabetic foot ulcer microbiome is associated with clinical factors. Diabetes 62(3):923–930

Ghannoum MA, Jurevic RJ, Mukherjee PK, Cui F, Sikaroodi M, Naqvi A, Gillevet PM (2010) Characterization of the oral fungal microbiome (mycobiome) in healthy individuals. PLoS Pathog 6(1):e1000713

Ghisletti S, Meda C, Maggi A, Vegeto E (2005) 17β-estradiol inhibits inflammatory gene expression by controlling NF-κB intracellular localization. Mol Cell Biol 25(8):2957–2968

Gilbert JA, Blaser MJ, Caporaso JG, Jansson JK, Lynch SV, Knight R (2018) Current understanding of the human microbiome. Nat Med 24(4):392

Gill SR, Pop M, DeBoy RT, Eckburg PB, Turnbaugh PJ, Samuel BS et al (2006) Metagenomic analysis of the human distal gut microbiome. Science 312(5778):1355–1359

Gomez A, Espinoza JL, Harkins DM, Leong P, Saffery R, Bockmann M et al (2017) Host genetic control of the oral microbiome in health and disease. Cell Host Microbe 22(3):269–278

Gong Z, Luna Y, Yu P, Fan H (2014) Lactobacilli inactivate chlamydia trachomatis through lactic acid but not H2O2. PLoS One 9(9):e107758

Graves DT, Corrêa JD, Silva TA (2019) The oral microbiota is modified by systemic diseases. J Dent Res 98(2):148–156

Grice EA (2015) The intersection of microbiome and host at the skin interface: genomic-and metagenomic-based insights. Genome Res 25(10):1514–1520
Grice EA, Kong HH, Conlan S, Deming CB, Davis J, Young AC et al (2009) Topographical and temporal diversity of the human skin microbiome. Science 324(5931):1190–1192
Grice EA, Segre JA (2011) The skin microbiome. Nat Rev Microbiol 9(4):244–253
Gross EL, Leys EJ, Gasparovich SR, Firestone ND, Schwartzbaum JA, Janies DA et al (2010) Bacterial 16S sequence analysis of severe caries in young permanent teeth. J Clin Microbiol 48 (11):4121–4128
Haffajee AD, Teles RP, Socransky SS (2006) The effect of periodontal therapy on the composition of the subgingival microbiota. Periodontol 42(1):219–258
Hajishengallis G, Lambris JD (2010) Crosstalk pathways between toll-like receptors and the complement system. Trends Immunol 31(4):154–163
He J, Li Y, Cao Y, Xue J, Zhou X (2015) The oral microbiome diversity and its relation to human diseases. Folia Microbiol 60(1):69–80
Hijazi K, Lowe T, Meharg C, Berry SH, Foley J, Hold GL (2015) Mucosal microbiome in patients with recurrent aphthous stomatitis. J Dent Res 94(3_suppl):87S–94S
Hintao J, Teanpaisan R, Chongsuvivatwong V, Ratarasan C, Dahlen G (2007) The microbiological profiles of saliva, supragingival and subgingival plaque and dental caries in adults with and without type 2 diabetes mellitus. Oral Microbiol Immunol 22(3):175–181
Holz C, Alexander C, Balcke C, Moré M, Auinger A, Bauer M et al (2013) Lactobacillus paracasei DSMZ16671 reduces mutans streptococci: a short-term pilot study. Probiot Antimicrob Proteins 5(4):259–263
Hooper LV, Stappenbeck TS, Hong CV, Gordon JI (2003) Angiogenins: a new class of microbicidal proteins involved in innate immunity. Nat Immunol 4(3):269–273
Hooper SJ, Crean SJ, Fardy MJ, Lewis MA, Spratt DA, Wade WG, Wilson MJ (2007) A molecular analysis of the bacteria present within oral squamous cell carcinoma. J Med Microbiol 56 (12):1651–1659
Hujoel PP, Hujoel MLA, Kotsakis GA (2018) Personal oral hygiene and dental caries: A systematic review of randomised controlled trials. Gerodontology 35(4):282–289
Iebba V, Totino V, Gagliardi A, Santangelo F, Cacciotti F, Trancassini M et al (2016) Eubiosis and dysbiosis: the two sides of the microbiota. New Microbiol 39(1):1–12
Kang D, Shi B, Erfe MC, Craft N, Li H (2015) Vitamin B12 modulates the transcriptome of the skin microbiota in acne pathogenesis. Sci Transl Med 7(293):293ra103
Keijser BJF, Zaura E, Huse SM, Van der Vossen JMBM, Schuren FHJ, Montijn RC et al (2008) Pyrosequencing analysis of the oral microflora of healthy adults. J Dent Res 87(11):1016–1020
Kilian M (2018) The oral microbiome–friend or foe? Eur J Oral Sci 126:5–12
Kim YJ, Choi YS, Baek KJ, Yoon SH, Park HK, Choi Y (2016) Mucosal and salivary microbiota associated with recurrent aphthous stomatitis. BMC Microbiol 16(1):57
Kobayashi T, Glatz M, Horiuchi K, Kawasaki H, Akiyama H, Kaplan DH et al (2015) Dysbiosis and Staphylococcus aureus colonization drives inflammation in atopic dermatitis. Immunity 42 (4):756–766
Kojima Y, Ohshima T, Seneviratne CJ, Maeda N (2016) Combining prebiotics and probiotics to develop novel synbiotics that suppress oral pathogens. J Oral Biosci 58(1):27–32
Kong HH (2011) Skin microbiome: genomics-based insights into the diversity and role of skin microbes. Trends Mol Med 17(6):320–328
Kong HH, Oh J, Deming C, Conlan S, Grice EA, Beatson MA et al (2012) Temporal shifts in the skin microbiome associated with disease flares and treatment in children with atopic dermatitis. Genome Res 22(5):850–859
Koren O, Spor A, Felin J, Fåk F, Stombaugh J, Tremaroli V et al (2011) Human oral, gut, and plaque microbiota in patients with atherosclerosis. Proc Natl Acad Sci 108(Suppl 1):4592–4598
Krasse P, Carlsson B, Dahl C, Paulsson A, Nilsson A, Sinkiewicz G (2006) Decreased gum bleeding and reduced gingivitis by the probiotic Lactobacillus reuteri. Swed Dent J 30(2):55–60

Kumar PS, Griffen AL, Barton JA, Paster BJ, Moeschberger ML, Leys EJ (2003) New bacterial species associated with chronic periodontitis. J Dent Res 82(5):338–344
Kumar PS, Griffen AL, Moeschberger ML, Leys EJ (2005) Identification of candidate periodontal pathogens and beneficial species by quantitative 16S clonal analysis. J Clin Microbiol 43 (8):3944–3955
Lai Y, Di Nardo A, Nakatsuji T, Leichtle A, Yang Y, Cogen AL et al (2009) Commensal bacteria regulate toll-like receptor 3–dependent inflammation after skin injury. Nat Med 15(12):1377
Lajoie J, Juno J, Burgener A, Rahman S, Mogk K, Wachihi C et al (2012) A distinct cytokine and chemokine profile at the genital mucosa is associated with HIV-1 protection among HIV-exposed seronegative commercial sex workers. Mucosal Immunol 5(3):277–287
Łaniewski P, Gomez A, Hire G, So M, Herbst-Kralovetz MM (2017) Human three-dimensional endometrial epithelial cell model to study host interactions with vaginal bacteria and Neisseria gonorrhoeae. Infect Immun 85(3):e01049–e01016
Larsen FJ, Ekblom B, Sahlin K, Lundberg JO, Weitzberg E (2006) Effects of dietary nitrate on blood pressure in healthy volunteers. N Engl J Med 355:2792–2793
Lassalle F, Spagnoletti M, Fumagalli M, Shaw L, Dyble M, Walker C et al (2018) Oral microbiomes from hunter-gatherers and traditional farmers reveal shifts in commensal balance and pathogen load linked to diet. Mol Ecol 27(1):182–195
Lax S, Smith DP, Hampton-Marcell J, Owens SM, Handley KM, Scott NM et al (2014) Longitudinal analysis of microbial interaction between humans and the indoor environment. Science 345 (6200):1048–1052
Lepp PW, Brinig MM, Ouverney CC, Palm K, Armitage GC, Relman DA (2004) Methanogenic archaea and human periodontal disease. Proc Natl Acad Sci 101(16):6176–6181
Ley RE, Turnbaugh PJ, Klein S, Gordon JI (2006) Human gut microbes associated with obesity. Nature 444(7122):1022–1023
Leyden JJ, McGinley KJ, Mills OH, Kligman AM (1975) Propionibacterium levels in patients with and without acne vulgaris. J Investig Dermatol 65(4):382–384
Lim Y, Totsika M, Morrison M, Punyadeera C (2017) Oral microbiome: a new biomarker reservoir for oral and oropharyngeal cancers. Theranostics 7(17):4313
Lomholt HB, Kilian M (2010) Population genetic analysis of Propionibacterium acnes identifies a subpopulation and epidemic clones associated with acne. PLoS One 5(8):e12277
Lu M, Xuan S, Wang Z (2019) Oral microbiota: a new view of body health. Food Sci Human Wellness 8(1):8–15
Marchini L, Campos MS, Silva AM, Paulino LC, Nobrega FG (2007) Bacterial diversity in aphthous ulcers. Oral Microbiol Immunol 22(4):225–231
Markopoulos AK (2012) Current aspects on oral squamous cell carcinoma. Open Dent J 6:126
Marples MJ (1965) The ecology of the human skin. Springer, Cham
Marsh PD (2006) Dental plaque as a biofilm and a microbial community–implications for health and disease. BMC Oral Health 6(1):S14
Mastromarino P, Di Pietro M, Schiavoni G, Nardis C, Gentile M, Sessa R (2014) Effects of vaginal lactobacilli in chlamydia trachomatis infection. Int J Med Microbiol 304(5–6):654–661
Morrison C, Fichorova RN, Mauck C, Chen PL, Kwok C, Chipato T et al (2014) Cervical inflammation and immunity associated with hormonal contraception, pregnancy, and HIV-1 seroconversion. JAIDS J Acquired Immune Def Synd 66(2):109–117
Myles IA, Williams KW, Reckhow JD, Jammeh ML, Pincus NB, Sastalla I et al (2016) Transplantation of human skin microbiota in models of atopic dermatitis. JCI insight 1(10):e86955
Nagy KN, Sonkodi I, Szöke I, Nagy E, Newman HN (1998) The microflora associated with human oral carcinomas. Oral Oncol 34(4):304–308
Näse L, Hatakka K, Savilahti E, Saxelin M, Pönkä A, Poussa T et al (2001) Effect of long-term consumption of a probiotic bacterium, Lactobacillus rhamnosus GG, in milk on dental caries and caries risk in children. Caries Res 35(6):412–420

Newman L, Rowley J, Vander Hoorn S, Wijesooriya NS, Unemo M, Low N et al (2015) Global estimates of the prevalence and incidence of four curable sexually transmitted infections in 2012 based on systematic review and global reporting. PLoS One 10(12):e0143304

Nguyen-Hieu T, Khelaifia S, Aboudharam G, Drancourt M (2013) Methanogenic archaea in subgingival sites: a review. APMIS 121(6):467–477

Nodake Y, Matsumoto S, Miura R, Honda H, Ishibashi G, Matsumoto S et al (2015) Pilot study on novel skin care method by augmentation with Staphylococcus epidermidis, an autologous skin microbe–A blinded randomized clinical trial. J Dermatol Sci 79(2):119–126

Paetzold B, Willis JR, de Lima JP, Knödlseder N, Brüggemann H, Quist SR et al (2019) Skin microbiome modulation induced by probiotic solutions. Microbiome 7(1):1–9

Palmer CN, Irvine AD, Terron-Kwiatkowski A, Zhao Y, Liao H, Lee SP et al (2006) Common loss-of-function variants of the epidermal barrier protein filaggrin are a major predisposing factor for atopic dermatitis. Nat Genet 38(4):441–446

Paternoster L, Standl M, Waage J, Baurecht H, Hotze M, Strachan DP, Curtin JA (2015) The EArly genetics and Lifecourse epidemiology (EAGLE) eczema consortium multi-ancestry genome-wide association study of 21,000 cases and 95,000 controls identifies new risk loci for atopic dermatitis. Nat Genet 47:1449–1456

Perry RJ, Peng L, Barry NA, Cline GW, Zhang D, Cardone RL et al (2016) Acetate mediates a microbiome–brain–β-cell axis to promote metabolic syndrome. Nature 534(7606):213–217

Petersilka GJ, Ehmke B, Flemmig TF (2002) Antimicrobial effects of mechanical debridement. Periodontol 28(1):56–71

Peterson J, Garges S, Giovanni M, McInnes P, Wang L, Schloss JA et al (2009) The NIH human microbiome project. Genome Res 19(12):2317–2323

Pihlstrom BL, Michalowicz BS, Johnson NW (2005) Periodontal diseases. Lancet 366 (9499):1809–1820

Preshaw PM, Alba AL, Herrera D, Jepsen S, Konstantinidis A, Makrilakis K, Taylor R (2012) Periodontitis and diabetes: a two-way relationship. Diabetologia 55(1):21–31

Presti RM, Handley S, Droit L, Ghannoum M, Jacobson M, Shiboski CH et al (2018) Alterations in the oral microbiome in HIV-infected participants after ART administration are influenced by immune status. AIDS 32(10):1279

Pucar A, Milasin J, Lekovic V, Vukadinovic M, Ristic M, Putnik S, Kenney EB (2007) Correlation between atherosclerosis and periodontal putative pathogenic bacterial infections in coronary and internal mammary arteries. J Periodontol 78(4):677–682

Pushalkar S, Ji X, Li Y, Estilo C, Yegnanarayana R, Singh B et al (2012) Comparison of oral microbiota in tumor and non-tumor tissues of patients with oral squamous cell carcinoma. BMC Microbiol 12(1):144

Ramirez JH, Parra B, Gutierrez S, Arce RM, Jaramillo A, Ariza Y, Contreras A (2014) Biomarkers of cardiovascular disease are increased in untreated chronic periodontitis: a case control study. Aust Dent J 59(1):29–36

Ramsey SD, Newton K, Blough D, McCulloch DK, Sandhu N, Reiber GE, Wagner EH (1999) Incidence, outcomes, and cost of foot ulcers in patients with diabetes. Diabetes Care 22 (3):382–387

Ray K (2017) Gut microbiota: Oral microbiome could provide clues to CRC. Nat Rev Gastroenterol Hepatol 14(12):690

Reznik DA (2005) Oral manifestations of HIV disease. Topics HIV Med Publ Int AIDS Soc U S A 13(5):143–148

Ricklin D, Hajishengallis G, Yang K, Lambris JD (2010) Complement: a key system for immune surveillance and homeostasis. Nat Immunol 11(9):785

Roberfroid MB, Bornet F, Bouley CE, Cummings JH (1995) Colonic microflora: nutrition and health. Summary and conclusions of an international Life Sciences Institute (ILSI)[Europe] workshop held in Barcelona, Spain. Nutr Rev 53(5):127–130

Roth RR, James WD (1988) Microbial ecology of the skin. Ann Rev Microbiol 42(1):441–464

Sakamoto M, Huang Y, Ohnishi M, Umeda M, Ishikawa I, Benno Y (2004) Changes in oral microbial profiles after periodontal treatment as determined by molecular analysis of 16S rRNA genes. J Med Microbiol 53(6):563–571
Sällberg M (2009) Oral viral infections of children. Periodontol 49(1):87–95
Scharschmidt TC, Fischbach MA (2013) What lives on our skin: ecology, genomics and therapeutic opportunities of the skin microbiome. Drug Discov Tod Dis Mech 10(3–4):e83–e89
Schauber J, Svanholm C, Termen S, Iffland K, Menzel T, Scheppach W et al (2003) Expression of the cathelicidin LL-37 is modulated by short chain fatty acids in colonocytes: relevance of signalling pathways. Gut 52(5):735–741
Scully C, Greenman J (2008) Halitosis (breath odor). Periodontol 48(1):66–75
Segre JA (2006) Epidermal barrier formation and recovery in skin disorders. J Clin Invest 116 (5):1150–1158
Selwitz RH, Ismail AI, Pitts NB (2007) Dental caries. Lancet 369(9555):51–59
Seoudi N, Bergmeier LA, Drobniewski F, Paster B, Fortune F (2015) The oral mucosal and salivary microbial community of Behcet's syndrome and recurrent aphthous stomatitis. J Oral Microbiol 7(1):27150
Serour F, Samra Z, Kushel Z, Gorenstein A, Dan M (1997) Comparative periurethral bacteriology of uncircumcised and circumcised males. Sex Transm Infect 73(4):288–290
Socransky SS, Haffajee AD, Cugini MA, Smith C, Kent RL Jr (1998) Microbial complexes in subgingival plaque. J Clin Periodontol 25(2):134–144
Spach DH, Stapleton AE, Stamm WE (1992) Lack of circumcision increases the risk of urinary tract infection in young men. JAMA 267(5):679–681
Stamatas GN, Nikolovski J, Luedtke MA, Kollias N, Wiegand BC (2010) Infant skin microstructure assessed in vivo differs from adult skin in organization and at the cellular level. Pediatr Dermatol 27(2):125–131
Turnbaugh PJ, Ley RE, Hamady M, Fraser-Liggett CM, Knight R, Gordon JI (2007) The human microbiome project. Nature 449(7164):804–810
Turnbaugh PJ, Ley RE, Mahowald MA, Magrini V, Mardis ER, Gordon JI (2006) An obesity-associated gut microbiome with increased capacity for energy harvest. Nature 444(7122):1027
Ursell LK, Haiser HJ, Van Treuren W, Garg N, Reddivari L, Vanamala J et al (2014) The intestinal metabolome: an intersection between microbiota and host. Gastroenterology 146(6):1470–1476
Valensi P, Girod I, Baron F, Moreau-Defarges T, Guillon P (2005) Quality of life and clinical correlates in patients with diabetic foot ulcers. Diabetes Metab 31(3):263–271
Vanhatalo A, Blackwell JR, L'Heureux JE, Williams DW, Smith A, van der Giezen M et al (2018) Nitrate-responsive oral microbiome modulates nitric oxide homeostasis and blood pressure in humans. Free Radic Biol Med 124:21–30
Wade WG (2013) The oral microbiome in health and disease. Pharmacol Res 69(1):137–143
Wang J, Gao Y, Zhao F (2016) Phage–bacteria interaction network in human oral microbiome. Environ Microbiol 18(7):2143–2158
Wang L, Clavaud C, Bar-Hen A, Cui M, Gao J, Liu Y et al (2015) Characterization of the major bacterial–fungal populations colonizing dandruff scalps in Shanghai, China, shows microbial disequilibrium. Exp Dermatol 24(5):398–400
Wang X, Du L, You J, King JB, Cichewicz RH (2012) Fungal biofilm inhibitors from a human oral microbiome-derived bacterium. Org Biomol Chem 10(10):2044–2050
Webb AJ, Patel N, Loukogeorgakis S, Okorie M, Aboud Z, Misra S, Rashid R, Miall P, Deanfield J, Benjamin N, MacAllister R, Hobbs AJ, Ahluwalia A (2008) Acute blood pressure lowering, vasoprotective, and antiplatelet properties of dietary nitrate via bioconversion to nitrite. Hypertension 51:784–790
Welch JLM, Rossetti BJ, Rieken CW, Dewhirst FE, Borisy GG (2016) Biogeography of a human oral microbiome at the micron scale. Proc Natl Acad Sci 113(6):E791–E800
Wira CR, Fahey JV, Sentman CL, Pioli PA, Shen L (2005) Innate and adaptive immunity in female genital tract: cellular responses and interactions. Immunol Rev 206(1):306–335

Wiswell TE, Roscelli JD (1986) Corroborative evidence for the decreased incidence of urinary tract infections in circumcised male infants. Pediatrics 78(1):96–99

Woo SB, Challacombe SJ (2007) Management of recurrent oral herpes simplex infections. Oral Surg Oral Med Oral Pathol Oral Radiol Endodontol 103:S12–Se1

Zarco MF, Vess TJ, Ginsburg GS (2012) The oral microbiome in health and disease and the potential impact on personalized dental medicine. Oral Dis 18(2):109–120

Zaura E, Nicu EA, Krom BP, Keijser BJ (2014) Acquiring and maintaining a normal oral microbiome: current perspective. Front Cell Infect Microbiol 4:85

Zevin AS, Xie IY, Birse K, Arnold K, Romas L, Westmacott G et al (2016) Microbiome composition and function drives wound-healing impairment in the female genital tract. PLoS Pathog 12 (9):e1005889

Zhao H, Chu M, Huang Z, Yang X, Ran S, Hu B et al (2017) Variations in oral microbiota associated with oral cancer. Sci Rep 7:11773

A Systematic Review on Crosstalk Between Microbiome and Immune System

3

Y. Aparna, K. Anuradha, S. Anju, and J. Sarada

Abstract

The microbiome includes a diverse group of microorganisms that inhabit a particular niche. The microbes that inhabit humans and exhibit beneficial or symbiotic association are termed as normal flora, most of which secrete metabolites like vitamins and other growth factors which supplement the basic nutritional requirements for the body. The interspecies balance that is observed among the diverse groups of microbes is termed eubiosis and any deviation from this state is termed dysbiosis. It was hypothesized that there exists a crosstalk between the immune system and the microbiome of the host. These organisms play an important role in the induction and education of immune cells. This bilateral interaction plays a crucial role in maintaining human health by striking a fine balance between tolerances exhibited to normal flora and evoking responses against invading pathogens. Recent studies on the discovery of Pattern recognition receptors (PRRs), Toll-like receptors (TLRs), C-type lectin receptors, etc. expressed by a variety of cells are known to play a central role in immune surveillance mechanisms and are involved in maintaining homeostasis between the microbiome and immune responses. Many studies have proven that disruption of this balance paves the way for the development of pathogenesis in many disease conditions like allergy, autoimmune disorders, asthma, etc. The present review highlights the role of the human microbiome and its interaction with the immune system concerning its beneficial aspects and disease pathogenesis.

Keywords

Microbiome · Immune system · Eubiosis · Dysbiosis · Autoimmunity

Y. Aparna (✉) · K. Anuradha · S. Anju · J. Sarada
Department of Microbiology, Bhavans Vivekananda College of Science, Humanities, and Commerce, Sainikpuri, Telangana, India

P. V. Bramhachari (ed.), *Microbiome in Human Health and Disease*,
https://doi.org/10.1007/978-981-16-3156-6_3

3.1 Introduction

Humans are made up of both microscope and macroscopic structures. It is estimated that the total number of microbes far exceeds the human cells inside the body. They are known to express unique genes which are far different from the host genome and exist as complex communities in humans. Microbes that inhabit humans include bacteria, virus, fungi, and protozoa which are known to control host physiology.

A recent development in immunology has paved the way for better understanding the role of microorganisms in induction, education, and functioning of the immune system in humans. It is understood that commensals in the human body play a fundamental role in educating and training the immune system. Many studies prove that microbes play a remarkable role in controlling many disease conditions in the body. Different parts of the body inhabit different types of microflora depending on the type of environment they are exposed to. The highest microbial load was observed in the colon. The type of microbes varies from human to human. Despite the variation, 90% of gut microflora belong to Gram-Negative bacteria like *Bacteroides* and Gram-Positive Bacteria like *Firmicutes* (Sender et al. 2016).

Microbiota forms a symbiotic relationship with their host. This mutualistic relationship between the microbes and the host co-evolved thereby integrating signaling and sensing pathways to ensure its survival in a microbial predominated world. This dynamic interaction is necessary to maintain homeostasis in individuals. If this homeostasis is disturbed, it leads to dysbiosis causing diseases in humans. The microbiome composition in an individual is greatly influenced mainly by two factors, termed as genetic factors and immunological factors. They are also influenced by environmental diet and hygiene. All these factors together shape the microbiome in an individual (Mezouar et al. 2018). Recent advances in identifying and characterizing the microbial flora reveal the importance of microbes in maintaining a healthy steady state in humans.

The term microbiome includes not only bacteria but also fungal communities. Both bacteria and fungi form essential parts of the microbiome. It is estimated that nearly 400 species are found in humans (Halwachs et al. 2017). A shift in these bacterial and fungal compositions results in disease conditions like inflammatory bowel disease, cystic fibrosis in humans (Kim et al. 2015).

It was believed earlier that host-microbe interaction has evolved for metabolic and nutritional requirements which led to a mutualistic interaction between microorganisms and humans. It was observed now that there is an interaction between the microbiome and the immune system. The natural interface between microorganisms and the immune system is the gastrointestinal tract and the mucosal cell surfaces. The immune system can differentiate between the invading pathogens and the commensals residing in the gut. This led to the evolution of the immune system. The commensals protect the body from invading pathogens by consuming the nutrients that are required for the growth of pathogens. Also, the commensals secrete secondary metabolites which inhibit the growth of invading pathogens. This competitive exclusion plays an important role in the immunomodulatory mechanism (Corr et al. 2007). Many studies suggest the role of commensals in shaping the immune system. It was identified that they play a major role in the maturation of Mucus-Associated Lymphoid Tissue (MALT).

On the other hand, the immune system controls the microbiome by spatial segregation which is noted especially in the intestine. This type of stratification is essential to protect the beneficial microbiome in the body. For example in the colon, a load of microbes is more. A thick impermeable mucus layer made of O-glycosylated MUC2 mucin which is rich in lectin-like proteins such as ZG 16 and beta-defensins keeps away the microbiome (Bergstorm et al. 2016).

3.2 Role of Microbiota in Hematopoiesis

Though there is a clear anatomical separation between the microbiota and the human immune system, there is very clear evidence which marks the dispersion of detectable commensal metabolites in the tissues following the colonization by microbes. These metabolites reach circulation and show a profound impact in tuning the host immune system. Few experimental pieces of evidence state that commensals in the bloodstream help in bringing the steady-state hematopoiesis (Maslowski et al. 2009; Shi et al. 2011). Recognition of commensal-derived products by TLRs is important to maintain bone marrow myeloid cells. TLR components like MyD88 and TICAM1 specifically are known to mediate steady-state granulopoiesis, though it is not very clear which microbial product is responsible for this activation. Such results indicate the coevolution of the host and the normal flora, where the host is dependent on the microbial-derived signals to maintain vigilance on invading pathogens (Balmer et al. 2014).

The process of hematopoiesis and education of hematopoietic stem cells are influenced by the products of bacterial metabolism in mammals. The gut microbiome helps in breaking down of dietary components like fiber which are indigestible. The net result of this digestion is the production of short-chain fatty acids (SCFA). These products are produced uniquely by bacterial fermentation in the intestine (Cummings et al. 1987). These SCFAs serve as energy sources for enterocytes. They not only can activate G-protein-coupled receptors which are expressed on hematopoietic cells and epithelial cells but also inhibit Histone deacetylase (HDAC) in the gut leading to altering the gene expression of local macrophages (Macia et al. 2015; Maslowski et al. 2009; Chang et al. 2014; Singh et al. 2014). Research on SCFA which is produced by the metabolites derived from commensals led to alterations in the process of hematopoiesis which are identified by increased production of macrophages and dendritic cell precursors (Trompette et al. 2014). Also, neutrophil aging is known to be promoted by commensal microbiota by tonic sensing of TLR ligands (Zhang et al. 2015).

Several findings state that Gut microbiota is involved in promoting the process of hematopoiesis and controlling central immunity. The key findings of this study are gnotobiotic and oral antibiotic-treated mice were susceptible to *Listeria monocytogenes* due to defects in myelopoiesis. But when they are re-colonized by complex microbiota, it restores the defects in myelopoiesis thereby resisting the infection with *Listeria monocytogenes.* These findings are promising as they demonstrate the evolutionary connection between the microbes and the host. They are known to promote and maintain both embryonic-derived myeloid cells and hematopoietic stem cells thereby striking a perfect balance under steady-state conditions. The absence of commensals also reduced the population of neutrophils, monocytes, and macrophages in germ-free animals. This study states that regulation of hematopoiesis and shaping of immunity are largely mediated by gut microflora (Khosravi et al. 2014).

Commensal bacteria are reported to influence T-helper type 2 (TH2) cytokine-dependent inflammation leading to allergic disorders. Alteration of commensal bacteria by oral antibiotic treatment in mice resulted in elevated serum IgE which in turn exaggerated the basophil production resulting in hyperimmunoglobulinemia E syndrome in mice. This clearly states commensal bacteria-derived signals play a vital role in regulating basophil hematopoiesis (Hill et al. 2012).

Nevertheless, it was clearly understood that microbial signals that are transmitted to the neonate during gestation, through the placenta, breast milk, and during the early developmental stages shape the immune repertoire in the bone marrow. Components of microbiota like SCFAs, MAMPs, LPS, Peptidoglycans, etc. reach circulation, access the distal sites like bone marrow, and regulate the proliferation and differentiation of HSCs. These responses protect the neonate from invading pathogens by reducing the susceptibility and modulating the inflammatory responses (McCoy and Thomson 2018).

Dysbiosis, a term that is generally referred to as imbalance of gut microbiome is associated with suppression of hematopoiesis. This is observed in conditions like inflammatory bowel disease (IBD), obesity, anemia, nutritional disorders, and altered hematological abnormalities both in humans and in mice. A model was proposed on how microbial products regulate the cascade of reactions and regulate hematopoiesis. Briefly, MAMP's and other microbial products activate the TLR pathway while Diaminopimelic acid (DAP) activates the NOD1 pathway in bone marrow stromal cells which are MyD88-dependent. These two pathways have a common signaling molecule, i.e., Tumor necrosis factor receptor-associated factor 3 (TRAF3) which in turn signals IRF 3. This signaling cascade induces type I interferon which activates STAT1, which is known to induce the process of hematopoiesis (Yan et al. 2018) (Fig. 3.1).

3.3 Microbiome and Innate Immune Responses

The immunity that is conferred by an individual right from birth is termed innate immunity. The understanding of host-microbial interactions has revolutionized the field of immunology. The past two decades have brought two Paradigm shifts that enable us to understand the role of microbes in conferring immunity. The major discoveries that led a strong impact in this field include the discovery of Pattern Recognition Receptors (PRR's) present in the host which are involved in identifying the conserved molecular structures in microorganisms. Some of the main families in

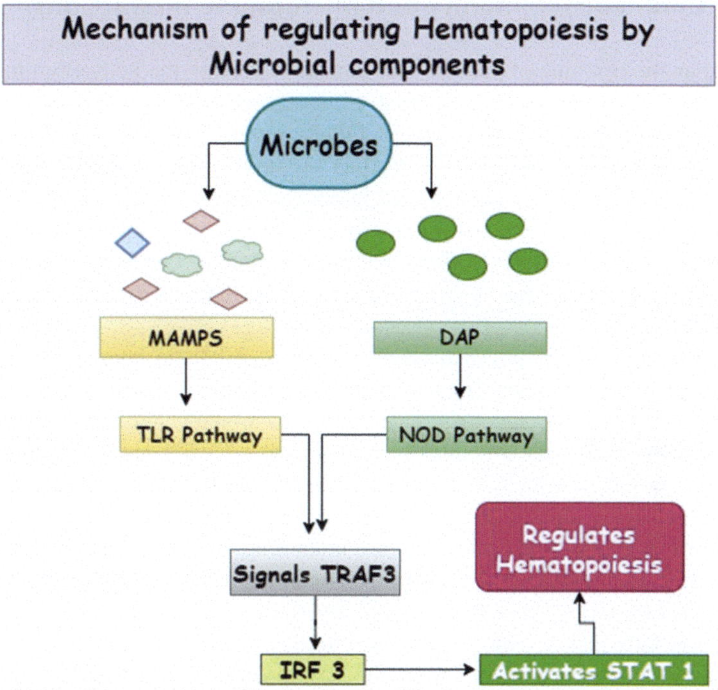

Fig. 3.1 Mechanism describing the activation of cascade of events for regulating the process of hematopoiesis induced by microbial components

PRR's include Toll-like receptors (TLR's), Nucleotides-binding oligomerization (NOD)-like receptors, C-type lectin receptors, RIG-I-like receptors, AIM2-like receptors, OAS-like receptors. These receptors act as continuous surveillance systems for invading microorganisms. Another breakthrough in the past two decades is the characterization of a culture-independent microbiome.

The different innate barriers include Anatomical barriers like skin and mucous membrane that line the gastrointestinal tract, respiratory tract, genitourinary tract, physiological barriers, cellular barriers, inflammatory barriers, etc. The diversity of microorganisms far exceeds the total number of cells in the body. The immune system plays a crucial role in shaping the microbial communities that are tolerated by the host. This bilateral interaction plays a pivotal role in maintaining human health. Innate immunity and microbiota are interdependent. Let us look at the processes in detail.

3.4 Cutaneous Immunity and Microbiome Interactions

Skin acts as the first line of defense mechanism protecting the body from invading foreign pathogens and toxic substances. Skin is colonized by a diverse group of microorganisms, which includes bacteria, fungi, viruses, mites, etc. These organisms live as symbionts on the skin protecting the body from harmful pathogenic microorganisms. They are also involved in educating the T-cells thereby striking a delicate balance between the host immunity and the microorganisms. Disruption of this balance leads to skin infections and skin disorders. Cutaneous immunity is brought both by innate and adaptive immune responses which are modulated by the microbiota.

Skin acts as a strong physical barrier because of its top layer epidermis, the Stratum corneum. This layer consists of enucleated cells known as Keratinocytes which are also known as squames. These squames are embedded in the lipid layer forming bricks and mortar structures in the epidermis. Cutaneous invaginations and appendages in the skin include sweat glands, sebaceous glands, and hair follicles, which attract many microorganisms. The characteristic body odor is due to the presence of different types of microbes inhabiting the skin. Sebaceous glands support the growth of facultative anaerobes like *Propionibacterium acne*, which is a commensal bacteria. These bacteria hydrolyse triglycerides present in the sebum and release fatty acids which contribute to the acidic pH on the skin. Many pathogens like *Staphylococcus aureus* and *Streptococcus pyogenes* are inhibited because of this acidic pH. Host factors like age, location, and gender significantly contribute to the variation of microbial flora on the skin. Environmental factors like occupation, usage of antibiotics, and clothing choice also significantly influence the type of skin microbiota. The use of cosmetics, soaps, and moisturizers are the most potential factors that influence the diversity of microbes in an individual. Molecular analysis and genomic approaches revealed a much greater diversity of organisms. Four predominant phyla on most of the human skins are identified as *Actinobacteria, Firmicutes, Bacteroidetes,* and *Proteobacteria.* They are also found in the inner mucosal surfaces underlying the skin. Metagenomic analysis revealed *Staphylococcus* and *Corynebacterium spps* to inhabit mostly in the moist areas like umbilicus, axillary vault, side of the groin, sole inner elbow, etc. Temporal variation of the skin microbiome depends on the sampled site (Grice and Segre 2011).

Skin not only acts as a physical barrier but it also acts as an immunological barrier. The immune responses generated during infectious conditions significantly modulate the commensal microbiota colonizing the skin. Pattern Recognition Receptors (PRR's) such as Toll-Like Receptors (TLR's), Mannose receptors, and NOD-like receptors are continuously involved in recognizing Pathogen-Associated Molecular Patterns (PAMP's) which include LPS layer, nucleic acids, mannans, a peptidoglycan layer, teichoic acids, etc. This identification results in the activation of keratinocytes which initiates the innate immune response resulting in the secretion of cytokines, chemokines, and antimicrobial peptides. They also assist in lysing bacteria, fungi, and enveloped viruses. Recent shreds of evidence state that Langerhans cells, a set of dendritic cells found in the skin promote tolerance to self-antigens and

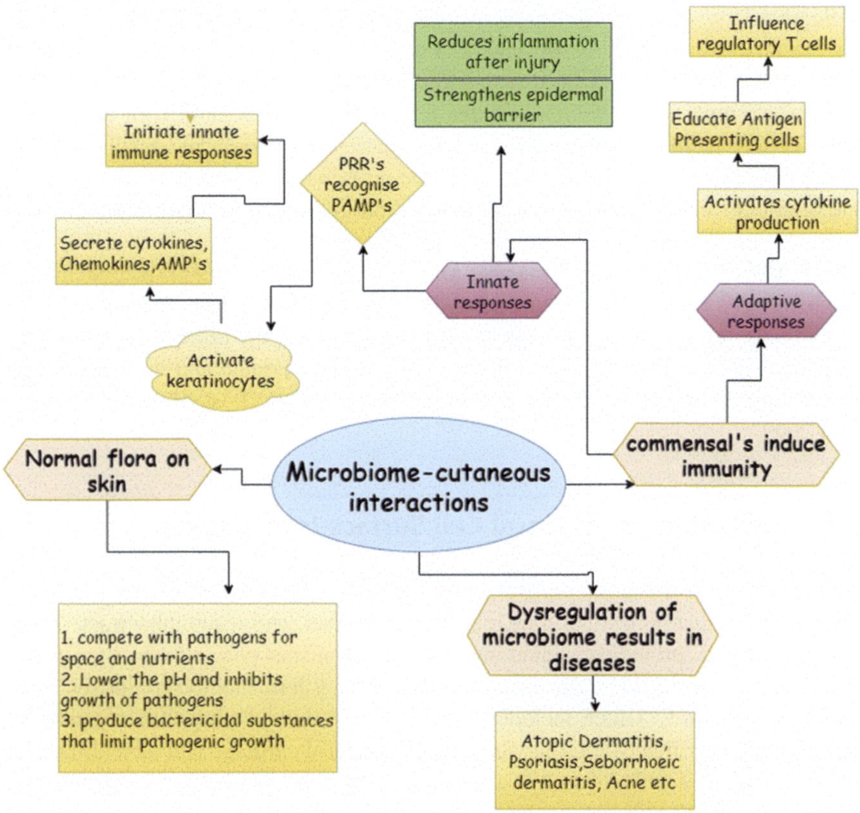

Fig. 3.2 Microbiome interactions with skin and their role in cutaneous immunity

commensal bacteria on the skin by inducing regulatory T cells in a steady-state (Seneschal et al. 2012; Romani et al. 2012). Though the mechanism is not clear they are known to be involved in the induction of immune tolerance which is achieved by the combined recognition of PAMP's by PRR's (Strober 2004; Fukao and Koyasu 2003). Many reports specify the significant role played by microbes in inducing immune responses, for example, *Staphylococcus epidermidis* is known to modulate the innate immune response by producing phenol soluble modulins (PSM's) that selectively inhibit skin pathogens like *Staphylococcus aureus* and Group A *Streptococcus.* They are also known to assist the host AMP's by enhancing lytic activity. Recent studies also reported that commensal-induced TLR signaling is essential to repair the damaged cells postinfection. Lipoteichoic acid-mediated crosstalk between TLR2 and TLR3 inhibits skin inflammation (Fig. 3.2).

On the other hand, dysregulation of skin immune response leads to several skin disorders like Atopic Dermatitis (AD), Psoriasis, and other disorders. Skin disorders are known to be associated with microorganisms inhabiting the skin like Seborrhoeic dermatitis, a hyperproliferative disorder that affects the scalp. The causative agent is

Malassezia spp, cultured from infective sites. They are also found on healthy skin. The factors responsible for turning it into a pathogen depend on age, genetic, environmental factors and need to be studied in more detail (Gupta et al. 2004). Another example is *Propionibacterium acne* which is associated with acne especially in teenagers. This organism secretes lipases, proteases, and hyaluronidases which injure the tissue lining. It damages the pilosebaceous unit, activates classical and alternate complement pathways, induces proinflammatory cytokines, and also activates many neutrophil chemotactic factors (Webster and Leyden 1980; Kim 2005). Atopic Dermatitis (AD) is also known to be associated with pathogens like *Staphylococcus aureus*. Sometimes the commensals like *Staphylococcus epidermidis* can turn into opportunistic pathogens in immunocompromised individuals. Most frequent cases related to this are nosocomial and increased levels of antibiotic resistance which are confined specifically to the host (Uckay et al. 2009; Otto 2009).

3.5 Microbiome–Mucosal Cell Surface Interactions

Mucous linings form another anatomical barrier apart from the skin. They are found mainly in the gastrointestinal tract, genitourinary tract, and respiratory tract. These linings act as a protective shield from microorganisms entering the host tissue (McGuckin et al. 2011). The intestinal mucosal system mainly comprises three lymphoid structures, which include Peyer's patches, lamina propria, and epithelial cells. Respiratory mucosal linings include submucosal cartilage and adventitia. The gut microbiome is influenced by various factors which include diet, environmental factors, exogenous substrates which are the key regulatory factors. Changes in the diet like consumption of high fat and high sugar diet changed the composition and diversity of normal flora in the intestine.

Mucosal epithelial cells produce peptides that have an antimicrobial property that limits the growth of commensal bacteria by exerting their enzymatic action on cell walls (Hooper and Macpherson 2010). An example of the mucosal antimicrobial peptide is RegIIIγ (lectin) which is produced soon after birth. This lectin protein forms a "demilitarized zone" forming a physical barrier between microbiota and host intestine. The production of this protein is controlled by the MyD88 pathway which is highly regulated (Cash et al. 2006; Vaishnava et al. 2011).

In addition to these inflammasomes, a special type of cells perform pleiotropic immune functions and activate inflammatory caspases. NLRs (NOD-like receptors) assemble to form multiprotein complexes and are found more in these inflammasomes. One such studied example is NLRP6 inflammasome. It regulates microbiome composition and is involved in maintaining homeostasis. The signaling of this NLRP6 is modulated by metabolites derived from microbiota and is involved in regulating epithelial IL 18 secretion and AMP expression (Levy et al. 2015).

The majority of antimicrobial peptides were produced by Paneth cells, which are secretory cells present in small intestine mucosa (Bevins and Salzman 2011). In addition to this pancreatic acini secretions are also vital in maintaining intestinal homeostasis which is demonstrated with increased mortality in mice due to reduced secretions of pancreatic acini (Ahuja et al. 2017). Bacteria like *Bacteroides fragilis* is a well-studied commensal as it produces Polysaccharide A (PSA), a single molecule involved in promoting symbiosis and educating the host immune system. PSA is identified by TLR2/TLR1 in association with Dectin-1 and regulates intestinal immunity by controlling Treg cell differentiation thereby conferring intestinal immunity and bringing homeostasis.

Recent reports state that in a steady state, commensal microbiota present in the intestine signals for the production of Interleukin-17 (Th 17) and Interferon-γ (Th 1) (Gaboriau-Routhiau et al. 2009). The constant signaling by microbes is helping in maintaining homeostasis. The categorization of intestinal bacteria also depends on secretory immunoglobulin, i.e., IgA. IgA specific for the intestinal microbiome is produced by B cells with the help of dendritic cells and T-cell interactions. This IgA is specifically produced against commensal-derived antigens. If these commensals cross the intestinal barrier, the phagocytic cells like macrophages and dendritic cells engulf them, transport them to MALT tissue, induce B cells for IgM production. These reactions prevent the adherence of commensals to epithelial cell surfaces thereby striking a balance between commensal community and host immune responses. However, it was noted that IgA lacks classical memory and hence changes with antigen encounters (Hapfelmeier et al. 2010).

Commensal containment is attributed to discrete pathways in the host in addition to nonspecific and broad modes. For example, bacteria belonging to the *Alcaligens* genus will proliferate more in the presence of IL_22 which is produced by Intestinal Lymphoid cells. These works state that microbes and host immunity might have coevolved together to promote the maintenance of selective species in an ecological niche (Qiu et al. 2013; Sonnenberg et al. 2012).

3.6 Microbiome and Adaptive Immunity

In addition to innate immune responses, recent advances in research demonstrated the role of commensals in promoting adaptive immunity. They are involved in training the immune cells and acting as adjuvants in whole. GALT (Gut-Associated Lymphoid Tissue) with Lymph Nodes act as major sites of antigen priming in adaptive responses. This is well studied in a process termed "homeostatic immunity", which is defined as a process in which the development and establishment of adaptive responses with the response to microbiota in absence of inflammation. The best-studied example of homeostatic immunity is IgA responses. This isotype is the most abundant secretory form which is produced in mammals and plays a pivotal role in shaping the initial interactions with microbes. Secretory IgA is referred to as both innate and adaptive as it is produced by both T-independent and T-dependent mechanisms (Kawamoto et al. 2014; Sutherland et al. 2016). IgA produced by the T Cell-dependent mechanism plays a significant role in shaping the microbiome. The mutualistic relationship between IgA and microbiota results in maintaining a balanced and diversified microbiome. This facilitates the expansion of Foxp3+ T cells which are involved in maintaining a regulatory loop.

Advances in sequencing techniques revealed that the microbiome contains nearly 3X10(Bevins and Salzman 2011) genes which indicate that human microbiota has coevolved with the host (Qin et al. 2010). Many bacteria are known to exhibit mutualism and aid in maintaining homeostasis. In an experiment, gnotobiotic mice were monocolonized with *Bacteroides thetaiotaomicron*. This bacteria colonizes the intestines of both humans and mice and is involved in altering host genes which shown an impact on the absorption of nutrients, angiogenesis, maturation of immune cells, etc. while monocolonization with other bacteria like *E.Coli* and *Bifidobacterium* showed different effects. These data reveal that different bacterial species colonizing the human body have varying functions and need to be investigated in detail for their role in host immunity (Hooper et al. 2001).

3.7 Microbiome and Cell-Mediated Immunity

Recent studies demonstrated the role of the microbiome in cell-mediated immune responses fostering the development of different T-cell subsets. Nearly more than half of genes of microbiome colonized are known to be associated with immune responses when experimented with GF mice. This data reveals the role of microbes in shaping immunity. Monocolonization with Segmented filamentous bacteria (*Clostridia*-related organism) produced multiple T-cell lineages specifically in the induction of T17 cells. They augmented mRNA transcripts, RegIII-γ in ileal cells. Up-regulation of RegIII-γ increased IL-17 production. The SFB monocolonized mice are resistant to infections caused by *Citrobacter rodentium* by increasing the production of cytokines like IL-22, IL-17, and IL-23 which are associated with increased Th17. This data suggests the protective role of SFB in maintaining the health of epithelial cells (Gaboriau-Routhiau et al. 2009).

Many studies have reported the role of bacteria in the induction of Regulatory T cells like CD4 + Foxp3+ and CD4+ IL10+ in the intestine. In an experiment, 46 different strains of *Clostridium* are reconstituted in GF mice, which induced production of CD4+ Foxp3+ cells through TGF-β activator production only in the colon and cecum where the introduced clostridium strains inhabit more. They also induced the production of IL-10 under inflammatory conditions. When tested with *Lactobacillus* and SFB, the levels of Treg production were minimal which indicate the specific responses to clostridia alone. Interestingly they are also resistant to DSS and oxazolone-induced colitis indicating the significant role of clostridia in maintaining mucosal homeostasis (Atarashi et al. 2011).

Also, Effector T-cell production and Regulatory T-cell production were dynamically maintained by commensal bacteria in the gut by TLR9-dependent pathway. The process is known to be mediated by production by SCFAs, colonization of segmented Filamentous bacteria, and other related organisms through signal recognition by TLR5 and TLR9 (Hall et al. 2008).

3.8 Microbiome and Humoral Immunity

Mucosal IgA binds to epithelial serves and confers protection to the host against any invading bacteria. It also regulates the composition of the microbiota, controls the expression of genes by microbes inhabiting the intestine. For instance, the commensal bacterium *Bacteroides thetaiotaomicron* will not trigger the process of inflammation in the gut. IgA after affinity maturation and somatic hypermutation selects for specific components in microbes which results in increasing the diversity of the microbiome and bringing mutualism between commensals and host. IgA-producing plasma cells are generated by T-dependent and T-independent mechanisms along with help of epithelial cells, macrophages, dendritic cells, and innate lymphoid cells (ILCs). Levels of IgA-producing plasma cells from Peyer's patches and lamina propria showed significant variation in production when comparative studies are done with normal mice and germ-free mice which is very less in later cases. In an interesting study bacteria, *Sutterella* species inversely correlated with IgA production in feces in contrast to the above statement. In Rheumatoid arthritis, an autoimmune disorder, these bacteria degrade IgA and J chain peptide in IgA that is required for its stability in the lumen. This study states that the anatomical location of bacteria also determines the stability of IgA responses.

Concerning T Cell-dependent responses, the bacteria SFB and *Mucispirillum* in intestinal epithelium elicits T Cell-dependent IgA responses. SFB induces Th17 and follicular T-helper cells resulting in high-affinity IgA responses (Fig. 3.3). Mice lacking TCR chains β and δ also produce IgA specific to commensal bacteria indicating T cell-independent mechanisms (Fig. 3.4). IgA-producing B Cell clones persist for longer periods. The repertoire of IgA changes dynamically with the increasing diversity of commensal bacteria. The composition of microbiota changes concerning various factors like diet, exercise, environmental conditions, etc. (Honda and Littman 2016).

3.9 Microbiome and Immune System Dysbiosis

In general, genetically susceptible individuals' dysbiosis between the microbiome and immune system leads to the development of different types of immune-mediated diseases. Two research studies state that dysregulation of microbes and immune system results in many multifactorial diseases like neurodegenerative diseases, some of the well-studied diseases are highlighted in the study (Table 3.1).

Inflammatory bowel disease (IBD) is a chronic inflammatory disorder. Many studies have shown evidence of the role of the gut microbiome in the pathogenesis of inflammatory bowel disease. This may be due to a reduction in the number of various bacteria taxa like *Firmicutes, Lactobacillus, Clostridia,* etc., and increased diversity of *Enterobacteriaceae* members. The main reasons identified were alterations in the microbes-associated metabolites, breakdown of the intestinal mucosal barrier, tissue injury, mutations in NOD2 gene, autophagy-related ATG16L1, suppression of pro-inflammatory bacteria like *Bacteriodes vulgatus*, enhanced colonization of

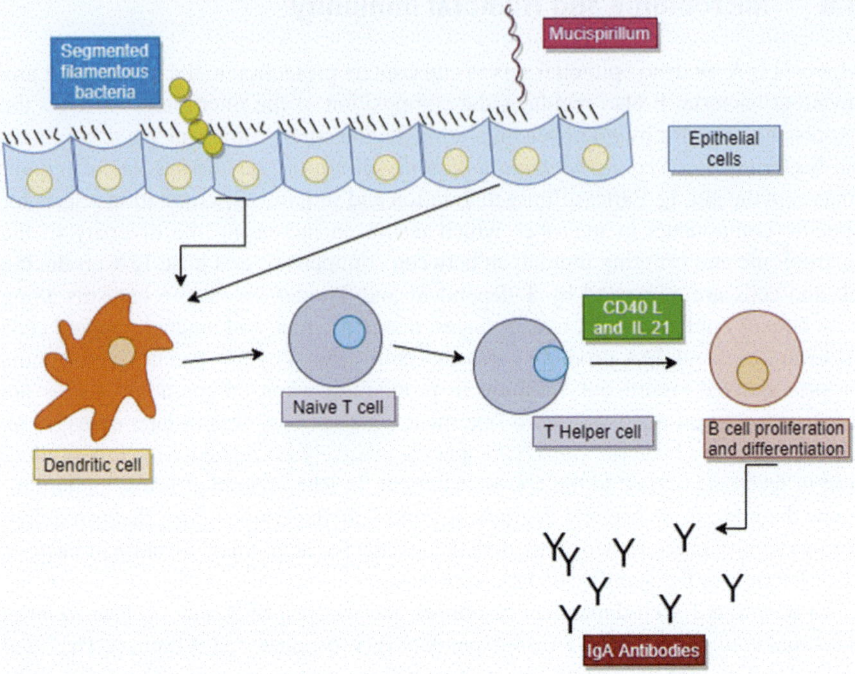

Fig. 3.3 T Cell-dependent pathway

bacteria like *Akkermansia muciniphila,* etc. However, the detailed mechanism was yet not clear (Zheng et al. 2020).

Rheumatoid arthritis (RA) which involves joints, Synovial inflammation, bone cartilage destruction is also known to be implicated in the microbiome along with environmental and genetic factors. Different bacteria like *Prevotella copri, Collinsella, Eggerthella, faecalibacterium, Lactobacillus salivarius* are proved to be associated with Rheumatoid arthritis condition. Microbe-derived metabolites like short-chain fatty acid derivatives are observed to be interacting with immune pathways which are implicated in RA condition. Further research is required to understand the details of microbiome alterations in RA condition (Zheng et al. 2020).

Another disease identified with microbiome dysbiosis is cardiometabolic disease. This is characterized by chronic low-grade inflammation, which is found to be the root cause of many metabolic disorders like diabetes mellitus, obesity, nonalcoholic fatty liver disease, atherosclerosis, etc. Recent pieces of evidence stated that microbiome-derived metabolites are fueling the process of inflammation by crossing the gut barrier. It was noted that TLRs present in the liver identify bacterial ligands initiating the inflammatory cascades. Activation of these cascades contributes to NAFLD and nonalcoholic steatohepatitis (NASH). In the case of obesity, tryptophan metabolites produced by microbes are known to modulate adipose tissue

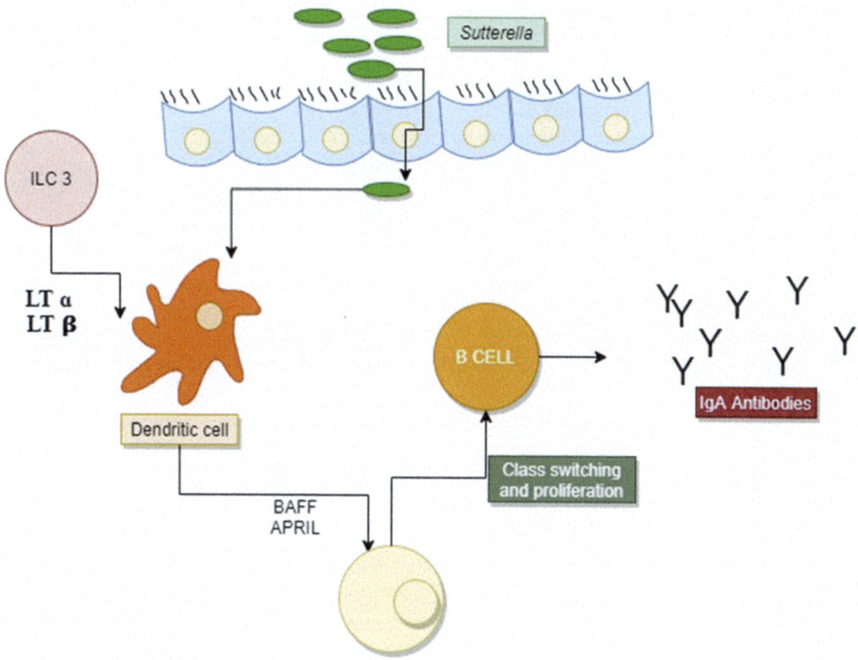

Fig. 3.4 T cell-independent pathway

inflammation which is mediated by the miR-181 family of microRNAs. In atherosclerosis heart disease metabolite TMAO produced by gut microbiota up-regulates macrophage scavenger receptors CD36 and SR-A1 which results in the accumulation of cholesterol (Virtue et al. 2019; Truax et al. 2018; Koeth et al. 2019; Wang et al. 2011).

A direct link is also found between gut microbiota and cancer immune surveillance mechanisms. Presence of *Fusobacterium nucleatum* in colorectal cancers is involved in the inhibition of NK cells which has tumor suppressor function. Another example is the Pancreatic adenocarcinoma (PDAC) study in which *Gammaproteobacteria* and intratumor microbiota promote carcinogenesis in both humans and mice via TLRs and T-cell energy. In contrast, presence of more commensals like *Clostridium, Bifidobacterium longum, Collinsella aerofaciens, and Enterococcus faecium* favor T cell-mediated responses to anti PD 1 therapy in metastasized melanoma (Zheng et al. 2020).

3.10 Concluding Remarks

The present review is focused on the current knowledge of crosstalk between the microbiome and immune reactions with a special focus on eubiosis and dysbiosis. As discussed in the paper, majority of immune responses are controlled either

Table 3.1 Microorganisms associated with some major chronic diseases

S. No.	Disease	Microorganisms associated with disease	References
1	Inflammatory bowel disease	Increased diversity of *Enterobacteriaceae, Akkermansia Muciniphila,* etc. and decreased number of *Firmicutes, Lactobacillus, Clostridia, Bacteroides vulgates, Bilophila wadsworthia, E.Coli, Bacteroidetes*	Zheng et al. (2020), Devkota et al. (2012)
2	Cystic fibrosis	*Pseudomonas,Staphylococcus*	Blainey et al. (2012)
3	Rheumatoid arthritis	*Prevotella copri, Collinsella, Eggert Hella, faecalibacterium, Lactobacillus salivarius*	Scher et al. (2013)
4	Cardiovascular disease	*Candida, campylobacter, Shigella sps*	Jin et al. (2019)
5	Cancer	*Fusobacterium nucleatum, Helicobacter pylori* in colorectal cancers; *Gammaproteobacteria in Pancreatic adenocarcinoma,* Bacteroides fragilis in tumorogenesis	Atherton and Blaser (2009)
6	Type II diabetes	*E.Coli, clostridia*	Karlsson et al. (2013), Qin et al. (2010)
7	Obesity	*Firmicutes*	Turnbaugh et al. (2006)
8	Crohn's disease	Adherent invasive *E.Coli, Yersinia, Clostridium difficile*	Issa et al. (2008)

directly or indirectly by microbiome. The discovery of PAMPs, TLRs, and other novel molecular determinants helped in expanding the knowledge on interactions of microbes with immunity. Microbes play a crucial role right from the production of immune cells by hematopoiesis to a well-developed acquired immunity in humans. There is growing interest in the field as is evidenced by the findings. Though there are many research works focused on studying interactions between microbiome and immunity, the study is still in its infancy as many mechanisms remain unexplored concerning their regulation via integration of environmental, diet, microbiology, and immunological approaches. Moreover, less is known about the interactions of commensal bacteria, virus, fungi, and protozoa cooperate among themselves and influence each other in bringing homeostatic immunity. Many diseases known today are known to be associated with microorganisms, their products, and metabolites which are termed as dysbiosis. Modulation of microbial derivatives or molecules may provide a novel way for combating antibiotic resistance and also inflammation-induced diseases in humans.

Acknowledgments There is no financial support for the present review paper. The authors thank the Principal and Management of Bhavan's Vivekananda College of Science, Humanities, and Commerce for their support and encouragement.

Conflicts of Interest The authors claim that there are no conflicts of interest.

References

Ahuja M, Schwartz DM, Tandon M, Son A, Zeng M, Swaim W et al (2017) Orai1-mediated antimicrobial secretion from pancreatic acini shapes the gut microbiome and regulates gut innate immunity. Cell Metab 25(3):635–646

Atarashi K, Tanoue T, Shima T, Imaoka A, Kuwahara T, Momose Y et al (2011) Induction of colonic regulatory T cells by indigenous Clostridium species. Science 331(6015):337–341

Atherton JC, Blaser MJ (2009) Coadaptation of helicobacter pylori and humans: ancient history, modern implications. J Clin Invest 119(9):2475–2487

Balmer ML, Schürch CM, Saito Y, Geuking MB, Li H, Cuenca M, Macpherson AJ (2014) Microbiota-derived compounds drive steady-state granulopoiesis via MyD88/TICAM signaling. J Immunol 193(10):5273–5283

Bergström JH, Birchenough GM, Katona G, Schroeder BO, Schütte A, Ermund A et al (2016) Gram-positive bacteria are held at a distance in the colon mucus by the lectin-like protein ZG16. Proc Natl Acad Sci 113(48):13833–13838

Bevins CL, Salzman NH (2011) Paneth cells, antimicrobial peptides and maintenance of intestinal homeostasis. Nat Rev Microbiol 9(5):356–368

Blainey PC, Milla CE, Cornfield DN, Quake SR (2012) Quantitative analysis of the human airway microbial ecology reveals a pervasive signature for cystic fibrosis. Sci Transl Med 4 (153):153ra130

Cash HL, Whitham CV, Behrendt CL, Hooper LV (2006) Symbiotic bacteria direct expression of an intestinal bactericidal lectin. Science 313(5790):1126–1130

Chang PV, Hao L, Offermanns S, Medzhitov R (2014) The microbial metabolite butyrate regulates intestinal macrophage function via histone deacetylase inhibition. Proc Natl Acad Sci 111 (6):2247–2252

Corr SC, Li Y, Riedel CU, O'Toole PW, Hill C, Gahan CG (2007) Bacteriocin production as a mechanism for the anti-infective activity of Lactobacillus salivarius UCC118. Proc Natl Acad Sci 104(18):7617–7621

Cummings J, Pomare EW, Branch WJ, Naylor CP, Macfarlane GT (1987) Short chain fatty acids in human large intestine, portal, hepatic and venous blood. Gut 28(10):1221–1227

Devkota S, Wang Y, Musch MW, Leone V, Fehlner-Peach H, Nadimpalli A et al (2012) Dietary-fat-induced taurocholic acid promotes pathobiont expansion and colitis in Il10−/− mice. Nature 487(7405):104–108

Fukao T, Koyasu S (2003) PI3K and negative regulation of TLR signaling. Trends Immunol 24 (7):358–363

Gaboriau-Routhiau V, Rakotobe S, Lécuyer E, Mulder I, Lan A, Bridonneau C, Eberl G (2009) The key role of segmented filamentous bacteria in the coordinated maturation of gut helper T-cell responses. Immunity 31(4):677–689

Grice EA, Segre JA (2011) The skin microbiome. Nat Rev Microbiol 9(4):244–253

Gupta AK, Batra R, Bluhm R, Boekhout T, Dawson TL Jr (2004) Skin diseases associated with Malassezia species. J Am Acad Dermatol 51(5):785–798

Hall JA, Bouladoux N, Sun CM, Wohlfert EA, Blank RB, Zhu Q et al (2008) Commensal DNA limits regulatory T-cell conversion and is a natural adjuvant of intestinal immune responses. Immunity 29(4):637–649

Halwachs B, Madhusudhan N, Krause R, Nilsson RH, Moissl-Eichinger C, Högenauer C et al (2017) Critical issues in mycobiota analysis. Front Microbiol 8:180

Hapfelmeier S, Lawson MA, Slack E, Kirundi JK, Stoel M, Heikenwalder M, Geuking MB (2010) Reversible microbial colonization of germ-free mice reveals the dynamics of IgA immune responses. Science 328(5986):1705–1709

Hill DA, Siracusa MC, Abt MC, Kim BS, Kobuley D, Kubo M et al (2012) Commensal bacteria-derived signals regulate basophil hematopoiesis and allergic inflammation. Nat Med 18 (4):538–546
Honda K, Littman DR (2016) The microbiota in adaptive immune homeostasis and disease. Nature 535(7610):75–84
Hooper LV, Macpherson AJ (2010) Immune adaptations that maintain homeostasis with the intestinal microbiota. Nat Rev Immunol 10(3):159–169
Hooper LV, Wong MH, Thelin A, Hansson L, Falk PG, Gordon JI (2001) Molecular analysis of commensal host-microbial relationships in the intestine. Science 291(5505):881–884
Issa M, Ananthakrishnan AN, Binion DG (2008) Clostridium difficile and inflammatory bowel disease. Inflamm Bowel Dis 14(10):1432–1442
Jin M, Qian Z, Yin J, Xu W, Zhou X (2019) The role of intestinal microbiota in cardiovascular disease. J Cell Mol Med 23(4):2343–2350
Karlsson FH, Tremaroli V, Nookaew I, Bergström G, Behre CJ, Fagerberg B et al (2013) Gut metagenome in European women with normal, impaired and diabetic glucose control. Nature 498(7452):99–103
Kawamoto S, Maruya M, Kato LM, Suda W, Atarashi K, Doi Y et al (2014) Foxp3+ T cells regulate immunoglobulin a selection and facilitate diversification of bacterial species responsible for immune homeostasis. Immunity 41(1):152–165
Khosravi A, Yáñez A, Price JG, Chow A, Merad M, Goodridge HS, Mazmanian SK (2014) Gut microbiota promote hematopoiesis to control bacterial infection. Cell Host Microbe 15 (3):374–381
Kim J (2005) Review of the innate immune response in acne vulgaris: activation of toll-like receptor 2 in acne triggers inflammatory cytokine responses. Dermatology 211(3):193–198
Kim SH, Clark ST, Surendra A, Copeland JK, Wang PW, Ammar R et al (2015) Global analysis of the fungal microbiome in cystic fibrosis patients reveals loss of function of the transcriptional repressor Nrg1 as a mechanism of pathogen adaptation. PLoS Pathog 11(11):e1005308
Koeth RA, Lam-Galvez BR, Kirsop J, Wang Z, Levison BS, Gu X et al (2019) L-carnitine in omnivorous diets induces an atherogenic gut microbial pathway in humans. J Clin Invest 129 (1):373–387
Levy M, Thaiss CA, Zeevi D, Dohnalova L, Zilberman-Schapira G, Mahdi JA et al (2015) Microbiota-modulated metabolites shape the intestinal microenvironment by regulating NLRP6 inflammasome signaling. Cell 163(6):1428–1443
Macia L, Tan J, Vieira AT, Leach K, Stanley D, Luong S et al (2015) Metabolite-sensing receptors GPR43 and GPR109A facilitate dietary fibre-induced gut homeostasis through regulation of the inflammasome. Nat Commun 6(1):1–15
Maslowski KM, Vieira AT, Ng A, Kranich J, Sierro F, Yu D, Mackay CR (2009) Regulation of inflammatory responses by gut microbiota and chemoattractant receptor GPR43. Nature 461 (7268):1282–1286
McCoy KD, Thomson CA (2018) The impact of maternal microbes and microbial colonization in early life on hematopoiesis. J Immunol 200(8):2519–2526
McGuckin MA, Lindén SK, Sutton P, Florin TH (2011) Mucin dynamics and enteric pathogens. Nat Rev Microbiol 9(4):265–278
Mezouar S, Chantran Y, Michel J, Fabre A, Dubus JC, Leone M et al (2018) Microbiome and the immune system: From a healthy steady-state to allergy-associated disruption. Hum Microbiom J 10:11–20
Otto M (2009) Staphylococcus epidermidis—the 'accidental' pathogen. Nat Rev Microbiol 7 (8):555–567
Qin J, Li R, Raes J, Arumugam M, Burgdorf KS, Manichanh C et al (2010) A human gut microbial gene catalogue established by metagenomic sequencing. Nature 464(7285):59–65
Qiu J, Guo X, Zong-ming EC, He L, Sonnenberg GF, Artis D, Zhou L (2013) Group 3 innate lymphoid cells inhibit T-cell-mediated intestinal inflammation through aryl hydrocarbon receptor signaling and regulation of microflora. Immunity 39(2):386–399

Romani N, Brunner PM, Stingl G (2012) Changing views of the role of Langerhans cells. J Investig Dermatol 132(3):872–881
Scher JU, Sczesnak A, Longman RS, Segata N, Ubeda C, Bielski C et al (2013) Expansion of intestinal Prevotella copri correlates with enhanced susceptibility to arthritis. elife 2:e01202
Sender R, Fuchs S, Milo R (2016) Revised estimates for the number of human and bacteria cells in the body. PLoS Biol 14(8):e1002533
Seneschal J, Clark RA, Gehad A, Baecher-Allan CM, Kupper TS (2012) Human epidermal Langerhans cells maintain immune homeostasis in skin by activating skin resident regulatory T cells. Immunity 36(5):873–884
Shi C, Jia T, Mendez-Ferrer S, Hohl TM, Serbina NV, Lipuma L, Pamer EG (2011) Bone marrow mesenchymal stem and progenitor cells induce monocyte emigration in response to circulating toll-like receptor ligands. Immunity 34(4):590–601
Singh N, Gurav A, Sivaprakasam S, Brady E, Padia R, Shi H et al (2014) Activation of Gpr109a, receptor for niacin and the commensal metabolite butyrate, suppresses colonic inflammation and carcinogenesis. Immunity 40(1):128–139
Sonnenberg GF, Monticelli LA, Alenghat T, Fung TC, Hutnick NA, Kunisawa J, Tardif MR (2012) Innate lymphoid cells promote anatomical containment of lymphoid-resident commensal bacteria. Science 336(6086):1321–1325
Strober W (2004) Epithelial cells pay a toll for protection. Nat Med 10(9):898–900
Sutherland DB, Suzuki K, Fagarasan S (2016) Fostering of advanced mutualism with gut microbiota by immunoglobulin a. Immunol Rev 270(1):20–31
Trompette A, Gollwitzer ES, Yadava K, Sichelstiel AK, Sprenger N, Ngom-Bru C et al (2014) Gut microbiota metabolism of dietary fiber influences allergic airway disease and hematopoiesis. Nat Med 20(2):159–166
Truax AD, Chen L, Tam JW, Cheng N, Guo H, Koblansky AA et al (2018) The inhibitory innate immune sensor NLRP12 maintains a threshold against obesity by regulating gut microbiota homeostasis. Cell Host Microbe 24(3):364–378
Turnbaugh PJ, Ley RE, Mahowald MA, Magrini V, Mardis ER, Gordon JI (2006) An obesity-associated gut microbiome with increased capacity for energy harvest. Nature 444(7122):1027
Uckay I, Pittet D, Vaudaux P, Sax H, Lew D, Waldvogel F (2009) Foreign body infections due to Staphylococcus epidermidis. Ann Med 41(2):109–119
Vaishnava S, Yamamoto M, Severson KM, Ruhn KA, Yu X, Koren O, Hooper LV (2011) The antibacterial lectin RegIIIγ promotes the spatial segregation of microbiota and host in the intestine. Science 334(6053):255–258
Virtue AT, McCright SJ, Wright JM, Jimenez MT, Mowel WK, Kotzin JJ et al (2019) The gut microbiota regulates white adipose tissue inflammation and obesity via a family of microRNAs. Sci Transl Med 11(496):eaav1892
Wang Z, Klipfell E, Bennett BJ, Koeth R, Levison BS, DuGar B et al (2011) Gut flora metabolism of phosphatidylcholine promotes cardiovascular disease. Nature 472(7341):57–63
Webster GF, Leyden JJ (1980) Characterization of serum-independent polymorphonuclear leukocyte chemotactic factors produced by Propionibacterium acnes. Inflammation 4(3):261–269
Yan H, Baldridge MT, King KY (2018) Hematopoiesis and the bacterial microbiome. Blood 132 (6):559–564
Zhang D, Chen G, Manwani D, Mortha A, Xu C, Faith JJ et al (2015) Neutrophil ageing is regulated by the microbiome. Nature 525(7570):528–532
Zheng D, Liwinski T, Elinav E (2020) Interaction between microbiota and immunity in health and disease. Cell Res 30:492–506

Diversity and Dynamics of the Gut Microbiome and Immune Cells

4

Prerna Pathak

Abstract

Human microbiome consists of multiple species out of which most of them reside in gut. Gut microbiota is most complex and dynamic in terms of species diversity and therefore regulates the host homeostasis. The intricate relation between the gut microbiota and host is crucial for host functioning. Dysbiosis in microbiota affects myriads of processes which result in multiple diseases such as IBD, type 1 diabetes, and rheumatoid arthritis, etc. This chapter highlights the role of gut microbiota in innate and adaptive immune system development and further explains how alteration in microbiota leads to dysbiosis which makes host susceptible to several diseases.

Keywords

Gut microbiota · Autoimmunity · Innate immunity · Adaptive immunity · T1DM · Rheumatoid arthritis · Systemic lupus erythematosus

4.1 Introduction to Gut Microbiota

Microorganisms are the part of normal human microbiota, over the period of time a symbiotic relationship leads to their colonization in the nasal tract, oral cavity, skin, respiratory, and genitourinary tract (Opazo et al. 2018). Apart from microorganisms, human microbiota comprises protozoans, fungi, archaea as well as viruses (Neish 2009; Sekirov et al. 2010; Sonnenburg and Bäckhed 2016), this collective colonization is termed as gut microbiota. The gastro intestinal tract (GIT) of humans is an intricate open system which harbors 10^{14} microorganisms (Seksik and Landman

P. Pathak (✉)
Jawaharlal Nehru University, New Delhi, India

P. V. Bramhachari (ed.), *Microbiome in Human Health and Disease*,
https://doi.org/10.1007/978-981-16-3156-6_4

2015). Recent advances in deep sequencing technology bring insight about the genome of gut microbiome, it encodes 3.3 billion genes; this surpasses the number of human gene by 100-fold (Dwivedi et al. 2017), therefore it is also termed as "human second genome" (Xu et al. 2019). Human health is governed by the microbes in the gut microbiota which have both favorable as well as pathogenic effects. GIT is the largest region which gets exposed with the external habitat and comprises two-third of the total human microbiome (Virili et al. 2018).

The diversity and the dynamics of the gut microbiota are governed by several factors like age, gender, health alignments, immunity, genetics of the host, geographical changes, lifestyle, and treatments (Ardissone et al. 2014; De Martino et al. 2004). Metagenomics study reveals that the most of the species of the microbiome are missing in the same person at a same time frame; however, microbiota of healthy individuals shows abundant of some species over others (Human Microbiome Project Consortium 2012; Qin et al. 2010). In fact, there is diversity in the microbiota of the gut on the bases of the types of cells; cells of mucus layer, intestinal lumen, and epithelial cells show diverse microhabitats (Sekirov et al. 2010).

Development of human microbiome starts even before the birth as it is revealed by the study on microbiome composition of the placenta (Aagaard et al. 2014). Moreover the study performed on the first stool of the infants shows the presence of 30 genera which are normal inhabitant of amniotic fluid, oral, and vaginal cavity (Ardissone et al. 2014; Clemente et al. 2012; De Martino et al. 2004). It is believed that this is occurred due to mother to child transmission at the time of pregnancy (Lagier et al. 2012). The microbiota of the infants depends on the mode of delivery such as caesarean section born infants have microbiota similar to that of skin while vaginally born infants have vaginal microbiota (Dominguez-Bello et al. 2010). Pregnancy period and the initial few months after the delivery are crucial for the development of the microbiota which further influences immune homeostasis (Gordon Jeffrey et al. 2012).

Breast feeding is another decisive factor that links with microbiome and immunity development (Stewart et al. 2018). Human milk contains ~10^9 bacterial cells/L (Endesfelder et al. 2014), apart from its nutritional value it possesses various bioactive and immunological molecules which govern the microbiome and intestine maturation of the infants. Studies illustrate that immunological components of the human milk such as sIgA, lysozyme, complex lipids, alpha lactalbumin, and lactoferrin impart protection to the infants (Gordon Jeffrey et al. 2012). Initiation and development of microbiota during infant stage impact the health and immunity during adulthood (Ranucci et al. 2017), any perturbation in this development may lead to negative consequences (Fulde et al. 2018). The gut microbiome keeps on evolving from infants to early childhood in a phased manner (Xu et al. 2019).

The mutual relationship between host and gut microbiota plays significant beneficial role. It impacts the development of the gut by influencing proliferation of epithelial cells and host cells apoptosis. Short-chain fatty acids which are the by-product of polysaccharide fermentation mediate interactions between host cells and gut microbiota (Lazar et al. 2018). Apart from maintaining gastrointestinal homeostasis, gut microbiota play role in the development of components of immune

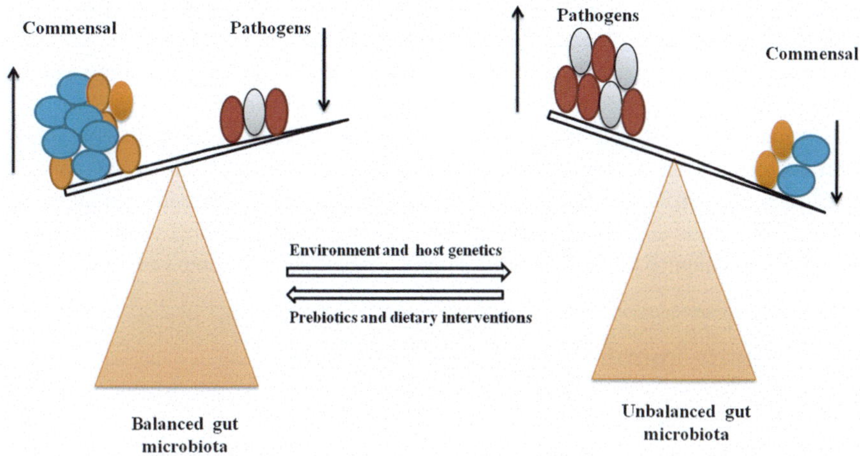

Fig. 4.1 Gut microbiota in normal and diseased conditions

system, synthesis of vitamins like B-complex, folic and biotin, detoxification of xenobiotic compounds, maintaining nutritional homeostasis (Gérard 2013). Among all the known members of gut microbiota, bacteria are most explored. They are classified into three categories, aerobic, facultative anaerobic, and obligate anaerobic bacteria out of which obligate anaerobic dominates them all (Fig. 4.1).

The interspecies balance is pivotal for the proper functioning of the body; this balance is termed as eubiosis. Any imbalance to eubiosis is termed as dysbiosis could lead to plethora of diseases which further affects multiple organs (Clemente et al. 2012). Hereafter this chapter will focus on the role of gut microbiota in the development of immune system and disease.

4.2 Interactions Between Gut Microbiota and Immune System

The symbiotic relationship between the microbiome and the human gut is beneficial for both of them as human gut acts as nutrient source as well as provides breeding habitat to microflora; in return gut micobiota helps in vitamin synthesis, gut development, and forms mucosal barrier as a defense mechanism (Berg et al. 2015). The human mucosa is the largest and the most exposed component to the external environment. Eubiosis is pivotal for maintaining host homeostasis and its defense. The relevance of gut microbiota came into light after the studies performed on germ-free (GF) mice revealed that it produces relatively reduced amount of Intraepithelial lymphocytes (IELs) (Bandeira et al. 1990), IgA-secreting plasma cells (Crabbé et al. 1968), Tregs cells (Ostman et al. 2006), and Angiogenin-4 (Ang4) (Hooper et al. 2003).

Moreover in GF mice, the germinal center of Peyer's patches is smaller compare to conventional mice (McDermott and Huffnagle 2014). Another study illustrates that concentration of IgA in the feces enhanced reasonably after prebiotics treatment; however, expression of Peyer's patches and pro-inflammatory factor of mesenteric lymph nodes reduced considerably (Carasi et al. 2015). All together these results pinpoint the role of gut microbiota in immunity development as well as eubiosis.

Immune system works basically by recognizing and eliminating the pathogen from our system. The intestinal immune homeostasis is crucial for both host as well as trillions of microbes within the system. This immune homeostasis is a combined effect of innate and adaptive immunity. Several innate and adaptive responses which are crucial in shaping the intestinal microbiota are explained in next section.

4.3 Innate Immunity and Gut Microbiota

Gut-Associated Lymphoid Tissues (GALTs) are present throughout the intestine; they are part of mucosa-associated lymphoid tissues (MALTs) (Brandtzaeg et al. 2008). The innate immune cells of the GALTs are involved in presenting antigen to activate adaptive immune response after recognizing pathogens in a nonspecific manner (Jiao et al. 2020). GALTs have dual function of immune tolerance and immune homeostasis. GALTs include following components; Peyer's patches, isolated lymphoid follicles (ILFs), crypt patches, M cells, appendix, and mesenteric lymph nodes (mLNs) (Brandtzaeg et al. 2008; Mowat 2003). M cells are involved in delivering intestinal antigen to GALTs (Mabbott et al. 2013). Several studies pin point the role of gut microbiota in shaping GALTs. Lymphoid tissue inducer (LTi) cells are involved in formation of secondary lymphoid organs of gut such as Peyer's patches, mLNs, and ILFs (Adachi et al. 1997; Mebius et al. 1997).

Pattern-recognition receptors (PRRs) are crucial for innate immune response; they sense pathogen through specific structures. The pathogen-associated molecular patterns (PAMPs) of the intestinal microorganisms are recognized by PRRs and lead to the development of ILFs. Mouse deficient in PRRs shows defects in ILFs development. PRRs are of several types depending on their location, ligand specificity, and functions, several PRR-related molecules are involved in mechanism like toll-like receptors 2 (TLR2) (Round et al. 2011), myeloid differentiation primary response 88 protein (MyD88) (Medzhitov et al. 1998; Wesche et al. 1997), nucleotide-binding oligomerization domain 1/2 (NOD 1/2) (Bouskra et al. 2008; Clarke et al. 2010; Petnicki-Ocwieja et al. 2009), and TIR domain-containing adaptor protein inducing interferon-β (TRIF) (Bouskra et al. 2008). PRR-PAMP recognition plays pivotal role during host defense response as well as structural development of GALTs.

Toll-like receptors (TLRs) are another crucial member of innate immune system, they recognize specific region in pathogens and start immune response (Rakoff-Nahoum et al. 2004). They are also involved in balancing microbiota composition (Larsson et al. 2012; Wen et al. 2008). Several studies report that TLR5-deficient mice show compositional changes in microbiota (Vijay-Kumar et al. 2010). These changes may further lead to the development of spontaneous colitis, metabolic syndrome, and obesity; this highlights the role of TLR5 in maintaining gut microbiota composition (Carvalho et al. 2012; Chassaing et al. 2014; Chassaing et al. 2014; Vijay-Kumar et al. 2010).

The nucleotide-binding oligomerization domain-like receptors or Nod-like receptors (NLRs) are intracellular stress sensors of pathogen-associated molecular patterns (PAMPs) associated with cellular stress. Like TLRs, they are also involved in maintaining microbiota composition as it is shown by the study where NOD1/2-deficient mice show altered composition of microbiota (Bouskra et al. 2008; Couturier-Maillard et al. 2013; Petnicki-Ocwieja et al. 2009). Paneth cells which are present in small intestine have enhanced expression of NOD2 protein; upon getting exposed to pathogens this protein induces multiple responses which include cytokines production, autophagy initiation, generation of antimicrobial peptides, and intracellular vesicle trafficking, thus impacts the composition of the microbiota (Couturier-Maillard et al. 2013; Nigro et al. 2014; Ramanan et al. 2014). These members of NLR family elicit immune response after recognizing bacterial peptidoglycans, viruses, and parasites.

Innate immune system functions by initiating response upon sensing the metabolic state of the gut microbiota. Evidence from all the studies reveals the role of innate immune system in governing composition of microbiota (Levy et al. 2015), as mice deficient in *NOD2* (Couturier-Maillard et al. 2013; Petnicki-Ocwieja et al. 2009; Ramanan et al. 2014), *NLRP6* (Elinav et al. 2011), and *TLR5* (Vijay-Kumar et al. 2010), leads to dysbiosis. It is therefore believed that sensors of innate immune system work by promoting the growth of beneficial microorganisms as well as maintaining the stable microbiota. Additionally, there are several other molecules which are involved in maintaining composition of microbiota such as any alterations in paneth cells, which produce antimicrobial peptides (AMPs) lead to dysbiosis (Salzman et al. 2010; Salzman and Bevins 2013). Moreover altered AMP expression brings about alterations in spatial organization of microbiota; RegIIIγ deficiency causes colonization of microorganism in the inner mucus layer, which is devoid of microorganism in normal condition (Vaishnava et al. 2011).

4.4 Adaptive Immunity and Gut Microbiota.

4.4.1 T cells

T cells are the crucial member of adaptive immune system; they are of two types $CD4^+$ T cells and $CD8^+$ T cells. $CD4^+$ T cells are present in lamina propria of intestine and upon activation it differentiates into following subtypes T helper 1 (Th1), Th2, Th17, or regulatory T cell (Treg). The balanced expression of these subtypes is important determinant factor of human health. Each subtype has different functions like Th1 cells play important role during host defense against microorganism, while Th2 cells remove parasitic infections (Wu and Eric 2012). Unregulated Th responses lead to autoimmune diseases and allergic reactions. $CD8^+$ T cells are present in intraepithelial compartment of the gut. GF mice show less number of $CD8^+$ T cells with reduction in their cytotoxicity indicates the role of microbiota in monitoring $CD8^+$ T cells and its function (Helgeland et al. 2004; Imaoka et al. 1996; Kawaguchi-Miyashita et al. 1996).

Several studies revealed that mice deficient in adaptive immune system display changes in their microbiota, this suggest the role of adaptive immune system in balancing microbiota composition (Kato et al. 2014; Zhang et al. 2015). Mice lacking T cells also show altered microbiota, it is believed that T cells regulate microbiota by triggering expression of AMP; however, there is no direct evidence to this. Rather the prime mechanism by which T cells regulate microbiota is by influencing B cells to produce secretory IgA (Kato et al. 2014). Tregs and Th17 cells are known to be involved in intestinal IgA production, there are reports which suggest that Tregs assist B cells in IgA production (Tsuji et al. 2009). Another study points out the role of Th17 cells in antigen-specific IgA production upon immunization with cholera toxin (Hirota et al. 2013). All together these studies brought insight about the role of T cells in development of microbiota.

4.4.2 B Cells

Another crucial molecule of adaptive immune system which governs the composition of intestinal microbiota is immunoglobulin A (IgA). It is produced by plasma cells into the intestinal lumen where it attaches to microbes as well as microbial components. This generates a physical barrier which averts detrimental interactions with immune system (Pabst 2012). IgA maintains the eubiosis by two proposed mechanism, first by inhibiting the growth or inflammatory effects of microorganisms and secondly by preserving the diversity of healthy microbiota (Palm et al. 2015). Mice deficient in Activation-Induced Cytidine Deaminase (AID) show defect in class switching or somatic hypermutation, this defect is reversed by substituting IgA; this highlights the role of IgA in forging microbiota (Fagarasan et al. 2002; Suzuki et al. 2004), IgA known to play crucial part in forging microbiota during development, as its deficiency fails to curb proteobacteria during microbiota maturation (Mirpuri et al. 2014). IgA is also known to suppress the inflammatory response there by promoting the mutualism between host and microbiota (Peterson et al. 2007). The phenomenon of immune exclusion is mainly involved in suppressing inflammatory response by prohibiting microorganisms from approaching mucosal epithelium (Corthésy 2013). Moreover alternate mechanism by which IgA directly suppresses inflammatory responses is by coating microorganisms gene expression (Cullender et al. 2013). Studies are still at their infancy about how IgA arbitrates these responses.

4.5 Role of Gut Microbiota in Disease Development

The impact of gut microbiota on innate and adaptive immune system is already discussed above, any dysbiosis in microbiota leads to critical diseases. The modification in the eubiotic state of microbiota triggers myriad of diseases such as type I diabetes, nonalcoholic fatty liver disease, rheumatoid arthritis, obesity, cancer, etc., however how these alterations cause these diseases is still ambiguous. The symbiosis between host and microbiota is crucial for homeostasis. Together, this chapter reveals the crosstalk between host, microbiota, and environmental cues which leads to these pathophysiologies.

4.6 Type 1 Diabetes

Type 1 diabetes (T1DM) was previously called as juvenile-onset diabetes; it is a chronic disease which is linked with high mortality at premature stage (Lazar et al. 2018). It is occurred due to inability of pancreatic β cells to produce insulin due to autoimmune obliteration (Aathira and Jain 2014). Normally this disease occurs during early stage of life but there are reports which reveal that 50% of T1DM occurs in individuals older than 20 year. Several factors govern the development of this disease such as diet, genetics, and gut microbiota (Pociot and Lernmark 2016; Rewers and Ludvigsson 2016; Todd et al. 2007). Data from various studies show that the composition of gut microbiota varies between healthy individuals and individuals with T1DM. Bio-Breeding (BB) rat and nonobese diabetic (NOD) mouse bear alike attribute with that of human disease (Pearson et al. 2016). The composition of gut microbiota in Bio-Breeding diabetes-prone (BB-DP) rats is altered strikingly before and after the outbreak of T1DM (Brugman et al. 2006). Consistent with this, the composition of gut microbiota is altered between healthy and T1DM individuals (Han et al. 2018).

Large amount of research in this direction highlights the role of gut microbiota in TIDM development by regulating immune responses. The outer membrane component of gram-negative bacteria lipopolysaccharide (LPS) or endotoxin is crucial in enhancing the proinflammatory cytokines and damaging the function of pancreatic β cells (Allin et al. 2015), which further leads to diabetes (Pussinen et al. 2011). Another study showed that circulating LPS is higher in T1DM individuals compared to that of healthy individuals (Devaraj et al. 2009). Additionally it is considered that LPS is derived from gut microbiota. Therefore, it is considered as a link between gut microbiota and TIDM (Han et al. 2018). Any change in gut microbiota causes LPS and fatty acids leakage by damaging the mucosal barrier, this leads to simultaneous induction of *TLR4* which results in metabolic inflammation (Velloso et al. 2015). Studies performed in NOD mouse lacking *TLR4* show the increased rate of T1DM development (Gülden et al. 2013). Moreover, a study in NOD mice lacking MyD88 reveals that the T1DM development is not their when raised under specific-pathogen-free (SPF) condition; however, NOD mice lacking MyD88 shows increased development of T1DM under GF conditions (Wen et al. 2008). MyD88 deficiency changes the composition of gut microbiota and leads to T1DM by regulating host immune response (Wen et al. 2008).

4.7 Rheumatoid Arthritis

Rheumatoid arthritis (RA) is an autoimmune disorder which is caused by obliteration of bone and cartilage which leads to development of pain and swelling in and around the joints of body. It is very frequent disease affecting 1% of total population and is quite common among females.

RA correlates with the inflammatory responses caused by $CD4^+$ Th1 and Th17 cells and any variation in these responses leads to advancement of RA (Xu et al. 2019). Study on collagen induction arthritis (CIA) mice model shows that gut microbiota plays crucial role in impacting susceptibility to arthritis. The gut microbiota of CIA-susceptible and CIA-resistant mice was altered (Liu et al. 2016). It is also reported that the GF mice having microbiota of CIA-susceptible mice show increased initiation of RA than those having microbiota of CIA-resistant mice (Liu et al. 2016).

4.8 Inflammatory Bowel Disease (IBD)

IBD is a gastrointestinal disorder in which structure of mucosa gets altered, composition of gut microbiota changes along with some systemic deformity (Mulder et al. 2014). Depending on the symptoms and intestinal localization it is mainly divided in two forms, Crohn's disease (CD) and ulcerative colitis (UC) (Mulder et al. 2014; Wijmenga 2005). IBD is a progressive disease and its frequency increased worldwide (Chow et al. 2009; Kaplan 2015; Wang et al. 2010). Recent research points out the role of gut microbiota in development of IBD, Th17, and Treg cells harmony is crucial for intestinal homeostasis. Study on segmented filamentous bacteria (SFB) in mice reveals the inflation of Th1 and Th17 cytokines (Gaboriau-Routhiau et al. 2009; Ivanov et al. 2009; Lee and Mazmanian 2014).

Several studies suggested that IBD patients with dysbiosis show alteration in their stool microbiome as well as loss of beneficial microorganisms compared to that of healthy individuals (Moustafa et al. 2018). The cause and development of IBD is linked with dysbiosis in various reports (Abu-Shanab and Quigley 2010; Casén et al. 2015; Huttenhower et al. 2014; Marchesi et al. 2007; Tamboli et al. 2004; Wright et al. 2015; Zhang et al. 2007). In IBD patients, number of commensal bacteria such as Firmicutes and Bacteroides are comparatively less in number; however, bacteria of family *Enterobacteriaceae* are higher in number (Bien et al. 2013; Hedin et al. 2014; Li et al. 2015; Mondot et al. 2011; Nguyen 2011). Another study reports the interaction between reduced gut diversity and disease onset in individuals with CD (Gevers et al. 2014). The studies performed in CD and UC individuals reveal decrease in *Clostridium* groups of bacteria and increase in Proteobacteria (Frank et al. 2007; Macpherson et al. 2000; Sartor 2008), as well as compelling reduction of commensal bacterial species belong to genera *Bacteriodes*, *Lactobacillus*, and *Eubacterium* (Nemoto et al. 2012; Sha et al. 2013).

4.9 Celiac Disease (CD)

It is a chronic disorder of the digestive system. Gut microbiota plays pivotal role in onset of this disease, any alteration in the composition of gut microbiota leads to the CD (Festi et al. 2014). It is hypothesized that gut microbiota have pathogenic role in CD development (Collado et al. 2007). Studies show that species like *Streptococcus mutans* and *Streptococcus anginosus* were present in less number in CD patients compared to that of healthy individuals (Lazar et al. 2018). Galactoside 2-alpha-L-fucosyltransferase2 is an enzyme which is encoded by *FUT2* gene; monitors the expression of ABH blood group antigens in intestinal mucus as well as other secretions. Study conducted in Fut2-deficient mice illustrates higher susceptibility to *Candida albicans* colonization compared to that of control mice, this leads to induction of CD (Lazar et al. 2018). *Bifidobacterium* spp. of bacteria are commensal of gut and provide protection against pathogens, any changes in the microbiome due to mutation in *FUT2* gene lead to development of CD (Nagao-Kitamoto et al. 2016).

4.10 Systemic Lupus Erythematosus (SLE)

SLE as the name suggests is a systemic, chronic, and inflammatory autoimmune disease of ambiguous mechanism mainly defines by inflammation at multiple site of the body (Paglia et al. 2017). Manifestation of this disease is indicated by upsurge of *Bacteroides* phyla and reduction in *Firmicutes* (Hevia et al. 2014). The composition of microbiota is believed to be crucial, as any modification in microbiota is associated with onset of SLE. There are several reports which explained the link between dysbiosis and SLE development. Recent study shows that microbiomes of SLE patients of northeastern China have higher number of Proteobacteria and lesser number of Ruminococcaceae (Wei et al. 2019).

4.11 Cancer

The major factor which is pivotal for cancer development is chronic inflammation. Inflammation accelerates the tumor development and hastens invasion and metastasis. Inflammatory cytokines cause damage in the DNA, any changes in the methylation of DNA induce inflammation-associated cancers (Nagao-Kitamoto et al. 2016). Development of cancer is not associated with change in single entity; however, it is linked with dysbiosis of entire microbiome. During dysbiosis, there is alteration in the bacterial populations with upsurge in tumor-inducing species and decrease in commensal species (Lazar et al. 2018). During inflammation, there is increased alteration of microbiota which assists bacterial translocation into the neoplastic tissue and leads to expression of inflammatory cytokines which in turn causes tumor growth (Grivennikov et al. 2012). The microbiota of the colon induces colorectal cancer by triggering immune response of Th17 cells (Wu et al. 2009).

The balance between microbiota and host defense mechanism is crucial for the development of colorectal cancer.

4.12 Conclusion and Future Perspectives

The diversity and dynamics of human microbiome are intriguing and therefore there is increased research in this direction in the past decade. Eubiosis of the microbiome is crucial for the host functions. The role of the gut microbiota in the development of the host immune system and autoimmune disorder is already evident from the research. Host genetics and environmental cues are pivotal in shaping the gut microbiota. Any modulation in gut microbiota increases the prospect of autoimmune disease.

This chapter highlights the association between the gut microbiota with immune system and disease development. The identification and characterization of crucial microorganism and their mechanism of action will allow us to understand their contribution in disease development and progression, as well as lead the way for the development of novel strategies which can prevent disease development. Apart from this, human microbiome can also be used to identify gut-associated disease as change in gut microbiota is a hallmark in various gut-linked diseases. Considering the fact that microbiome can change upon dietary changes this could be used to customize diets which can reshape microbiota and its function in order to prevent disease. This knowledge can be used in future for accurate and efficient treatment of patients. Further recognition of unique symbiotic microorganism which prevents aggregation of disease causing bacteria and boosts host immunity will pave a way in development of medicine which can reverse the defects caused by dysbiosis.

Acknowledgments I gratefully acknowledge the University Grants Commission (UGC), Government of India for awarding Junior and Senior Research Fellowship.

Conflict of Interest Author has no conflict of interest.

References

Aagaard K et al (2014) The placenta harbors a unique microbiome. Sci Transl Med 6(237):237ra65

Aathira R, Jain V (2014) Advances in management of Type 1 diabetes mellitus. World J Diabetes 5 (5):689–696

Abu-Shanab A, Quigley EMM (2010) The role of the gut microbiota in nonalcoholic fatty liver disease. Nat Rev Gastroenterol Hepatol 7(12):691–701

Adachi S, Yoshida H, Kataoka H, Nishikawa S (1997) Three distinctive steps in Peyer's patch formation of murine embryo. Int Immunol 9(4):507–514

Allin KH, Nielsen T, Pedersen O (2015) Mechanisms in endocrinology: gut microbiota in patients with type 2 diabetes mellitus. Eur J Endocrinol 172(4):R167–R177

Ardissone AN et al (2014) Meconium microbiome analysis identifies Bacteria correlated with premature birth. PLoS One 9(3):e90784

Bandeira A et al (1990) Localization of gamma/Delta T cells to the intestinal epithelium is independent of Normal microbial colonization. J Exp Med 172(1):239–244

Benoit C, Ley RE, Gewirtz AT (2014) Intestinal epithelial cell toll-like receptor 5 regulates the intestinal microbiota to prevent low-grade inflammation and metabolic syndrome in mice. Gastroenterology 147(6):1363–1377

Berg D, Clemente JC, Colombel J-F (2015) Can inflammatory bowel disease be permanently treated with short-term interventions on the microbiome? Exp Rev Gastroenterol Hepatol 9 (6):781–795

Bien J, Palagani V, Bozko P (2013) The intestinal microbiota Dysbiosis and Clostridium Difficile infection: is there a relationship with inflammatory bowel disease? Ther Adv Gastroenterol 6 (1):53–68

Bouskra D et al (2008) Lymphoid tissue genesis induced by commensals through NOD1 regulates intestinal homeostasis. Nature 456(7221):507–510

Brandtzaeg P, Kiyono H, Pabst R, Russell MW (2008) Terminology: nomenclature of mucosa-associated lymphoid tissue. Mucosal Immunol 1(1):31–37

Brugman S et al (2006) Antibiotic treatment partially protects against type 1 diabetes in the bio-breeding diabetes-prone rat. Is the gut Flora involved in the development of type 1 diabetes? Diabetologia 49(9):2105–2108

Carasi P et al (2015) Impact of kefir derived Lactobacillus Kefiri on the mucosal immune response and gut microbiota. J Immunol Res 2015:361604

Carvalho FA et al (2012) Transient inability to manage Proteobacteria promotes chronic gut inflammation in TLR5-deficient mice. Cell Host Microbe 12(2):139–152

Casén C et al (2015) Deviations in human gut microbiota: a novel diagnostic test for determining Dysbiosis in patients with IBS or IBD. Aliment Pharmacol Ther 42(1):71–83

Chassaing B et al (2014) AIEC Pathobiont instigates chronic colitis in susceptible hosts by altering microbiota composition. Gut 63(7):1069–1080

Chow DKL et al (2009) Long-term follow-up of ulcerative colitis in the Chinese population. Am J Gastroenterol 104(3):647–654

Clarke TB et al (2010) Recognition of peptidoglycan from the microbiota by Nod1 enhances systemic innate immunity. Nat Med 16(2):228–231

Clemente JC, Ursell LK, Parfrey LW, Knight R (2012) The impact of the gut microbiota on human health: an integrative view. Cell 148(6):1258–1270

Collado MC, Calabuig M, Sanz Y (2007) Differences between the fecal microbiota of coeliac infants and healthy controls. Curr Issues Intest Microbiol 8(1):9–14

Corthésy B (2013) Multi-faceted functions of secretory IgA at mucosal surfaces. Front Immunol 4:185

Couturier-Maillard A et al (2013) NOD2-mediated Dysbiosis predisposes mice to transmissible colitis and colorectal Cancer. J Clin Invest 123(2):700–711

Crabbé PA, Bazin H, Eyssen H, Heremans JF (1968) The Normal microbial Flora as a major stimulus for proliferation of plasma cells synthesizing IgA in the gut. The germ-free intestinal tract. Int Arch Allergy Appl Immunol 34(4):362–375

Cullender TC et al (2013) Innate and adaptive immunity interact to quench microbiome flagellar motility in the gut. Cell Host Microbe 14(5):571–581

De Martino SJ et al (2004) Peripartum Bacteremias due to Leptotrichia Amnionii and Sneathia Sanguinegens, rare causes of fever during and after delivery. J Clin Microbiol 42 (12):5940–5943

Devaraj S, Dasu MR, Park SH, Jialal I (2009) Increased levels of ligands of toll-like receptors 2 and 4 in type 1 diabetes. Diabetologia 52(8):1665–1668

Dominguez-Bello MG et al (2010) Delivery mode shapes the acquisition and structure of the initial microbiota across multiple body habitats in newborns. Proc Natl Acad Sci U S A 107 (26):11971–11975

Dwivedi M, Ansarullah, Radichev I, Kemp EH (2017) Alteration of immune-mechanisms by human microbiota and development and prevention of human diseases. J Immunol Res 2017:6985256
Elinav E et al (2011) NLRP6 Inflammasome regulates colonic microbial ecology and risk for colitis. Cell 145(5):745–757
Endesfelder D et al (2014) Compromised gut microbiota networks in children with anti-islet cell autoimmunity. Diabetes 63(6):2006–2014
Fagarasan S et al (2002) Critical roles of activation-induced cytidine deaminase in the homeostasis of gut flora. Science 298(5597):1424–1427
Festi D et al (2014) Gut microbiota and metabolic syndrome. World J Gastroenterol 20 (43):16079–16094
Frank DN et al (2007) Molecular-phylogenetic characterization of microbial community imbalances in human inflammatory bowel diseases. Proc Natl Acad Sci U S A 104 (34):13780–13785
Fulde M et al (2018) Neonatal selection by toll-like receptor 5 influences long-term gut microbiota composition. Nature 560(7719):489–493
Gaboriau-Routhiau V et al (2009) The key role of segmented filamentous bacteria in the coordinated maturation of gut helper T-cell responses. Immunity 31(4):677–689
Gérard P (2013) Metabolism of cholesterol and bile acids by the gut microbiota. Pathogens 3 (1):14–24
Gevers D et al (2014) The treatment-naive microbiome in new-onset Crohn's disease. Cell Host Microbe 15(3):382–392
Gordon Jeffrey I, Dewey KG, Mills DA, Medzhitov RM (2012) The human gut microbiota and undernutrition. Sci Transl Med 4(137):137ps12
Grivennikov SI et al (2012) Adenoma-linked barrier defects and microbial products drive IL-23/IL-17-mediated tumour growth. Nature 491(7423):254–258
Gülden E et al (2013) Toll-like receptor 4 deficiency accelerates the development of insulin-deficient diabetes in non-obese diabetic mice. PLoS One 8(9):e75385
Han H et al (2018) Gut microbiota and type 1 diabetes. Int J Mol Sci 19(4):251–259
Hedin CR et al (2014) Altered intestinal microbiota and blood T-cell phenotype are shared by patients with Crohn's disease and their unaffected siblings. Gut 63(10):1578–1586
Helgeland L et al (2004) Microbial colonization induces Oligoclonal expansions of intraepithelial CD8 T cells in the gut. Eur J Immunol 34(12):3389–3400
Hevia A et al (2014) Intestinal dysbiosis associated with systemic lupus erythematosus. mBio 5(5): e01548–e01514
Hirota K et al (2013) Plasticity of Th17 cells in Peyer's patches is responsible for the induction of T cell-dependent IgA responses. Nat Immunol 14(4):372–379
Hooper LV, Stappenbeck TS, Hong CV, Gordon JI (2003) Angiogenins: a new class of Microbicidal proteins involved in innate immunity. Nat Immunol 4(3):269–273
Human Microbiome Project Consortium (2012) Structure, function and diversity of the healthy human microbiome. Nature 486(7402):207–214
Huttenhower C, Kostic AD, Xavier RJ (2014) Inflammatory bowel disease as a model for translating the microbiome. Immunity 40(6):843–854
Imaoka A et al (1996) Proliferative recruitment of intestinal intraepithelial lymphocytes after microbial colonization of germ-free mice. Eur J Immunol 26(4):945–948
Ivanov II et al (2009) Induction of intestinal Th17 cells by segmented filamentous Bacteria. Cell 139 (3):485–498
Jiao Y, Wu L, Huntington ND, Zhang X (2020) Crosstalk between gut microbiota and innate immunity and its implication in autoimmune diseases. Front Immunol 11:282
Kaplan GG (2015) The global burden of IBD: from 2015 to 2025. Nat Rev Gastroenterol Hepatol 12(12):720–727
Kato LM, Kawamoto S, Maruya M, Fagarasan S (2014) The role of the adaptive immune system in regulation of gut microbiota. Immunol Rev 260(1):67–75

Kawaguchi-Miyashita M et al (1996) Development and Cytolytic function of intestinal intraepithelial T lymphocytes in antigen-minimized mice. Immunology 89(2):268–273
Lagier J-C et al (2012) Microbial Culturomics: paradigm shift in the human gut microbiome study. Clin Microbiol Infect 18(12):1185–1193
Larsson E et al (2012) Analysis of gut microbial regulation of host gene expression along the length of the gut and regulation of gut microbial ecology through MyD88. Gut 61(8):1124–1131
Lazar V et al (2018) Aspects of gut microbiota and immune system interactions in infectious diseases, immunopathology, and Cancer. Front Immunol 9:1830
Lee YK, Mazmanian SK (2014) Microbial learning lessons: SFB educate the immune system. Immunity 40(4):457–459
Levy M, Thaiss CA, Elinav E (2015) Metagenomic cross-talk: the regulatory interplay between Immunogenomics and the microbiome. Genome Med 7:120
Li J, Butcher J, Mack D, Stintzi A (2015) Functional impacts of the intestinal microbiome in the pathogenesis of inflammatory bowel disease. Inflamm Bowel Dis 21(1):139–153
Liu X et al (2016) Role of the gut microbiome in modulating arthritis progression in mice. Sci Rep 6:30594
Mabbott NA et al (2013) Microfold (M) cells: important Immunosurveillance posts in the intestinal epithelium. Mucosal Immunol 6(4):666–677
Macpherson AJ et al (2000) A primitive T cell-independent mechanism of intestinal mucosal IgA responses to commensal bacteria. Science 288(5474):2222–2226
Marchesi JR et al (2007) Rapid and noninvasive Metabonomic characterization of inflammatory bowel disease. J Proteome Res 6(2):546–551
McDermott AJ, Huffnagle GB (2014) The microbiome and regulation of mucosal immunity. Immunology 142(1):24–31
Mebius RE, Rennert P, Weissman IL (1997) Developing lymph nodes collect CD4+CD3- LTbeta+ cells that can differentiate to APC, NK cells, and follicular cells but not T or B cells. Immunity 7 (4):493–504
Medzhitov R et al (1998) MyD88 is an adaptor protein in the HToll/IL-1 receptor family signaling pathways. Mol Cell 2(2):253–258
Mirpuri J et al (2014) Proteobacteria-specific IgA regulates maturation of the intestinal microbiota. Gut Microbes 5(1):28–39
Mondot S et al (2011) Highlighting new phylogenetic specificities of Crohn's disease microbiota. Inflamm Bowel Dis 17(1):185–192
Moustafa A et al (2018) Genetic risk, Dysbiosis, and treatment stratification using host genome and gut microbiome in inflammatory bowel disease. Clin Transl Gastroenterol 9(1):e132
Mowat AMI (2003) Anatomical basis of tolerance and immunity to intestinal antigens. Nat Rev Immunol 3(4):331–341
Mulder DJ, Noble AJ, Justinich CJ, Duffin JM (2014) A tale of two diseases: the history of inflammatory bowel disease. J Crohns Colitis 8(5):341–348
Nagao-Kitamoto H, Kitamoto S, Kuffa P, Kamada N (2016) Pathogenic role of the gut microbiota in gastrointestinal diseases. Intest Res 14(2):127–138
Neish AS (2009) Microbes in gastrointestinal health and disease. Gastroenterology 136(1):65–80
Nemoto H et al (2012) Reduced diversity and imbalance of fecal microbiota in patients with ulcerative colitis. Dig Dis Sci 57(11):2955–2964
Nguyen GC (2011) Editorial: bugs and drugs: insights into the pathogenesis of inflammatory bowel disease. Am J Gastroenterol 106(12):2143–2145
Nigro G et al (2014) The cytosolic bacterial peptidoglycan sensor Nod2 affords stem cell protection and links microbes to gut epithelial regeneration. Cell Host Microbe 15(6):792–798
Opazo MC et al (2018) Intestinal microbiota influences non-intestinal-related autoimmune diseases. Front Microbiol 9:432
Ostman S et al (2006) Impaired regulatory T-cell function in germ-free mice. Eur J Immunol 36 (9):2336–2346

Pabst O (2012) New concepts in the generation and functions of IgA. Nat Rev Immunol 12 (12):821–832
Paglia L, Concetta GM et al (2017) One year in review 2017: systemic lupus erythematosus. Clin Exp Rheumatol 35(4):551–561
Palm NW, de Zoete MR, Flavell RA (2015) Immune-microbiota interactions in health and disease. Clin Immunol 159(2):122–127
Pearson JA, Susan Wong F, Wen L (2016) The importance of the non obese diabetic (NOD) mouse model in autoimmune diabetes. J Autoimmun 66:76–88
Peterson DA, McNulty NP, Guruge JL, Gordon JI (2007) IgA response to symbiotic Bacteria as a mediator of gut homeostasis. Cell Host Microbe 2(5):328–339
Petnicki-Ocwieja T et al (2009) Nod2 is required for the regulation of commensal microbiota in the intestine. Proc Natl Acad Sci U S A 106(37):15813–15818
Pociot F, Lernmark Å (2016) Genetic risk factors for type 1 diabetes. Lancet 387 (10035):2331–2339
Pussinen PJ et al (2011) Endotoxemia is associated with an increased risk of incident diabetes. Diabetes Care 34(2):392–397
Qin J et al (2010) A human gut microbial gene catalogue established by metagenomic sequencing. Nature 464(7285):59–65
Rakoff-Nahoum S et al (2004) Recognition of commensal microflora by toll-like receptors is required for intestinal homeostasis. Cell 118(2):229–241
Ramanan D et al (2014) Bacterial sensor Nod2 prevents inflammation of the small intestine by restricting the expansion of the commensal Bacteroides Vulgatus. Immunity 41(2):311–324
Ranucci G et al (2017) Early-life intestine microbiota and lung health in children. J Immunol Res 2017:8450496
Rewers M, Ludvigsson J (2016) Environmental risk factors for type 1 diabetes. Lancet 387 (10035):2340–2348
Round JL et al (2011) The toll-like receptor 2 pathway establishes colonization by a commensal of the human microbiota. Science 332(6032):974–977
Salzman NH, Bevins CL (2013) Dysbiosis--a consequence of Paneth cell dysfunction. Semin Immunol 25(5):334–341
Salzman NH et al (2010) Enteric Defensins are essential regulators of intestinal microbial ecology. Nat Immunol 11(1):76–83
Sartor RB (2008) Microbial influences in inflammatory bowel diseases. Gastroenterology 134 (2):577–594
Sekirov I, Russell SL, Caetano L, Antunes M, Brett Finlay B (2010) Gut microbiota in health and disease. Physiol Rev 90(3):859–904
Seksik P, Landman C (2015) Understanding microbiome data: a primer for clinicians. Digest Dis 33 (Suppl 1):11–16
Sha S et al (2013) The biodiversity and composition of the dominant fecal microbiota in patients with inflammatory bowel disease. Diagn Microbiol Infect Dis 75(3):245–251
Sonnenburg JL, Bäckhed F (2016) Diet-microbiota interactions as moderators of human metabolism. Nature 535(7610):56–64
Stewart CJ et al (2018) Temporal development of the gut microbiome in early childhood from the TEDDY study. Nature 562(7728):583–588
Suzuki K et al (2004) Aberrant expansion of segmented filamentous Bacteria in IgA-deficient gut. Proc Natl Acad Sci U S A 101(7):1981–1986
Tamboli CP, Neut C, Desreumaux P, Colombel JF (2004) Dysbiosis in inflammatory bowel disease. Gut 53(1):1–4
Todd JA et al (2007) Robust associations of four new chromosome regions from genome-wide analyses of type 1 diabetes. Nat Genet 39(7):857–864
Tsuji M et al (2009) Preferential generation of follicular B helper T cells from Foxp3+ T cells in gut Peyer's patches. Science 323(5920):1488–1492

Vaishnava S et al (2011) The antibacterial lectin RegIIIgamma promotes the spatial segregation of microbiota and host in the intestine. Science 334(6053):255–258
Velloso LA, Folli F, Saad MJ (2015) TLR4 at the crossroads of nutrients, gut microbiota, and metabolic inflammation. Endocr Rev 36(3):245–271
Vijay-Kumar M et al (2010) Metabolic syndrome and altered gut microbiota in mice lacking toll-like receptor 5. Science 328(5975):228–231
Virili C et al (2018) Gut microbiota and Hashimoto's thyroiditis. Rev Endocr Metab Disord 19 (4):293–300
Wang YF, Ouyang Q, Ren Wei H (2010) Progression of inflammatory bowel disease in China. J Dig Dis 11(2):76–82
Wei F et al (2019) Changes of intestinal Flora in patients with systemic lupus erythematosus in Northeast China. PLoS One 14(3):e0213063
Wen L et al (2008) Innate immunity and intestinal microbiota in the development of type 1 diabetes. Nature 455(7216):1109–1113
Wesche H et al (1997) MyD88: an adapter that recruits IRAK to the IL-1 receptor complex. Immunity 7(6):837–847
Wijmenga C (2005) Expressing the differences between crohn disease and ulcerative colitis. PLoS Med 2(8):e230
Wright EK et al (2015) Recent advances in characterizing the gastrointestinal microbiome in Crohn's disease: a systematic review. Inflamm Bowel Dis 21(6):1219–1228
Wu H-J, Eric W (2012) The role of gut microbiota in immune homeostasis and autoimmunity. Gut Microbes 3(1):4–14
Wu S et al (2009) A human colonic commensal promotes colon tumorigenesis via activation of T-helper type 17 T-cell responses. Nat Med 15(9):1016–1022
Xu H et al (2019) The dynamic interplay between the gut microbiota and autoimmune diseases. J Immunol Res 2019:7546047
Zhang H et al (2015) Host adaptive immunity alters gut microbiota. ISME J 9(3):770–781
Zhang M et al (2007) Structural shifts of mucosa-associated lactobacilli and Clostridium Leptum subgroup in patients with ulcerative colitis. J Clin Microbiol 45(2):496–500

Overview on Human Gut Microbiome and its Role in Immunomodulation

5

Sudhakar Pola and Dhana Lakshmi Padi

Abstract

Microbes are first acquired from mother to child during pregnancy and they are involved in programming fetal immunity. Proper maintenance of these microbes in early infancy aids in adult life and, therefore, can be achieved by a healthy diet, exercise, and clean environment as it plays an essential role in shaping gut microbes. Microbes reside within different tissues and organs, and thus, they form a mutualistic relationship with the human body. The effects of microbes on the human body can be both beneficial and harmful. As definite changes in the microbiome of some specific microorganisms and alteration in the tumor-promoting and suppressing genes may lead to the development of cancer. Contrarily, they are also involved in boosting the immune system to fight against cancer. Therefore microbes are involved in modulating the immune responses and this can be triggered by microbial products (polysaccharides and formyl peptides), metabolites (short-chain fatty acids), immunotherapies (TLR agonists and immune checkpoint inhibitors), and drugs.

Keywords

Microbiome · Mutualistic relationship · Infancy · Cancer · Immunotherapy · Immunomodulation

S. Pola (✉) · D. L. Padi
Department of Biotechnology, Andhra University, Visakhapatnam, Andhra Pradesh, India
e-mail: sudhakar@andhrauniversity.edu.in

P. V. Bramhachari (ed.), *Microbiome in Human Health and Disease*,
https://doi.org/10.1007/978-981-16-3156-6_5

5.1 Human Gut Microbiome

There are trillions of microorganisms inhabited in the human body and they included bacteria, viruses, archaea, protists, and fungi (Backhed et al. 2005). Commensal microorganisms do not harm the human body even though they form a mutualistic relationship. Therefore, they colonize and form composite ecosystems in the areas such as our skin, mouth, and gut. In contrast to the human body, microorganisms encode many numbers of genes (Bhatt et al. 2017) and thus, their function and interaction with the host can be studied by many advanced genome sequencing technologies.

The human gut microbiome (Shreiner et al. 2015) plays a vital role in an individual's well-being; therefore, they help the host in maintaining many functions like homeostasis, metabolism, regulating nutrition and development and function of the innate and acquired immune system (Thaiss et al. 2016). A favorable environment and a proper diet are required for the remarkable growth of microbes. The right conditions for microbes help in the better development of good colonies; however, an uncertain environment may lead to the death of good microbes and cause the formation of harmful colonies.

Initially, human beings obtained microbes from their mothers during pregnancy; the maternal microbes from different sites such as the gut, skin, breast milk, and vagina help the infant in improving the immune system. In early infancy, if these microbes are nurtured properly, it helps in shaping up gut microbes, which in turn boosts the immune system. The alteration in the infant's gut microbiota may cause many disorders related to autoimmunity and inflammation (Sekirov et al. 2010; Amon and Sanderson 2017).

The microbes present in the human gut play a crucial role, as they interact with all the cells present in the human body. Therefore, gut mucosa has 60% of immune cells; thus, the mucosal immune system aid in resisting the attacks of harmful microbes and inflammatory reactions (Molloy et al. 2012), hence it acts as a defense barrier of our body. Pattern recognition receptors (PPRs) such as toll-like receptors (TLRs) help the innate immune system in recognizing the self and non-self-antigens and in turn, these TLRs recognize the pathogen-associated molecular patterns (PAMPs). These interactions activate the signaling cascade pathways, stimulate the effector responses, and thus produce cytokines, apoptotic factors, AMPs, and chemokines. Therefore, this mechanism is constructive in maintaining homeostasis and disease pathogenesis.

The human diet is linked with metabolic health; thus, gut microbes act as intermediator, which involves the conversion of metabolites. Gut microbiota helps the host in digesting dietary products such as non-digestible polysaccharides, complex proteins, aromatic amino acids, lipids, choline, and vitamins; hence the colonic microbes are involved in the degradation and fermentation of food particles. The main by-product produced after bacterial fermentation is short-chain fatty acids (SCFAs). Therefore, the metabolites are involved in the strengthening of the host-cell barrier against pathogens, maintaining controlled inflammation, and also regulating the mucosal immune responses (Belkaid and Hand 2014).

Sometimes, the toxic metabolites which are released into the human body may cause the onset of cancer (Garrett 2015). As many microbes may relocate to other parts of the body, there will be many chances of spreading these tumors (Rajagopala et al. 2017). The complications in the metabolic pathways may induce many diseases. For example, the difficulty in choline metabolism may lead to cardiovascular heart disease (Wang et al. 2011).

Homeostasis in the intestine is maintained by the gut microbial commensals. The homeostatic relationship between the host–microbe is generally disturbed due to the changes in food habits and the environment. The action of some medicines and antibiotics, exposure to pathogens, and also psychological disturbances may lead to the overall changes in the structure and activity of the gut microbes, which is known as dysbiosis (Shui et al. 2020). Some diseases like obesity, diabetes, liver problems, cancer, and even neurodegenerative diseases lead to modifications in the gut microbiota. The alterations in the gut microbes and their interaction with the human body may result in some illnesses and cause cancers. The gut microbes play a dual role in both tumor development and also in anti-cancer therapies (Cani 2018).

There are many numbers of research papers and articles were published on human gut microbes, as many studies were carried out on the host–gut microbiota interaction. Among these studies, the interactivity of human gut microbiota with the immune system and the discovery of novel immune therapeutics is one of the exciting topics in the last couple of years. Some immunotherapeutic treatments were in use to combat cancers such as immune checkpoint inhibitors (ICIs), activation and maturation of T- cells and B-cells, adoptive cell therapy (ACT), and vaccines. There are also some immunomodulatory drugs present that can elicit the anti-cancer immunity to kill the cancer cells.

5.2 Gut Microbes During Pregnancy and Early Infancy

Gut microbes are initially acquired from mother to child during delivery through breastfeeding, skin-to-skin contact, and delivery through the vaginal canal (Chu and Englund 2014). Gut microbial commensals in the early infants help in defining and anticipating the human's health status.

During pregnancy, a woman undergoes many physical and mental changes like gaining weight, hormonal, metabolic, and immune changes, and even mood swings. These changes may alter the gut microbes present in a pregnant woman; this may lead to some difficulties in the course of pregnancy. The poor maternal diet and alterations in the maternal gut microbiota throughout pregnancy may affect the infant's gut microbiota (Schwartz et al. 2012) and adult life (Kumbhare et al. 2020).

Maternal microbes present in the gut, breast milk, and vagina are involved in the development of fetal immunity before delivery. The gut microbes in the fetal intestine undergo colonization; however, the diet, antibiotics, and mode of birth during pregnancy impact the microbial colonization (La Rosa et al. 2014). The infant's gut microbial flora which is obtained from the mother helps in the

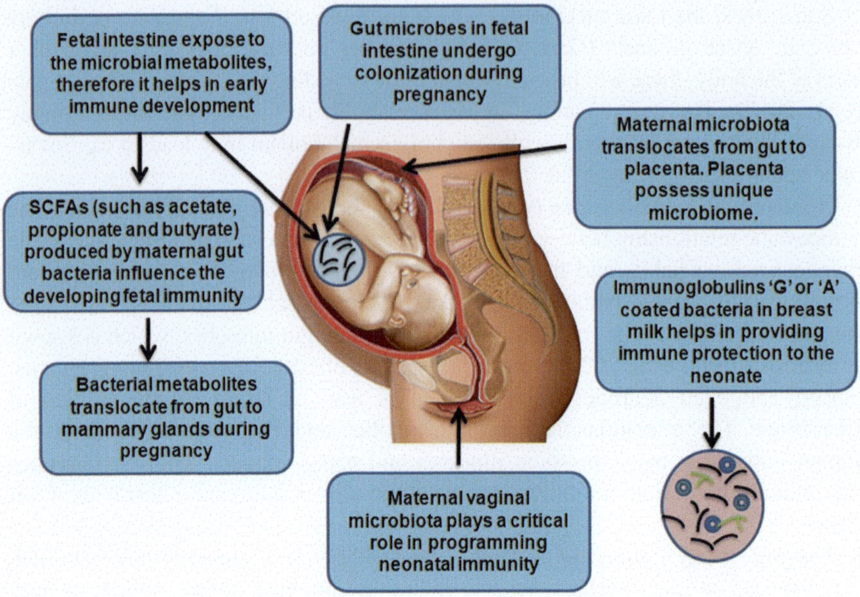

Fig. 5.1 Different mechanisms are showing the impact of maternal microbiota on a fetus during pregnancy

progression of the mucosal immune system; if that gut microbial environment gets disturbed, it may lead to mucosal infection and inflammation in childhood, and also its adverse effects continue to adulthood (Welliver and Ogra 2008) (Fig. 5.1).

Maternal microbial metabolites such as short-chain fatty acids (SCFAs) (Koh et al. 2016) help develop the fetal immune system. In the course of pregnancy, maternal bacterial metabolites translocate from gut to mammary glands and there the bacteria present in the breast milk are coated with immunoglobulins, IgG or IgA; therefore, this helps the neonate to get the proper immune protection. Maternal immunoglobulin G is also actively moved across the placenta and hence provides passive immunity to the neonate.

The uterus has been considered sterile, but likely susceptible to be affected by vaginal bacteria, therefore microbes are most probably seen at the endometrium. Thus these endometrial microbes help in assessing the success/failure of implantation during pregnancy (Moreno et al. 2016). Infants delivered vaginally have valuable maternal vaginal microbes such as *Bifidobacterium* and *Lactobacillus*, although cesarean section delivered infants are composed of *Staphylococcus spp.* (Dominguez-Bello et al. 2010) and also, they are more prone to type I diabetes by 20% (Charbonneau et al. 2016).

5.3 Shaping of Maternal Gut Microbes with Proper Diet during Pregnancy

The intake of fiber, prebiotic, and probiotic foods in the diet (De Filippo et al. 2010) during pregnancy helps in the increase of good microbes. The prenatal gut microbiota is stable and can grow better when good dietary food is introduced (Dawson et al. 2019). Not merely diet, but also external environmental factors play an essential role in shaping up gut microbes. High consumption of a fiber-rich diet assists the gut microbial commensals in maintaining the equilibrium in the intestine (Shui et al. 2020). Diets including high amounts of fruits, vegetables, and fibers are suitable for the enrichment of gut microbes, in contrast with a diet rich in fats, sugars, and animal protein (Gali 2015). Improper diet may cause many metabolic diseases like type 2 diabetes, thus focus on microbes that are linked to metabolism should be increased to overcome metabolic disorders (Sonnenburg and Backhed 2016).

Breastfeeding for the newly born child for at least four months to six months helps in the proper maintenance of gut microbes, as it contains carbohydrates, proteins, fats, nutrients, immunoglobulins, and endocannabinoids, thus it gives a complete supplement to the infant. Nevertheless, the early introduction of solid foods to the newly born child may lead to obesity (Koleva et al. 2015) and other disorders (Differding et al. 2020). Therefore, this leads to a drastic change in the microbiota composition in an infant's gut.

However, various prenatal, neonatal, and postnatal factors affect the infant's gut microbes, which are obtained from the mother during pregnancy and also affect the early infancy and childhood (Fig. 5.2). Therefore, the various factors are maternal diet, vaginal infections, psychological effects, gestational period, host genetics, mode of delivery (vaginal/cesarean), diet (milk consumption/solid food introduction), and environmental factors (Christian Milani et al. 2017).

5.4 Gut Microbes: Onset of Cancer

Despite commensals, some pathogenic microbes are linked to human carcinogenesis. Gene mutation and alteration in the oncogenes and tumor suppressor genes and inflammation, which provides the tumor-promoting environment, may lead to the beginning of cancer (Dzutsev et al. 2015). The modifications in the gut microbiota and their interaction with the human body may cause many health implications, including cancers like colon cancer, pancreatic cancer, and lung cancer (Zheng et al. 2020).

There are a high number of microorganisms located in the human colon and they play a significant part in colorectal carcinoma (CRC). The patients suffering from adenoma or adenocarcinoma have a relatively high number of some specific microbes and thus they adhere to the colon cells, inhibit the tumor suppressor genes, and activate the oncogenes that cause colon cancer (Abu-Ghazaleh et al. 2020). Some studies, like metagenomics and transcriptomics, are practical to know

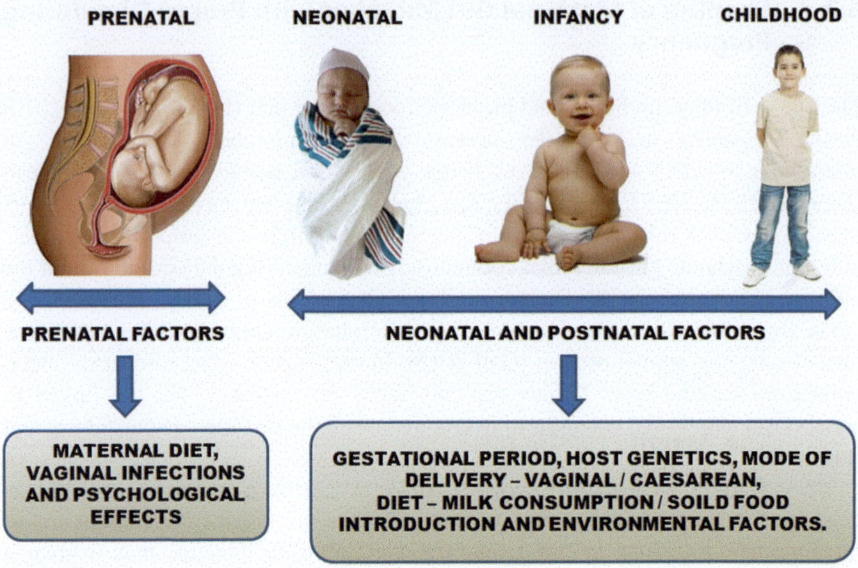

Fig. 5.2 Different prenatal, neonatal, and postnatal factors that are affecting the composition of gut microbes

the relation between *Fusobacterium nucleatum* and colon cancer (McCoy et al. 2013). The toxic metabolites produced during the degradation and fermentation of dietary products may also cause cancer. For instance, the fermentation of proteins releases toxic by-products like ammonia, amines, and branched-chain fatty acids, which causes the onset of colon cancer to the host.

The gut microbiota may also play a crucial role in pancreatic ductal adenocarcinoma (PDAC), human pancreatic diseases, pancreatitis (Akshintala et al. 2019), and lung cancers. The geographical and environmental factors affect the lung microbiome (Mao et al., 2018); therefore, the microbes present in the lungs are involved in the onset of lung cancer and lung diseases. Chronic obstructive pulmonary disease (COPD) is an inflammatory disease, which is persistent and causes when the respiratory tract is colonized by pathogenic microbes (Budden et al. 2016), whereas cystic fibrosis (CF) is caused by a gene mutation (Chmiel et al. 2014).

5.5 Immunomodulation

5.5.1 Immunomodulation by Microbial Components

The immune cells present in the gastrointestinal (GI) tract are continuously exposed to microbial antigens and thus the epithelial cells in the mucosa transcribe the requisite information to the immune cells which activates the host immune system. The PPR-mediated mechanisms through microbial components are helpful in the modulation of the immune system. The pathogen-associated molecular patterns (PAMPs) such as lipopolysaccharides, peptidoglycans, lipoteichoic acid, flagellin, and formyl peptides are recognized by pattern recognition receptors (PPRs) such as toll-like receptors (TLRs) and nod-like receptors (NLRs) (Osamu and Shizuo 2010).

TLRs have a significant role in maintaining homeostasis in the intestine and stimulating the inflammatory responses during pathogenesis (Valentini et al. 2014). These interactions activate the antigen-presenting cells (APCs) and dendritic cells (DCs) and then the priming of B- and T-cells occurs when they mediate with APCs at mesenteric lymph nodes. The naive T-cells are differentiated into $CD4^{+}$ T-cells and gut-associated lymphoid tissue (GALT) is stimulated by commensal bacteria, which induce B-cells to differentiate and also activate the production of IgA. The activity of the immune system can also be altered by circulating cytokines via soluble immunomodulatory factors (Khan et al. 2020).

The commensal microorganisms modulate epithelial immunity and thus it leads to the maturation and function of the mucosal immune system. Mucosal immune response arose from the three layers (epithelium, lamina propria, and muscularis mucosae) of the mucosa (Fig. 5.3a), as each layer composed of mucus proteins, mucins, immunoglobulins IgA, and lymphocytes (T- and B-cells and DCs), which helps in triggering the mucosal immune system to fight against pathogens (Shui et al. 2020).

Microbial products like polysaccharides and formyl peptides influence the PRR-mediated responses; this study helps to understand the host–microbe homeostasis. Polysaccharides and formyl peptides influence the innate and adaptive immune system (Fig. 5.3b). Therefore, they are useful immunomodulatory PAMPs, which are effective in treating autoimmune diseases, inflammation, neurodegenerative diseases, and cancers (Michelle and Wendy 2016).

5.5.2 Immunomodulation by Metabolites

Human dietary intake includes carbohydrates, proteins, lipids, and vitamins, which are degraded by gut microbes to produce metabolic products. Metabolism of dietary products yields energy and also maintains gut health. The main by-product, which is produced after microbial degradation, is SCFAs, which involve in modulating the immune system (Fig. 5.4).

The immune cells which are present in the epithelial layer of the mucosal interface when coming in contact with microbial metabolic products impact the immune responses and disease risks (Michelle and Wendy 2016). In this process, the involvement of host and microbes plays a critical role, as they permeate the metabolic products to get access to the epithelial layer of mucosa, which has immune cells.

Humans take carbohydrates as a diet; the majority of them include non-digestible polysaccharides, oligosaccharides, unabsorbed sugars, and plant fibers. These are resistant to the action of amylase; thus, colonic microbes play a vital role in degrading these substrates by producing various hydrolytic enzymes. Initially,

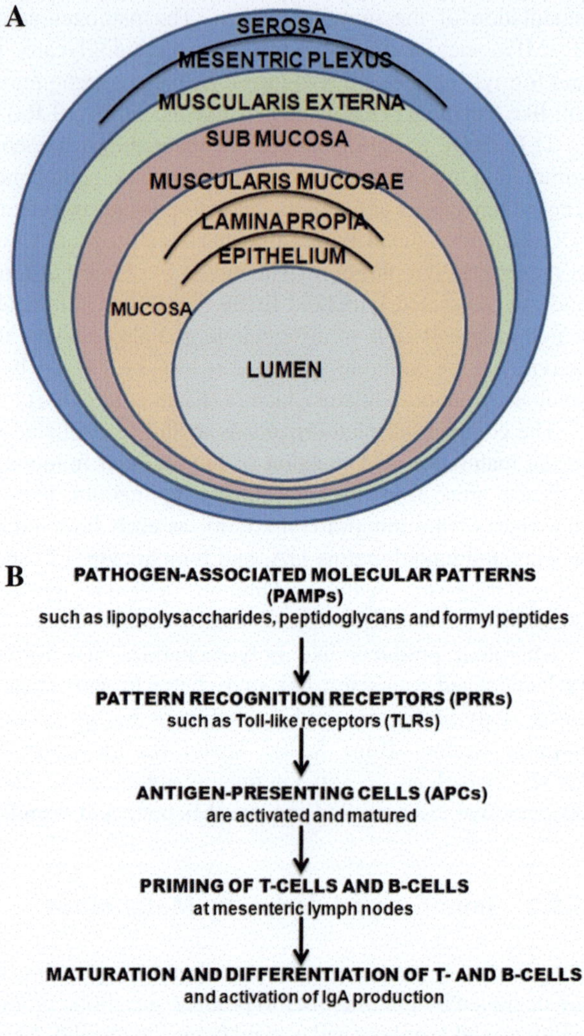

Fig. 5.3 (**a**) and 3(**b**) Different layers of the gastrointestinal tract (GI) and the modulation of the host immune system by microbial products

complex carbohydrates are degraded to polysaccharides and then polysaccharides to oligosaccharides; thus, these oligosaccharides undergo fermentation and produce by-products such as gases (like hydrogen, carbon dioxide, methane) and intraluminal solute (such as short-chain fatty acids (SCFAs).

SCFAs (such as acetate, butyrate, and propionate) act as immunopotentiators that enhance the production of antibodies. These metabolic products communicate with various immune cells present in the epithelial layer and they activate the immune system through many immune responses such as plasma B-cell proliferation, Treg development, macrophages, and dendritic cells (DCs) activity and the release of cytokines which involves reducing inflammation (Ziying Zhang et al. 2019). Butyrate strengthens the barrier between the blood–brain by connecting the cells, i.e., the

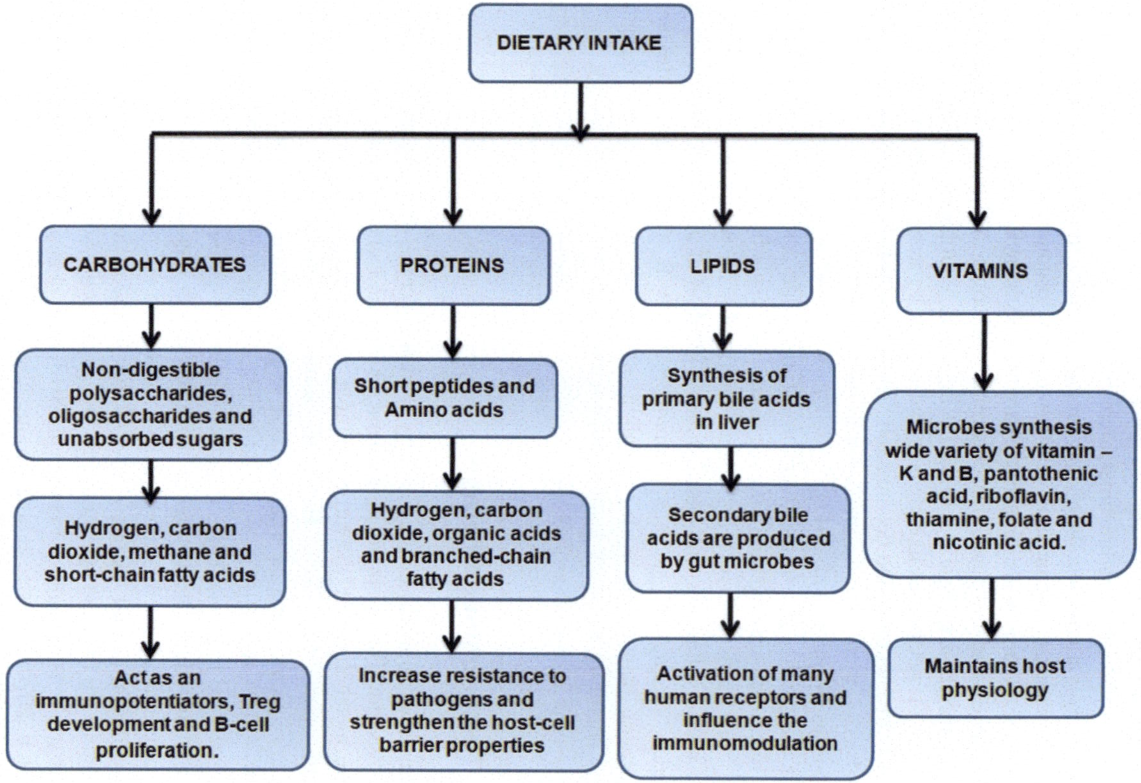

Fig. 5.4 Gut microbes are associated with the degradation of the dietary products such as carbohydrates, proteins, lipids, and vitamins by anaerobic degradation and fermentation; therefore, they maintain gut health by triggering the host immune system

gut–brain axis (Smith 2015). Lactate, another metabolite commonly found in the milk diet possesses many metabolic and immune properties. Lactic acid secreted by tumor cells helps in the development of pro-tumor immunity (Yan et al. 2018). Lactate creates a pro-tumor microenvironment, which helps in the increasing efficacy of tumor therapy (Feichtinger and Lang 2019).

Firstly, complex proteins are breakdown by microbes into short peptides and amino acids by various peptidases and proteases. After this, short peptides and amino acids undergo fermentation and produce by-products such as hydrogen, carbon dioxide, organic acids, branched-chain fatty acids (such as 2-methyl butyrate and isobutyrate), and trace amounts of amines, ammonia, and phenols. Aromatic amino acid metabolism is also carried by intestinal microbiota, as they produce various types of metabolites like phenols and indoles by transamination, decarboxylation, dehydrogenation, and deamination. These indoles help in strengthening the host-cell barrier, increasing the resistance to pathogens, and maintaining inflammatory responses (Monika et al. 2017).

Lipids are synthesized in the liver into primary bile acids; furthermore, these primary bile acids are degraded into secondary bile acids by gut microbes. Thus, secondary bile acids are involved in the activation of many receptors present in the host and it also influences immunomodulation (Ridlon et al. 2014). Moreover, a wide variety of vitamins such as vitamin-B, vitamin-K, riboflavin, pyridoxine, nicotinic acid, pantothenic acid, cobalamin, and folate are synthesized by gut microbes. Therefore, these are helpful in the maintenance of host physiology.

5.6 Tumor Immunotherapy

Tumor immunotherapy is an effective way to treat cancer as they directly target the immune system. Tumor microenvironment (TME) and chemotherapeutic drugs, which are metabolized by the gut microbiota, are helpful in the improvement of immunotherapies (Fig. 5.5). However, some immunosuppressive and anti-inflammatory factors such as arginase, TGF-B (Transforming growth factor-beta), and IL-10 (interleukin 10) are produced by macrophages, other myeloid and Treg cells are involved in preventing anti-cancer immune response (Dzutsev et al. 2015).

Many immunomodulatory drugs enhance anti-cancer immunity, namely, metformin, angiotensin receptor blockers (ARBs), anthracyclines, thalidomides, and statin (Maiko and Kawaguchi 2018). They modulate the anti-cancer immunity by strengthening the immune system to kill the cancer cells. Some chemotherapeutic drugs like cyclophosphamide (Viaud et al. 2013; Shui et al. 2020) and gemcitabine (Zhang et al. 2020; Banerjee et al. 2018) are also used for immunomodulation.

Some tumor immune-targeted treatments are useful to treat cancers, such as toll-like receptor (TLR) agonists, immune checkpoint inhibitors (ICI), adoptive T-cell therapy (ACT), and vaccines. The synthetic toll-like receptor (TLR) agonists are used as a vaccine for cancers. TLRs play a critical role in connecting adaptive and innate immunity. PPRs such as TLRs when recognizes pathogens via dendritic cells

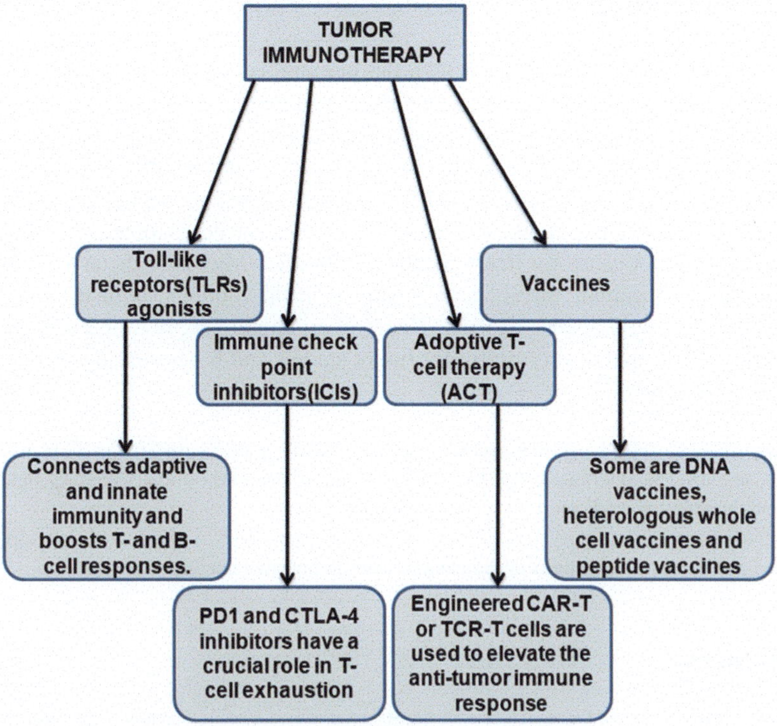

Fig. 5.5 Different types of tumor immune-targeted treatments to treat cancer by directly targeting the immune system

(DCs) involve boosting the T-cells and B-cells immune responses (Lazar et al. 2018).

PD-1 and its ligands PD-L1 and PD-L2 (Lee et al. 2016) are immune checkpoint inhibitors, which play a crucial role in T-cell exhaustion (Schildberg et al. 2016). These are usually expressed on antigen-presenting cells (APCs), which convert naive CD+ T-cells to regulatory T-cells.

Adoptive cell therapy (Cohen et al. 2017) includes the modification of cancer cells in vitro to make it more efficient for treating cancers by increasing their immune response. TCR-engineered T-cells or CAR-engineered T-cells are used to elevate the anti-tumor response.

Vaccines such as DNA vaccines, peptide vaccines, heterologous whole-cell vaccines, and immunotherapeutic drugs are used to treat some cancers (Thomas and Prendergast 2016). Vaccines are generally composed of antigens and adjuvants; they activate antigen-presenting cells, which in turn elevate the immune responses (Banchereau and Palucka 2018).

5.7 Conclusion

Gut microbes and their association with the immune system is one of the primary subjects to be focused. Gut microbes obtained from mother to child in the course of pregnancy play a critical role in an individual's entire life. Thus, shaping these microbes in early infancy is essential in combating many health issues and also cancers. Proper diet and environmental factors are essential in keeping gut microbes healthy. Focus on dietary intake should be increased as health is linked with the diet and this is the root cause for many metabolic diseases. Microbial dietary products, metabolites, and immunotherapies are involved in the modulation of the immune system and thus fight against cancers. Therefore, immunotherapy is an exciting and significant theme that is very useful for future studies and it has considerable scope for new researches.

Acknowledgments The authors **Sudhakar Pola** and **Dhana Lakshmi Padi** are grateful to the authorities of Andhra University Visakhapatnam, India for providing the necessary facilities to carry out the research work and for extending constant support.

Conflict of Interest The authors declare that there is no potential conflict of interest.

References

Abu-Ghazaleh N, Chua WJ, Gopalan V (2020) Intestinal microbiota and its association with colon cancer and red/processed meat consumption. J Gastroenterol Hepatol 36(1):75–88

Akshintala VS, Talukdar R, Singh VK, Goggins M (2019) The gut microbiome in pancreatic disease. Clin Gastroenterol Hepatol 17(2):290–295

Amon P, Sanderson I (2017) What is the microbiome? Arch Dis Child Educ Pract Ed 102 (5):257–260

Backhed F, Ley RE, Sonnenburg JL, Peterson DA, Gordon JI (2005) Host-bacterial mutualism in the human intestine. Nature 307:1915–1920

Banchereau J, Palucka K (2018) Immunotherapy: Cancer vaccines on the move. Nat Rev Clin Oncol 15(1):9–10

Banerjee K, Kumar S, Ross KA, Gautam S et al (2018) Emerging trends in the immunotherapy of pancreatic cancer. Cancer Lett 417:35–46

Belkaid Y, Hand TW (2014) Role of the microbiota in immunity and inflammation. Cell 157:121–141

Bhatt AP, Redinbo MR, Bultman SJ (2017) The role of the microbiome in cancer development and therapy. CA Cancer J Clin 67(4):326–344

Budden KF, Gellatly SL, Wood DLA, Cooper MA, Morrison M et al (2016) Emerging pathogenic links between microbiota and the gut-lung axis. Nat Rev Microbiol 15:55

Cani PD (2018) Human gut microbiome: hopes, threats and promises. Gut 67(9):1716–1725

Charbonneau MR, Blanton LV, DiGiulio DB et al (2016) A microbial perspective of human developmental biology. Nature 535:48–55

Chmiel JF, Aksamit TR, Chotirmall SH, Dasenbrook EC, Elborn JS et al (2014) Antibiotic Management of Lung Infections in cystic fibrosis. II. Non-tuberculous, mycobacteria, anaerobic Bacteria, and Fungi. Ann Am Thorac Soc 11:1298–1306

Chu HY, Englund JA (2014) Maternal immunization. Clin Infect Dis 59(4):560–568

Cohen JE, Merims S, Frank S, Engelstein R et al (2017) Adoptive cell therapy: past, present and future. Immunotherapy 9:183–196

Dawson SL, Craig JM, Clarke G, Mohebbi M et al (2019) Targeting the infant gut microbiota through a perinatal educational dietary intervention: protocol for a randomized controlled trial. JMIR Res Protoc 8(10):e14771

De Filippo C, Cavalieri D, Di Paola M, Ramazzotti M, Poullet JB et al (2010) Impact of diet in shaping gut microbiota revealed by a comparative study in children from Europe and rural Africa. Proc Natl Acad Sci U S A 107(33):14691–14696

Differding MK et al (2020) Timing of complementary feeding is associated with gut microbiota diversity and composition and short-chain fatty acid concentrations over the first year of life. BMC Microbiol 36(1):75–88

Dominguez-Bello MG, Costello EK, Contreras M et al (2010) Delivery mode shapes the acquisition and structure of the initial microbiota across multiple body habitats in newborns. Proc Natl Acad Sci U S A 107:11971–11975

Dzutsev A, Goldszmid RS, Viaud S, Zitvogel L, Trinchieri G (2015) The role of the microbiota in inflammation, carcinogenesis, and cancer therapy. Eur J Immunol 45:17–31

Feichtinger RG, Lang R (2019) Targeting L-lactate metabolism to overcome resistance to immune therapy of melanoma and other tumor entities. J Oncol 2019:2084195

Gali A (2015) The microbiome what we do and don't know. Nutr Clin Pract 30(6):734–746

Garrett WS (2015) Cancer and the microbiota. Science 348:80–86

Khan MAW, Ologun G, Arora R, McQuade JL, Wargo JA (2020) Gut microbiome modulates response to cancer immunotherapy. Dig Dis Sci 65(3):885–896

Koh A, De Vadder F, Kovatcheva-Datchary P, Backhed F (2016) From dietary fiber to host physiology: short-chain fatty acids as key bacterial metabolites. Cell 165:1332–1345

Koleva PT, Bridgman SL, Kozyrskyj AL (2015) The infant gut microbiome: evidence for obesity risk and dietary intervention. Nutrients 7(4):2237–2260

Kumbhare SV, Patangia DV, Mongad DS, Bora A et al (2020) Gut microbial diversity during pregnancy and early infancy: an exploratory study in the Indian population. FEMS Microbiol Lett 367(3):fnaa022

La Rosa PS, Warner BB, Zhou Y et al (2014) Patterned progression of bacterial populations in the premature infant gut. Proc Natl Acad Sci U S A 111:12522–12527

Lazar V, Ditu L-M, Pircalabioru GG, Gheorghe I et al (2018) Aspects of gut microbiota and immune system interactions in infectious diseases, immunopathology and cancer. Front Immunol 9:1830

Lee L, Gupta M, Sahasranaman S (2016) Immune checkpoint inhibitors: an introduction to the next-generation cancer immunotherapy. J Clin Pharmacol 56:157–169

Maiko M, Kawaguchi M (2018) Immunomodulatory effects of drugs for effective cancer immunotherapy. J Oncol 2018:8653489

Mao Q, Jiang F, Yin R, Wang J, Xia W et al (2018) Interplay between the lung microbiome and lung Cancer. Cancer Lett 415:40–48

McCoy AN, Araujo-Perez F, Azcarate-Peril A, Yeh JJ et al (2013) Fusobacterium is associated with colorectal adenomas. PLoS One 8(1):e53653

Michelle GR, Wendy SG (2016) Gut microbiota, metabolites and host immunity. Nat Rev Immunol 16(6):341–352

Milani C, Duranti S, Bottacini F, Casey E et al (2017) The first microbial colonizers of the human gut: composition, activities, and health implications of the infant gut microbiota. Microbiol Mol Biol Rev 81(4):e00036–e00017

Molloy MJ, Bouladoux N, Belkaid Y (2012) Intestinal microbiota: shaping local and systemic immune responses. Semin Immunol 24(1):58–66

Monika Y, Verma MK, Chauhan NS (2017) A review of metabolic potential of human gut microbiome in human nutrition. Arch Microbiol 200(2):203–217

Moreno I, Codoñer FM, Vilella F, Valbuena D, Martinez-Blanch JF et al (2016) Evidence that the endometrial microbiota has an effect on implantation success or failure. Am J Obstet Gynecol 215(6):684–703

Osamu T, Shizuo A (2010) Pattern recognition receptors and inflammation. Cell 140:805–820

Rajagopala SV, Vashee S, Lauren M et al (2017) The human microbiome and cancer. Cancer Prev Res 10(4):226–234
Ridlon JM, Kang DJ, Hylemon PB et al (2014) Bile acids and the gut microbiome. Curr Opin Gastroenterol 30:332–338
Schildberg F, Klein S, Freeman G, Sharpe A (2016) Coinhibitory pathways in the B7-CD28 ligand-receptor family. Immunity 44:955–972
Schwartz S, Friedberg I, Ivanov I, Davidson LA, Goldsby JS et al (2012) A metagenomic study of diet-dependent interaction between gut microbiota and host in infants reveals differences in immune response. Genome Biol 13:32
Sekirov I, Russell SL, Caetano Antunes ML, Finlay BB (2010) Gut microbiota in health and disease. Physiol Rev 90(3):859–904
Shreiner AB, Kao JY, Young VB (2015) The gut microbiome in health and in disease. Curr Opin Gastroenterol 31:69–75
Shui L, Yang X, Li J, Cheng Y et al (2020) Gut microbiome as a potential factor for modulating resistance to cancer immunotherapy. Front Immunol 10:2989
Smith PA (2015) Brain, meet gut. Nature 526:312–314
Sonnenburg JL, Backhed F (2016) Diet-microbiota interactions as moderators of human metabolism. Nature 535(7610):56–64
Thaiss CA, Zmora N, Levy M, Elinav E (2016) The microbiome and innate immunity. Nature 535:65–74
Thomas S, Prendergast GC (2016) Cancer vaccines: a brief overview. Methods Mol Biol 1403:755–761
Valentini M, Piermattei A, Di Sante G, Migliara G et al (2014) Immunomodulation by gut microbiota: role of toll-like receptor expressed by T cells. J Immunol Res 2014:586939
Viaud S, Saccheri F, Mignot G, Yamazaki T et al (2013) The intestinal microbiota modulates the anticancer immune effects of cyclophosphamide. Science 342:971–976
Wang Z, Klipfell E, Bennett BJ et al (2011) Gut flora metabolism of phosphatidylcholine promotes cardiovascular disease. Nature 472:57–63
Welliver RC, Ogra PL (2008) Effects of early environment on mucosal immunologic homeostasis, subsequent immune responses and disease outcome. Nestle Nutr Workshop Ser Pediatr Program 61:145–181
Yan J, Smyth MJ, Teng MWL (2018) Interleukin (IL)-12 and IL-23 and their conflicting roles in cancer. Cold Spring Harb Perspect Biol 10(7):a028530
Zhang X, Liu Q, Liao Q, Zhao Y (2020) Pancreatic cancer, gut microbiota and therapeutic efficacy. J Cancer 11(10):2749–2758
Zhang Z, Tang H, Chen P, Xie H, Tao Y (2019) Demystifying the manipulation of host immunity, metabolism, and extraintestinal tumors by the gut microbiome. Signal Transduct Target Ther 4:41
Zheng Y, Fang Z, Xue Y, Zhang J et al (2020) Specific gut microbiome signature predicts the early-stage lung cancer. Gut Microbes 2:1–13

Cancer Microbiome and Immunotherapy: Understanding the Complex Responses Between Microbes, Immunity, and Cancer

6

Kishore Kumar Godisela, Badithala Siva Sai Kiran, and Pallaval Veera Bramhachari

Abstract

The microbiome is the inherited substance of the numerous microscopic organisms that exist on and within the human body, including bacteria, protozoa, fungi, and viruses. The human gut microbiota is dominatingly made out of four groups of microbial phyla: Firmicutes, Bacteroides, Actinobacteria, and Proteobacteria; this entire microbiome assumes a significant job to the improvement of immunity. However, disruption of this homeostatic host–microorganism relationship can promote disease pathogenesis, such as in autoimmune diseases and cancer. Current investigations additionally show with the intention of the gut microbiome might influence the reaction to cancer treatment, by balancing the host cell inflammatory reaction. As the investigation of the microbiome is growing, various endeavors are being prepared to incline the range on the "ideal" microbiome region. In this regard, this chapter extensively discusses various types of cancers and the role of the microbiome in their treatment.

Keywords

Microbiome · Cancer · Microbial metabolites · Gastric malignancy · Fecal microbial transplantation

K. K. Godisela (✉)
Assistant Professor of Biochemistry, SRR Govt. Arts & Science College, Karimnagar, Telangana, India

B. S. S. Kiran
College of Pharmaceutical Sciences and Research, Krishna University, Machilipatnam, Andhra Pradesh, India

P. V. Bramhachari
Department of Biotechnology, Krishna University, Machilipatnam, Andhra Pradesh, India

P. V. Bramhachari (ed.), *Microbiome in Human Health and Disease*,
https://doi.org/10.1007/978-981-16-3156-6_6

6.1 Introduction

Humans are by most part microorganisms, more than a hundred trillion of them. Microscopic organisms predominate our human cells ten to one. The lion's share lives in our gut, particularly in the stomach-related organ. The microbiome is the inherited substance of the significant number of microscopic organisms—bacteria, protozoa, fungi, and viruses—that exist on and within the human body. The number of genes in all the microbes in one person's microbiome is two hundred times the number of genes in the person genome. The commensal microbiota in the host coexist in a symbiotic interaction, with every beneficial advantage and the wellness of the host could be considered as a superorganism. Much late research has centered on the bacterial part of the microbiota. By and large, a strong human body is contained roughly thirty trillion cells and is possessed by around thirty-nine trillion microscopic organisms (Sender et al. 2016). Microbial cells in the microbiome help us absorb nutrients, guide our immune system, protect us from other pathogenic microbes that cause infection, and manufacture supplements such as hormones, essential vitamins (riboflavin, thiamine, and B12), and other bioactive compounds (Vitamin K, which is required for blood clotting) that the host cannot obtain. (McFall-Ngai and Ruby 1991; Hooper et al. 2001; Lupp et al. 2012; Lepage et al. 2013). However, disturbance of this homeostatic host–microorganism relationship can promote disease pathogenesis, such as various autoimmune diseases (Frank et al. 2007; Paulos et al. 2007; Jenq et al. 2012).

6.2 Microbiome for Well-Being

The microbiome is fundamental for a person's turn of events, immunity, and nourishment. The microscopic organisms existing in and on us are not intruders except rather gainful settlers. Autoimmune disorders like hypoglycemia, rheumatoid arthritis, muscular dystrophy, multiple sclerosis, and fibromyalgia are related to brokenness in the microbiome. Infection-causing microorganisms build up over time, modifying gene activity and metabolic operations and ensuing in anomalous immune reactions contrary substances and tissues typically exiting in the body. Autoimmune disorders radiate an impression of being transferred in families not by DNA inheritance yet by gaining the family's microbiome.

6.3 The Microbiome and Immunity

The human gut microbiota is dominatingly made out of four groups of microbial phyla: Firmicutes, Bacteroides, Actinobacteria, and Proteobacteria; this entire microbiome assumes a significant job in the improvement of immunity (Dominguez-Bello et al. 2019). The microbiome is defined as the collective genomes of microorganisms within a network, whereas the term microbiota refers to the organisms as a whole. Inside a person, nearby trillion of microorganisms as various as human cells–which associate among the host continually at various locales (counting the epidermis and mucosal coverings, for example, the gastrointestinal tract) all through the turn of events (Fig. 6.1). In this manner, it is not amazing that they expect such a huge job in various host capacities including resistance (Morgan and Huttenhower 2012).

The crosstalk among microbiota and the immune framework at the level of the gut is basic and not just takes into consideration the resistance of commensal microscopic organisms and oral food antigens, yet furthermore enables the immune framework to perceive and assault shrewd microorganisms in this way forestalling bacterial intrusion and disease. Notwithstanding impacting restricted immune reactions, this microbiota likewise has more extensive impacts adding to intrinsic and adaptable immunity at different levels. This idea is upheld in preclinical models; germ-free mouse that needs intestinal microbiota is renowned to contain extreme imperfections in immunity, with a missing mucous coat, modified immunoglobulin A discharge, and diminished dimension and usefulness of Peyer's patches and depleting mesenteric lymph hubs (Johansson et al. 2015; Spiljar et al. 2017). Considering this, retarded immunomodulatory effects are surrendered by transferring the gut microbiome of wild-type mice to laboratory mice, and considerably after a few generations of reproduction, the effects stay. In this manner, it is favorable for combating viral diseases merely as mutagenic and inflammation-induced tumorigenesis (Rosshart et al. 2017).

The arrangement of the gut, which holds a mucosa involving a single epithelial layer comprised of intestinal epithelial cells (IEC) and intraepithelial lymphocytes, energizes this coordinated effort with the immune framework. The IECs have Paneth cells that discharge anti-bacterial polypeptides and goblet cells that emit mucus, which thus overlie the epithelial layer. Underneath the mucosal sheet conceals the lamina propria, a connexion tissue film enclosing Peyer's patches, and a large group of additional immune cells comprising antigen-presenting cells and native lymphoid cells, just as CD4+ and CD8+ B plus T cells. This gut-related lymphoid tissue speaks to the biggest segment of the immune organization inside the body and impacts immune reactions jointly local and systemically (Fig. 6.2).

Native immunity is fostered via appreciation of pathogen-linked molecular patterns (PAMPs) (namely, lipopolysaccharide as well as flagellin) by pattern recognition receptors (such as toll-like receptors existing on IECs as well as native immune performers inside the gut). Metabolites delivered by microorganisms perhaps likewise influence native immunity via the manufacture of short-chain fatty acids (SCFAs), which, within various main exercises, being appeared to expand immunity through IgA making by plasma cells (Pabst 2012). IgA works by effectively affecting bacterial virulence by freezing microbial attachment to epithelial cells; adhesion, trapping, and freezing microbial attachment to epithelial cells (Pabst 2012).

The role of the commensal microbiota in tweaking physiology becomes especially clear when routinely raised explicit sans pathogen (SPF) mice are compared to sans germ (GF, axenic) mice. The gut microbiota is personally associated with the turn of events and guidelines of the safe framework, particularly as for nearby

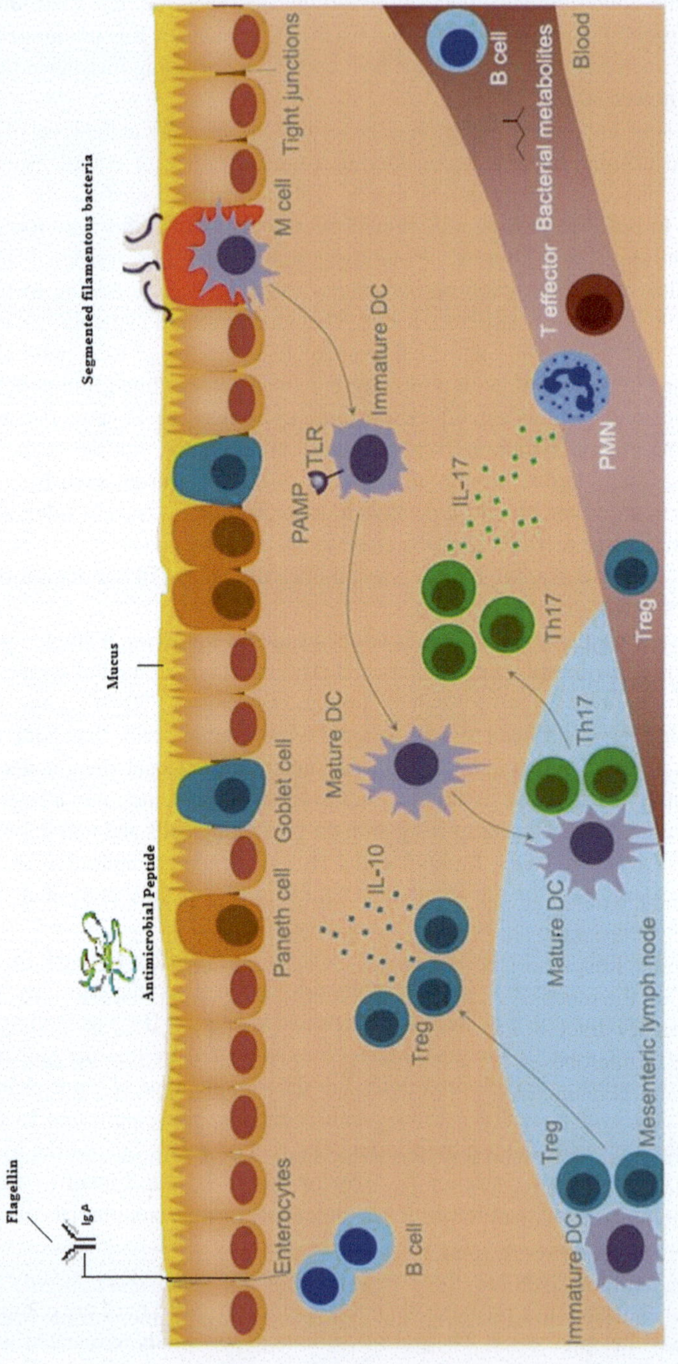

Fig. 6.1 The microbiome and immunity

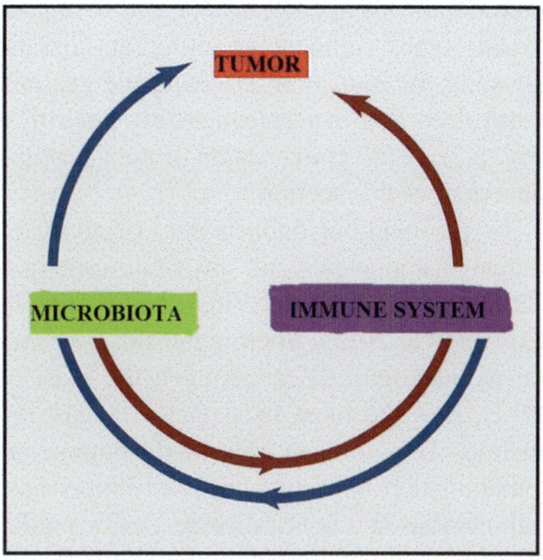

Fig. 6.2 The immune system, microbiota, and tumors are tightly associated

mucosal insusceptibility. Mainly focuses on GF mouse models as shown below: (1) littler mesenteric lymph hubs (MLH) and unusual higher endothelial arteries through poor lymphocyte restricting (Smith et al. 2007); (2) less and littler Peyer's patches which need germinal focuses (Lecuyer et al. 2014); and (3) absence of lymphoid follicles inside the intestinal lamina propria (LP), however nearness of early crypto patches which can form into useful secluded lymphoid follicles atop microbial habitation(Bouskra et al. 2008). Systemic native immune amelioration is additionally impacted by the symbiotic microbiota, with different paths of proof demonstrating stimulatory consequences for myelopoiesis at the degree of granulocyte-macrophage progenitors inside the bone marrow including in the periphery, just as on the capacity of dendritic cells, macrophages, and neutrophils (Gorjifard and Goldszmid 2016).

This chapter focuses on:

1. The connection between infection and malignant growth.
2. The inclination of malignancy patients to obtain infection disease(s);
3. The activity of the microbiota in malignant growth vulnerability;
4. Nitty Gritty of the microbiome in malignant growth treatment.

6.4 Infection and Malignant Growth

Human microbiome interruption is related to different kinds of cancer and for each report of the International Agency for Cancer Research (IARC), just 10 species were recognized as cancer-causing specialists to humans (de Martel et al. 2012). The intensity just as marks of immune reactions might be directed by the cross-reactivity among tumor and microbial antigens or by the incitement of example pattern recognition receptors (PRRs) via microbe-related antigens, Moreover, microorganisms have been appeared to impact carcinogenesis for various malignant growth types, influencing different host factors, directing disease trademarks, adjusting inflammation, affecting the genomic stability of host cells, and creating metabolites that can epigenetically control host gene expression (Bultman 2014) (Fig. 6.2). A full conversation of the job of the microbiota is as a matter of fact past the extent of this section

As per ongoing publications, oncoviruses are responsible for almost 12% of human malignancies and are significant factors in the actuation of oncogenesis. These are Epstein–Barr Virus, Human Papillomavirus, Hepatitis Virus, and HIV (Luo and Ou 2015). Microorganisms contamination of *H. pylori* infect over half of the world populace, rendering it the most regular disease on the planet (Camilo 2017). It is the most focal factor of gastric disease (GC) chance; because of its qualities, H. pylori can adjust to the outrageous acidic states of the stomach, to set up contamination and upset mucosal homeostasis of the host, bringing about gastric pathogenesis and at last disease. A few publications connect the nearness of fungal diseases with a higher danger of creating malignancy. A few strains of filamentous organisms, including *Aspergillus flavus, Aspergillus parasiticus, and Aspergillus nomius*, produce aflatoxins ready to incite liver malignancy and some can advance malignancy movement, such is the situation of expedient *Candida albicans* contaminations (Martins et al. 2008; Ramirez-Garcia et al. 2011).

6.5 The Inclination of Malignancy Patients to Obtain Infection Disease(s)

Microorganisms replicate quickly and stay alive in thickly involved biological niches. Supplement accessibility inside these small-scale natural surroundings is restricted; in this way, organisms go after assets, for example, amino acids sugars (Curtis et al. 2014), zinc (Gielda and DiRita 2012), iron (Deriu et al. 2013), and oxygen (Litvak et al. 2019) and anaerobic electron acceptors (Herp et al. 2019). A higher microbial decent variety suggests more organisms using an increasingly flexible pool of metabolites, representing a test for any bacterium to flourish. Any perturbation bringing about lost microbial burden or assorted variety destabilizes the microbial biological system, making an open door for strains with expanded wellness to multiply. Furthermore, it predisposes the host to diseases caused by either indigenous or exogenous pathogens. It makes the host more susceptible to infection from either indigenous or exogenous pathogens.

The elements of the immune system are sorted out to keep up parity amid the microbiota and to battle against the majority of the microbial attacks. Interestingly, in immune-compromised patients actuated by immunosuppressive medications, including malignant growth chemotherapy, these capacities are hindered and people are at an expanded danger of infection. However, these disease treatments change the associations flanked by the host and microbiota.

6.6 The Activity of the Microbiota in Malignant Growth Vulnerability

Microorganisms take an interest in host metabolic actions, and their metabolites can instigate inflammatory processes, meddling to be decided of tissue cell multiplication and demise. The emerging of malignant growth in certain areas could be actuated by microbial dissimilarities, for example, the more prominent powerlessness to disease in the large intestine because of the elevated microorganism thickness when contrasted with the tiny intestine (Burgess et al. 2014). Subsequently, the microbiota of all organs of the body is extraordinary, and their impact on irritation and carcinogenesis is likewise unmistakable in every organ. There is a developing admiration about the effect of microbiota at various anatomic locales on resistance and for various pathologic situations, including malignancy. This has been incredibly encouraged by the utilization of next-generation sequencing, which has extended our comprehension of the expansiveness and capacity of the microbiota ahead of conventional culturing techniques.

Dysbiosis because of malignancy prompted immunodeficiency, chemotherapy action regimens, or antibiotic employ, could likewise elevate the threat of bloodstream and *Clostridium difficile* contaminations, by upsetting the gut microbiome's capacity to oppose pathogen immigration or by debilitating the intestinal barricade. Tai and colleagues depicted that the recurrence pace of *Clostridium difficile* diseases in hospitalized oncologic kids was more than multiple times higher contrasted with those without malignancy (Tai et al. 2011). Wang and the team likewise discovered diminished diversity plus a bounty of the oral microbiome in patients by intense lymphoblastic leukemia (Wang et al. 2014). Another examination exhibited those pediatric oncology patients by intense lymphoblastic leukemia, beneath a few circles of treatment, demonstrated diminished microbial variety (Rajagopala et al. 2016).

The metabolic action of bacteria might actuate, impair or increase healing toxicity. An examination starting 1993 announced the limit of the microbiome to meddle through medicines in a gathering of Japanese who developed *Herpes zoster* while experiencing malignancy. *Herpes zoster* drug is changed by the typical stomach-related microbiota into a section that made the drug to treat malignant mortally poisonous. In this line of proof, Karin and collaborators expressed that a flawless microbiome is required for the fruitful control of tumor movement (Karin et al. 2014). Microbiota utilizes pivotal systems, for example, inflammation, metabolism, and genotoxicity to adjust cancer origination (Schwabe and Jobin 2013). Recent research also suggests that the gut microbiome's intention may influence the response to cancer treatment by balancing the inflammatory response of the host cell.

6.6.1 Pancreatic Cancer

Pancreatic cancer stays one of the main sources of cancer-related demise worldwide and has a poor forecast (Zambirinis et al. 2014). It is, indeed, a perceived inflammation determined malignancy, with enormous preclinical and clinical proof demonstrating that microscopic organisms likely impact this procedure, since they enact immune receptors, sustaining malignancy-related inflammation. In light of past research in the preceding decade, the gut microbiota could prompt modified adequacy of pharmacotherapeutics in malignant growth treatment (Thomas 2017). Although the express job of the microbiota in host immunity, particularly in the tumor-explicit tumor microenvironment (TME), stays muddled, the collaborations between tumor control and gut microbiota have gotten more interlaced than any other time in recent memory (Alexander et al. 2017; Yi et al. 2018). Microbial diversity changes, including bacterial species, for example, *Actinomycetes, Fusobacterium, Bacteroides, Neisseria, Streptococcus, Porphyromonas,* and *Bifidobacteria* have been related with pancreatic malignancy (Mitsuhashi et al. 2015; Ren et al. 2017). These microorganisms inside tumors may animate host immune reactions and create useful or problematic effects on anticancer treatment, as controlled by pharmacological systems, just as the significant reaction pathways (Cogdill et al. 2018).

6.6.2 Breast Cancer

As previously stated, danger may be linked to the unique microbial environment found in the tissue of the birthplace. An ongoing report announced that the microbes found in breast tissue differ with and without breast malignant growth. The researchers distinguished more significant levels of Comamonadaceae, Enterobacteriaceae, *Bacillus, Staphylococcus*, and Bacteroidetes in breast malignant sample plus, interestingly, healthy tissue uncovered more elevated levels of *Prevotella, Lactococcus, Streptococcus, Micrococcus,* and *Corynebacterium* species. Higher relative bounties of microorganisms that could cause DNA to injure in vitro were recognized in breast malignancy patients, just like a reduction in some lactic acid microscopic organisms, known for their advantageous well-being impacts, including anticarcinogenic properties. This examination brings up significant issues regarding the job of the mammary microbiome in balancing the danger of breast cancer development (Urbaniak et al. 2016).

This reality raises a significant issue: would women be able to modulate their breast microbiome to forestall elevated levels of *Escherichia coli* or *Staphylococcus aureus* colonization? A few examinations uncovered that drinking fermented food, for example, kefir, is related to a lower danger of breast malignancy. Different examinations utilizing living models indicated to orally ingested Lactobacillus may have a defensive job in opposition to breast malignancy improvement (Urbaniak et al. 2016). Considering this, forthcoming investigations are expected to exhibit the job of probiotics as a protective measure in opposition to breast malignancy. It was uncovered that a few antibiotics agents, for example, metronidazole, clarithromycin, and ciprofloxacin prompt a decrease within the biodiversity/plenitude of a few bacterial networks upsetting the balance of the gut microbiome, which comprises a higher danger of breast malignant growth (Jakobsson et al. 2010). Another investigation created by Goedert and collaborators distinguished a connection among fecal microbiome lesser assorted variety in postmenopausal female and breast cancer. This examination demonstrated that 87% of the females had estrogen receptor-positive tumors when contrasted with healthy manage females (Goedert et al. 2015).

6.6.3 3.3 Colorectal Cancer

Colorectal cancer (CRC), along with breast, prostate, and lung cancer, is one of the most well-known cancers. It likewise shows up more much of the time in males than females and, in the two sexes joined, shows the fourth most noteworthy death rate, after lung, breast, and prostate cancer (Beaulieu 2020). The human gastrointestinal microbiome assumes a significant function in managing immune standing and, in this way, is a further possible prescient biomarker used for CRC immunotherapy (Xu et al. 2020). Preclinical investigations have discovered that the degrees of explicit microbes are significantly elevated in tumors contrasted with those in close by usual tissue (Sivan et al. 2015). Bolster the speculation that a wellspring of between topic heterogeneity concerning programmed cell death protein1 (PD-L1) restorative efficiency might be the symphony of gastrointestinal microorganisms (Vetizou et al. 2015).

Additionally Vetizou et al. established that the efficacy of the cytotoxic T-lymphocyte-associated Ag 4 (CTLA-4) obstruction is impacted by the synthesis of *Bacteroides fragilis* and additionally *Bacteroides thetaiotaomicron* and *Burkholderiales* microbiota. To utilize the gastrointestinal microorganisms as a prescient biomarker in CRC, a superior comprehension of the useful function of microbiota is required, just like a progression of clinical examinations making an interpretation of preclinical outcomes to endorsed biomarkers (Vetizou et al. 2015).

6.6.4 Gastric Malignancy

Gastric malignancy is a major worldwide health issue, with more than one million new cases each year (Bray et al. 2018). Gastric chronic illness with Helicobacter pylori incites diminished acid discharge, which might animate the expansion of diverse microscopic organisms in the gastric mucosa. This change within the gastric microorganisms could prompt harm to the gastric mucosa and put into cancer. These outcomes are as per a preceding publication, indicating lesser bacterial variety in patients among gastric malignant contrasted and patients with non-atrophic gastritis (Aviles-Jimenez et al. 2014).

6.7 Nitty Gritty of **the** Microbiome in Malignant Growth Treatment

As of late reports exhibited in translational investigations that microscopic organisms inside the gut of malignancy patients can regulate retaliation to anticancer immunotherapy. In particular, by looking at the gut microorganisms of melanoma patients, huge contrasts were seen in the variety and make-up of patient gut microbiomes in responders to non-responders (Routy et al. 2018). There is developing proof for the influence of the gut microorganisms on the reactions to different types of malignant treatments; there are additionally a few components throughout the gut microscopic that may impact such responses. Nevertheless, it is turning out to be progressively certain to anticipate that there is a multifaceted bidirectional connection between the gut microscopic organisms and antimalignancy treatments; as well as, yet not restricted to, chemotherapy, physical radiation, molecularly focused on treatment, checkpoint inhibitors, cytokine-based treatment, receptive cell treatment (RCT), and operators that target inborn immunity (Table 6.1). A

Table 6.1 Multidimensional relationships between the gut microbiota, the tumor microenvironment, and systemic immunity

Treatment type	Tumor	Gut	Systemic response
Chemotherapy	Antitumor apoptosis impact by Th1 and Th17 effector cells	Gut dysbiosis brought about by changed proportion of gut microbiota	Bacterial translocation to lymphoid organs; increment of Th1 and Th17 subset expanded
Checkpoint and agonists	Develop dendritic cells initiate CD4+ and CD8+ T cells, diminished intratumoral FOXP3 Tregs, high intrinsic effector cells	Extension in CD11b DC actuation instigates IL-12 ward Th1 reaction in the lamina propria, decrease in incidentally determined colonic Tregs	Enlistment of CD4+ Th1 and CD8+ T cells
Cytokine therapy	Modifies resistant homeostasis in the tumor builds TNFα, apoptosis, neutrophil invade. Th1 and IL-17 increment	Treatment causes expanded intestinal harmfulness and gut shortening. Sodium take-up and increments in LPS cause a pathogenic inflammatory reaction	Systemic inflammatory reaction prompts proinflammatory cytokine reaction and an expansion in a supplement course enactment
Adoptive cell therapy	Multiplication of completely actuated T cells intercede tumor slaughtering and an expansion in cytokines IL-2,7, and 15	Lymphodepletion triggers intestinal injury prompting harm to gut epithelium and bacterial translocation to mesenteric lymph hubs	Increment in LPS serum level, cytokine sink, exhaustion of administrative T cells, MDSCs, NK cell, and B administrative cells

concise portrayal of the cooperation of organisms with every one of these helpful systems is incorporated later.

6.7.1 Chemotherapy

Chemotherapeutic regimens encompass to upset the intestinal boundary and advance antitumor adequacy in mice as well as humans. Firstly in mice and humans by way of sophisticated malignancies, secondly as in transplantable mice for CRC, cure with cyclophosphamide exhausted circulating T regulatory cells and supported Th1 and Th17 memory reactions (Ghiringhelli et al. 2007). Additionally, these progressions were related to the rebuilding of natural killer cell and T cell work plus the translocation of certain segments of the gut microorganisms to auxiliary lymphoid organs (Viaud et al. 2013). Furthermore, chemotherapeutic factors have been demonstrated, for example platinum-related oxaliplatin and cisplatin medicines, which can lead to explicit microorganisms (*L.johnsonii and E. hirae*) being translocalized into 2^0 lymphoid organs beginning with the intestinal lumen (Alexander et al. 2017).

This understanding showed that ideal chemotherapeutic adequacy was to some extent the consequence of undamaged microbes, which helped to make myeloid-derived usefulness in the tumor microenvironment (TME) (Iida et al. 2013). Taken jointly, these examinations bolster the idea that chemotherapeutics could differentially impact gut microbes in a huge amount of settings.

6.7.2 Checkpoint Inhibitors

Different examinations have demonstrated the powerful capacity of the microbes to control the antitumor reaction of the immune response framework with regard to checkpoint blockade (Bullman et al. 2017; Geva-Zatorsky et al. 2017; Zitvogel et al. 2017). Furthermore, the ability of gut microorganisms to potentiate the gut's reaction to the fringe is regulated to some extent, with specific bacterial populations seen as distinct from testee and non-testee patient cohorts. The immune reaction related with checkpoint restraint employing hostile to programmed cell death protein 1 (PD-1) has been additionally contemplated, whereby an improved cluster of differentiation 8+ T cell antitumor role has been appeared in malignancy, corresponding with the nearness of a "positive" microbial signature, include Bacteroidales and Burkholderiales (Vetizou et al. 2015). Moreover, key bits of knowledge into the basic and particular components engaged with against cytotoxic T-lymphocyte-associated protein 4 and hostile to PD-1 motivating interspecies populace dynamics may provide further understanding into the general instruments driving microscopic organisms interceded ameliorate of anticancer immunity. Take note that a recent study found that administering live microbial cultures to testees during immune checkpoint blockade can result in a stronger anticancer response than untreated testees, though these perceptions could be due to a combination of factors including diet, lifestyle, and well-being (Backhed et al. 2007).

6.7.3 Microbial Metabolites

Short-chain fatty acids (SCFA) are the microbial metabolites that secure intestinal mucosal obstruction and epithelial cells in opposition to death and harm, diminishing tissue penetrability and inflammation (Fukuda et al. 2011; Mathewson et al. 2016). Besides, SCFAs assume a crucial job in producing and controlling Treg lymphocytes restricting to inhabitant and systemic inflammation (a sign of malignant growth) (Arpaia et al. 2013). This may theoretically clarify the great impacts of abundant SCFA-producing microorganisms Blautia on account of hepatocarcinoma. Recent reports recommend that microbial metabolites intercede messages among the commensal microorganisms and the immune response, influencing the harmony among pro- and anti-inflammatory procedures.

Besides, a high-fiber diet, normal of veggie lover propensities, brings about increasing the SCFAs pool (Tomasello et al. 2016). The SCFAs are a subset of saturated fats, chiefly represented by butyrates, propionates, and acetic acid derivations, and they are the aftereffect of a multifaceted fermentation procedure of digestible to non-digestible polysaccharides. They are for the most part created in the proximal colon and need the presence of explicit microscopic organisms including especially Firmicutes and Bacteroides phylum (Tan et al. 2014). There is copious proof that an elevated level of SCFAs encourages healthy microorganisms and has different advantages, for example, mitigating and anti-tumorigenic impacts, just as antimicrobial activities.

6.7.4 Cytokine Therapy

Cytokine-based treatments (e.g., interleukin-2, Interferon-gamma) may likewise be affected by the microorganisms; particularly as gut bacterial dysbiosis is connected to distort immune activation which couples with irregular manufacture of provocative cytokines. Research connecting the individual gut microbiome to provocative cytokine making have uncovered the impact of bacterial commensals controlling fundamental immune action with means host–bacterial collaborations, for example, interferon α and interferon γ that partner with explicit bacterial metabolic pathways, for example, metabolisms of palmitoleic acid and tryptophan to tryptophol translation. Palmitoleic acid hampers the proinflammatory cytokine reaction of the host cell (Schirmer et al. 2016; Ter Horst et al. 2016). As an important element of the Human Functional Genomics Project (HFGP), they described the environmental, hereditary and bacterial factors associated with cytokine production in a cohort of five hundred healthy participants by various microbial stimuli.

6.7.5 Adoptive Cell Treatment

Adoptive cell treatments (ACT) are an additional hopeful way to deal with handling progressed oncological ailments. As of late research work, a rodent model of cervical malignancy, announced to adoptive T cell treatment viability was essentially affected by the make-up of the gut microorganisms when the mouse was medicated with versus without antibiotic agents or heterologous fecal transfer. The transloca-tion of gut microbes to the i.p. space, because of epithelial layer damage, can incite momentary contamination with the systemic rise of interleukins (IL-12) that influences the viability of adaptive cell treatment (Uribe-Herranz et al. 2018). Altogether, these findings feature the indispensable pretended by the gut microorganisms in regulating foundational plus anticancer adequacy of adoptive cell treatment.

6.8 Concluding Remarks

Gut microbiota, the "overlooked organ," influences different human physiological procedures prevalently resistance and it is still ineffectively investigated until this time. As the investigation of the microbiome is growing, various endeavors are being prepared to incline the range on the "ideal" microbiome region. Focusing on the gut and intracanceral microbiome might give already unexplored intend to develop a systemic treatment for various kinds of tumors. These possibly will go from straightforward nutritional changes to novel adoptive cell-based treatments. Fecal microbiota transplantation (FMT) is a clinical methodology that restores solid microorganisms in the colon by introducing stool by colonoscopy or bowel purge from a healthy person contributor. The relative simplicity with which the microbiota can be concentrated makes it particularly interesting, given that bacterial modulators are a useful extra for a surviving anti-tumor regime, especially as they have a significant impact on the effectors of the immune system. Anyway complications exist with so much methodology, as we do not yet have the foggiest idea about the synthesis of "ideal" gut microorganisms in the milieu of malignant growth cure, nor do we identify whether this would concern across medicines just as from corner to corner malignant types.

Furthermore, the decision of a specific methodology will be optional for thorough and cautious identifying in a clinical preliminary, as unquestionably there is an absence of accord in regards to which populations to target, how to focus on these, and what end-focuses to quantify. Evidently, a lot of research needs to be done to additionally explain the effect plus targetability of the gut microbiota on malignant growth and to completely fathom its whole potential. In any case, current endeavors will without a doubt end up being significant.

Commensal microbes in the gut lumen affect the immune system within the mucosa of the gut, exhausting mesenteric lymph nodes, and so on. The resistant framework in like manner can modify the gut microbiota. Goblet cells form a thick mucous defensive layer that protects the mucosa. Without germ creatures, this mucosal layer is severely deficient. Plasma cells in the lamina propria transmit IgA into the lumen of the gut. Paneth cells transmit different number of hostile to microbial peptides; their development is upgraded because of motioning from nearby insusceptible cells in light of the microbiota.

The microorganisms in the digestive tract can regulate tumor advancement through different various systems that can be predominantly separated into immediate and roundabout. The immediate components (blue line) are those in which microbial items legitimately advance tumor development. Of note, it has been indicated that a few inadequacies in the immune system can permit the development of certain procarcinogenic microbes. The backhanded instruments (red line) are those in which the microbes fundamentally cannot advance tumor commencement and development except if they cooperate with the invulnerable framework, which at last advances disease. At long last, it is additionally conceivable that inadequacies in explicit components of the immune response permit the extension of specific microorganisms, which, thus, initiate a protumorigenic immune reaction (blue + red line).

Acknowledgements The authors are grateful to their respective academic institutions for the support extended.

Conflict of Interest The authors declare that they have no competing interests.

References

Alexander JL, Wilson ID et al (2017) Gut microbiota modulation of chemotherapy efficacy and toxicity. Nat Rev Gastroenterol Hepatol 14(6):356–365

Arpaia N, Campbell C et al (2013) Metabolites produced by commensal bacteria promote peripheral regulatory T-cell generation. Nature 504(7480):451–455

Aviles-Jimenez F, Vazquez-Jimenez F et al (2014) Stomach microbiota composition varies between patients with non-atrophic gastritis and patients with intestinal type of gastric cancer. Sci Rep 4:4202

Backhed F, Manchester JK et al (2007) Mechanisms underlying the resistance to diet-induced obesity in germ-free mice. Proc Natl Acad Sci U S A 104(3):979–984

Beaulieu JF (2020) Colorectal Cancer research: basic, preclinical, and clinical approaches. Cancers 12(2):416

Bouskra D, Brezillon C et al (2008) Lymphoid tissue genesis induced by commensals through NOD1 regulates intestinal homeostasis. Nature 456(7221):507–510

Bray F, Ferlay J et al (2018) Global cancer statistics 2018: GLOBOCAN estimates of incidence and mortality worldwide for 36 cancers in 185 countries. CA Cancer J Clin 68(6):394–424

Bullman S, Pedamallu CS et al (2017) Analysis of fusobacterium persistence and antibiotic response in colorectal cancer. Science 358(6369):1443–1448

Bultman SJ (2014) Emerging roles of the microbiome in cancer. Carcinogenesis 35(2):249–255

Burgess SL, Buonomo E et al (2014) Bone marrow dendritic cells from mice with an altered microbiota provide interleukin 17A-dependent protection against Entamoeba histolytica colitis. mBio 5(6):e01817

Camilo V, Sugiyama T, Touati E (2017) Pathogenesis of helicobacter pylori infection. Helicobacter 22(Suppl 1):e12405

Cogdill AP, Gaudreau PO et al (2018) The impact of Intratumoral and gastrointestinal microbiota on systemic Cancer therapy. Trends Immunol 39(11):900–920

Curtis MM, Hu Z et al (2014) The gut commensal Bacteroides thetaiotaomicron exacerbates enteric infection through modification of the metabolic landscape. Cell Host Microbe 16(6):759–769

de Martel C, Ferlay J et al (2012) Global burden of cancers attributable to infections in 2008: a review and synthetic analysis. Lancet Oncol 13(6):607–615

Deriu E, Liu JZ et al (2013) Probiotic bacteria reduce salmonella typhimurium intestinal colonization by competing for iron. Cell Host Microbe 14(1):26–37

Dominguez-Bello MG, Godoy-Vitorino F et al (2019) Role of the microbiome in human development. Gut 68(6):1108–1114

Frank DN, Amand ALS et al (2007) Molecular-phylogenetic characterization of microbial community imbalances in human inflammatory bowel diseases. Proc Natl Acad Sci U S A 104 (34):13780–13785

Fukuda S, Toh H et al (2011) Bifidobacteria can protect from enteropathogenic infection through production of acetate. Nature 469(7331):543–547

Geva-Zatorsky N, Sefik E et al (2017) Mining the human gut microbiota for immunomodulatory organisms. Cell 168(5):928–943

Ghiringhelli F, Menard C et al (2007) Metronomic cyclophosphamide regimen selectively depletes CD4+CD25+ regulatory T cells and restores T and NK effector functions in end stage cancer patients. Cancer Immunol Immunother 56(5):641–648

Gielda LM, Dirita VJ (2012) Zinc competition among the intestinal microbiota. mBio 3(4):e00171–e00112

Goedert JJ, Jones G et al (2015) Investigation of the association between the fecal microbiota and breast cancer in postmenopausal women: a population-based case-control pilot study. J Natl Cancer Inst 107(8):djv147

Gorjifard S, Goldszmid RS (2016) Microbiota-myeloid cell crosstalk beyond the gut. J Leukoc Biol 100(5):865–879

Herp S, Brugiroux S et al (2019) Mucispirillum schaedleri antagonizes Salmonella virulence to protect mice against colitis. Cell Host Microbe 25(5):681–694

Hooper LV, Wong MH et al (2001) Molecular analysis of commensal host-microbial relationships in the intestine. Science 291(5505):881–884

Iida N, Dzutsev A et al (2013) Commensal bacteria control cancer response to therapy by modulating the tumor microenvironment. Science 342(6161):967–970

Jakobsson HE, Jernberg C et al (2010) Short-term antibiotic treatment has differing long-term impacts on the human throat and gut microbiome. PLoS One 5(3):e9836

Jenq RR, Ubeda C et al (2012) Regulation of intestinal inflammation by microbiota following allogeneic bone marrow transplantation. J Exp Med 209(5):903–911

Johansson ME, Jakobsson HE et al (2015) Normalization of host intestinal mucus layers requires long-term microbial colonization. Cell Host Microbe 18(5):582–592

Karin M, Jobin C et al (2014) Chemotherapy, immunity and microbiota--a new triumvirate? Nat Med 20(2):126–127

Lecuyer E, Rakotobe S et al (2014) Segmented filamentous bacterium uses secondary and tertiary lymphoid tissues to induce gut IgA and specific T helper 17 cell responses. Immunity 40 (4):608–620

Lepage P, Leclerc MC et al (2013) A metagenomic insight into our gut's microbiome. Gut 62 (1):146–158

Litvak Y, Mon KKZ et al (2019) Commensal Enterobacteriaceae protect against Salmonella colonization through oxygen competition. Cell Host Microbe 25(1):128–139

Luo GG, Ou JH (2015) Oncogenic viruses and cancer. Virol Sin 30(2):83–84

Lupp C, Skipper M et al (2012) Gut microbes and health. Nature 489(7415):219

Martins HM, Almeida I et al (2008) Interaction of wild strains of aspergilla with Aspergillus parasiticus ATCC15517 and aflatoxin production. Int J Mol Sci 9(3):394–400

Mathewson ND, Jenq R et al (2016) Corrigendum: gut microbiome-derived metabolites modulate intestinal epithelial cell damage and mitigate graft-versus-host disease. Nat Immunol 17 (10):1235

McFall-Ngai MJ, Ruby EG (1991) Symbiont recognition and subsequent morphogenesis as early events in an animal-bacterial mutualism. Science 254(5037):1491–1494

Mitsuhashi K, Nosho K et al (2015) Association of Fusobacterium species in pancreatic cancer tissues with molecular features and prognosis. Oncotarget 6(9):7209–7220

Morgan XC, Huttenhower C (2012) Chapter 12: human microbiome analysis. PLoS Comput Biol 8 (12):e1002808

Pabst O (2012) New concepts in the generation and functions of IgA. Nat Rev Immunol 12 (12):821–832

Paulos CM, Wrzesinski C et al (2007) Microbial translocation augments the function of adoptively transferred self/tumor-specific CD8+ T cells via TLR4 signaling. J Clin Invest 117 (8):2197–2204

Rajagopala SV, Yooseph S et al (2016) Gastrointestinal microbial populations can distinguish pediatric and adolescent acute lymphoblastic leukemia (ALL) at the time of disease diagnosis. BMC Genomics 17(1):635

Ramirez-Garcia A, Gallot N et al (2011) Molecular fractionation and characterization of a Candida albicans fraction that increases tumor cell adhesion to hepatic endothelium. Appl Microbiol Biotechnol 92(1):133–145

Ren Z, Jiang J et al (2017) Gut microbial profile analysis by MiSeq sequencing of pancreatic carcinoma patients in China. Oncotarget 8(56):95176–95191

Rosshart SP, Vassallo BG et al (2017) Wild mouse gut microbiota promotes host fitness and improves disease resistance. Cell 171(5):1015–1028

Routy B, Gopalakrishnan V et al (2018) The gut microbiota influences anticancer immunosurveillance and general health. Nat Rev Clin Oncol 15(6):382–396

Schirmer M, Smeekens SP et al (2016) Linking the human gut microbiome to inflammatory cytokine production capacity. Cell 167(7):1897

Schwabe RF, Jobin C (2013) The microbiome and cancer. Nat Rev Cancer 13(11):800–812

Sender R, Fuchs S et al (2016) Revised estimates for the number of human and Bacteria cells in the body. PLoS Biol 14(8):e1002533

Sivan A, Corrales L, Hubert N, Williams JB, Aquino-Michaels K, Earley ZM, Benyamin FW, Lei YM, Jabri B, Alegre ML et al (2015) Commensal Bifidobacterium promotes antitumor immunity and facilitates anti-PD-L1 efficacy. Science 350:1084–1089

Smith K, McCoy KD et al (2007) Use of axenic animals in studying the adaptation of mammals to their commensal intestinal microbiota. Semin Immunol 19(2):59–69

Spiljar M, Merkler D et al (2017) The immune system bridges the gut microbiota with systemic energy homeostasis: focus on TLRs, mucosal barrier, and SCFAs. Front Immunol 8:1353

Tai E, Richardson LC et al (2011) Clostridium difficile infection among children with cancer. Pediatr Infect Dis J 30(7):610–612

Tan J, McKenzie C et al (2014) The role of short-chain fatty acids in health and disease. Adv Immunol 121:91–119

Ter Horst R, Jaeger M et al (2016) Host and environmental factors influencing individual human cytokine responses. Cell 167(4):1111–1124

Thomas H (2017) Pancreatic cancer: Intra-tumour bacteria promote gemcitabine resistance in pancreatic adenocarcinoma. Nat Rev Gastroenterol Hepatol 14(11):632

Tomasello G, Mazzola M et al (2016) Nutrition, oxidative stress and intestinal dysbiosis: influence of diet on gut microbiota in inflammatory bowel diseases. Biomed Pap Med Fac Univ Palacky Olomouc Czech Repub 160(4):461–466

Urbaniak C, Gloor GB et al (2016) The microbiota of breast tissue and its association with breast Cancer. Appl Environ Microbiol 82(16):5039–5048

Uribe-Herranz M, Bittinger K et al (2018) Gut microbiota modulates adoptive cell therapy via CD8alpha dendritic cells and IL-12. JCI Insight 3(4):e94952

Vetizou M, Pitt JM et al (2015) Anticancer immunotherapy by CTLA-4 blockade relies on the gut microbiota. Science 350(6264):1079–1084

Viaud S, Saccheri F et al (2013) The intestinal microbiota modulates the anticancer immune effects of cyclophosphamide. Science 342(6161):971–976

Wang Y, Xue J et al (2014) Oral microbiota distinguishes acute lymphoblastic leukemia pediatric hosts from healthy populations. PLoS One 9(7):e102116

Xu S, Yin W et al (2020) Foes or friends? Bacteria enriched in the tumor microenvironment of colorectal Cancer. Cancers 12(2):372

Yi M, Jiao D et al (2018) Biomarkers for predicting efficacy of PD-1/PD-L1 inhibitors. Mol Cancer 17(1):129

Zambirinis CP, Pushalkar S et al (2014) Pancreatic cancer, inflammation, and microbiome. Cancer J 20(3):195–202

Zitvogel L, Daillere R et al (2017) Anticancer effects of the microbiome and its products. Nat Rev Microbiol 15(8):465–478

7 Dynamics of Respiratory Microbiome Profiles Contributes to Imbalance and Lung Dysbiosis in the Respiratory Tract

Harish Babu Kolla, Prakash Narayana Reddy, and Pallaval Veera Bramhachari

Abstract

Lungs and the other regions of respiratory tract are not sterile. Respiratory system consists of separate and balanced microbial communities called respiratory microbiome. Respiratory microbiome is the set of diverse microbial communities residing on the mucosal surface of respiratory tract. The presence of microbiota in respiratory tract is identified by 16S rRNA amplicon sequencing and the diversity of microbes in the respiratory tract was clearly understood. Microbiome equilibrium is maintained in healthy state of an individual. The composition of microbiome gets altered during disease. During health conditions the shape of microbiome is maintained by three detrimental factors: microbial immigration, emigration, and relative reproduction of microbes. Sometimes balance among these three factors is disturbed and the constancy of microbiota is affected. This condition is called as respiratory dysbiosis. There are certain factors such as smoking, antibiotics, chronic respiratory disorders (CRDs), and lung transplantation that contribute to imbalance and dynamic change of microbiome in respiratory tract. In this chapter, we discussed briefly the changes in microbial communities in response to these factors.

Keywords

Microbiome · Respiratory tract · Mucosal epithelium · Dysbiosis · Microbial diversity

H. B. Kolla · P. N. Reddy (✉)
Department of Biotechnology, Vignan's Foundation for Science Technology and Research (Deemed to be University), Vadlamudi, Guntur, Andhra Pradesh, India

P. V. Bramhachari
Department of Biotechnology, Krishna University, Machilipatnam, Andhra Pradesh, India

P. V. Bramhachari (ed.), *Microbiome in Human Health and Disease*, https://doi.org/10.1007/978-981-16-3156-6_7

7.1 Overview of Respiratory Microbiome

Respiratory system is not sterile and free from microbes. Lungs and other components of respiratory system have a separate microbial composition called respiratory microbiome. The microbiome helps in immune system regulation and protecting against several pathogenic microbes during a healthy state. This normal microbiome composition gets altered during disease conditions. When an individual is diseased, the normal microbiome of the respiratory systems gets slightly altered towards composition of pathogenic microbial communities probably Gammaproteobacteria, a class of bacteria that causes several respiratory system associated infections (RSAI) (Huffnagle et al. 2016). Because the respiratory tract is warm and moist, this environment is markedly convenient for microbes to grow. Till now there is no scientific evidence to show about sterile respiratory environment. Respiratory microbiome is different from the normal gut microbiota in its nature and dimension. The nature of microbiota in lungs is more dynamic than to GI tract. The movement of microbes from mouth to anus is unidirectional, whereas in the case of respiratory tract it is bidirectional. This is because of the variant environmental conditions in gastrointestinal and respiratory tracts contributing to distinguished microbial communities.

Presence and dynamics of respiratory microbiome is better explained with life in Antarctica as a model example. There might be a presence of small and constant microbial communities even in health conditions. If the environmental conditions in lung alter from optimal parameters, long-term microbial colonization takes place (Dickson et al. 2014a, 2015). Microbes in microbiome are characterized by 16SrRNA profiling. Microorganisms are identified, classified and the phylogenetic relationship among them was studied in human microbiome project (HMP). 16SrRNA amplicon sequencing from 300 adults in human microbiome project decoded that the nasal microbiota comprise species of *Staphylococcus*, *Corynibacterium*, *Propionibacterium*, and *Moraxella*. Further investigations of the project have revealed that the major organisms are *P. acnes*, *C. accolens*, *C. kroppenstedtii*, *S. aureus*, and *S. epidermidis* from the metagenomic studies (Human Microbiome Project Consortium 2012). Microbial communities of these bacteria were significantly different from the smokers and non-smokers (Charlson 2010). And from the reports of Brisgaard et al, it was known that the oral cavity also contains several hundreds of microbial communities in teeth, palates, tongue, and tonsils. Culture-independent studies have revealed that the occupants of oral cavity are the species of *Prevotella*, *Veillonella*, *Streptococcus*, *Haemophilus*, *Fusobacterium*, *Neisseria*, and *Corynebacteria* (Brisgaard et al. 2008). The respiratory microbiome is divided into upper and lower respiratory tract microbial communities. The upper airway tract initiates from buccal cavity, nose and terminates finally at the lungs. Mucus covers the upper respiratory tract. Lower airway tract continues to alveoli which are lined by a phospholipoprotein pulmonary surfactant that has bactericidal activity towards certain alien pathogenic bacteria. This property has made alveolar space distinct from the GI tract and upper respiratory tract. The surfactant layer of the alveoli contains free fatty acids along with sphingomyelins and primacy dipalmitoyl phosphatidylcholine, which are phosphatidylcholine-containing lipids. These are mainly responsible for fighting against the pathogens when entered into respiratory tract. Respiratory microbiome is a community of many microbes. Many research works with the involvement of metagenomics, bioinformatics, and metatranscriptomics approaches generated the data and information focusing on the bacterial communities in the respiratory tract. But there are also communities of several microbes such as virus (virome), fungi (mycobiome) that play an important role during healthy and diseased states. Even though the fungal communities were not characterized by culture methods, they are characterized at a molecular level. Microbiome of respiratory tract has an impact on the upshot of chronic respiratory disorders such as asthma, COPD, cystic fibrosis, bronchiectasis, etc. (Linh Nguyen et al. 2015).

7.2 Human Microbiome Project and Molecular Characterization of Respiratory Microbiome

Most of the microbiological works have focused on disease-causing pathogens and their mechanisms of pathogenesis. But the endogenous flora was poorly studied and understood. After the completion of the mega Human Genome Project, in 2001 Relman et al. called for the second project on human microbiota (Relman and Falkow 2001). This is a 5-year project made with a budget of nearly 150 million dollars by NIH. The main objective of this project is to understand the microbiome of human body and analyze the genomic data of the microbiota in human body. In this project, samples were collected from 250 volunteers and the genome of the microbial communities from the different regions of human body was sequenced (NIH HMP working group, The NIH Human microbiome project 2009). Microbial communities and their molecular characters were studied from oral, nasal, gut, airways in men and along with vagina in females. Before HMP, it was assumed that the human lung and the respiratory tract are sterile. Previously the source for microbiological research is culture-based studies. But for respiratory microbiome studies, sampling is the main challenge due to its low biomass and unavailable non-invasive sampling technologies. Bronchoscopy is one of the techniques used to study and examine the respiratory tract. But due to the high risk of pharyngeal contamination of samples associated with bronchoscopy, it is not being used for examining microbial communities in the respiratory tract (Dickson et al. 2014b, 2015). In HMP, microbes are characterized by culture-independent methods by sequencing the gene/genome of the microbes. The sequence data was assembled and the microbial interactions along with their evolutionary route and phylogenetic relationships are studied based on 16S rRNA analysis. HMP has two phases: the jumpstart phase and second phase. The jumpstart phase is the initial phase of HMP started in 2007. In this phase, four sequencing centers, the Board Institute, Baylor college of Medicine, Washington University School of medicine, and J. Craig Venter Institute, were supported by NIH. Nearly 500 bacterial reference genomes were sequenced up to July 2009 in the jumpstart phase. 16S rRNA sequencing of microbial communities was performed

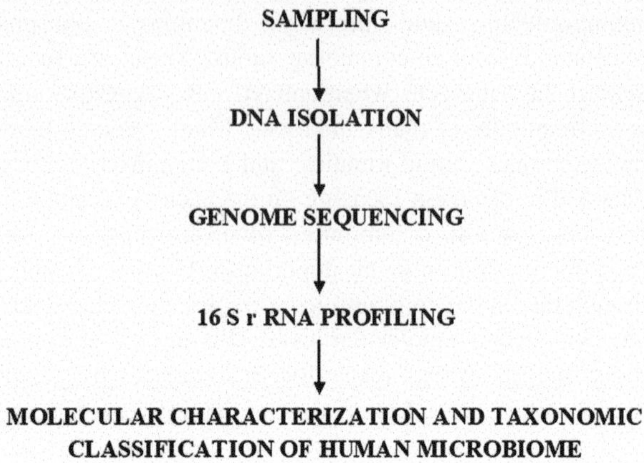

Fig. 7.1 Schematic representation of molecular characterization of normal human microbiome

from the four regions in males and five sites in females. Microbial communities of human microbiome are clustered into 5 phyla based on 16S rRNA sequence profile (Pace 1997; Woese and Fox 1977; Stahl et al. 1984) (Fig. 7.1). These include Firmicutes, Bacteroides, Proteobacteria, Actinobacteria, and Fusobacteria. The composition of these phyla varies in different regions in human body. For example, the resident microbes of oral cavity and lung are almost the same, but there is a significant difference in the proportion of bacteria in respiratory tract. Members of Actinobacteria are high in nasal tract than in oral cavity.

7.3 Determinants of Respiratory Microbiome

Three factors determine the shape of respiratory microbiome. These three are microbial immigration, elimination, and relative reproduction of microbial communities in airways. Microbiome composition was balanced by these three detrimental factors during health (Dickson et al. 2014c). In disease states, these factors are disturbed resulting in alteration of respiratory microbiome composition. Sources for immigration of microbes are microaspiration and inhalation of air (Lighthart 2000; Gleeson et al. 1997; Huxley et al. 1978; Quinn and Meyer 1929; Amberson 1954). Elimination of microbial communities was accomplished by innate and adaptive immune defenses, cough, and mucociliary clearances. During health, a balance between microbial immigration and emigration was established (Dickson et al. 2014a; Dickson and Huffnagle 2015; Venkataraman et al. 2015) (Fig. 7.2). Inflammation is an important factor that regulates reproduction of microbial communities in the respiratory tract and shapes immunity. Inflammation affects the growth conditions in respiratory tract such as temperature, pH, availability of nutrients and oxygen. The epithelial lining of the respiratory tract gets damaged during inflammation and bacteria adhere to the unshielded or exposed zones of respiratory lining. When bacteria binds to damaged cells in the respiratory lining, epithelia produce cytokines such as IL-25, IL-33, and thymic stromal lymphopoietin

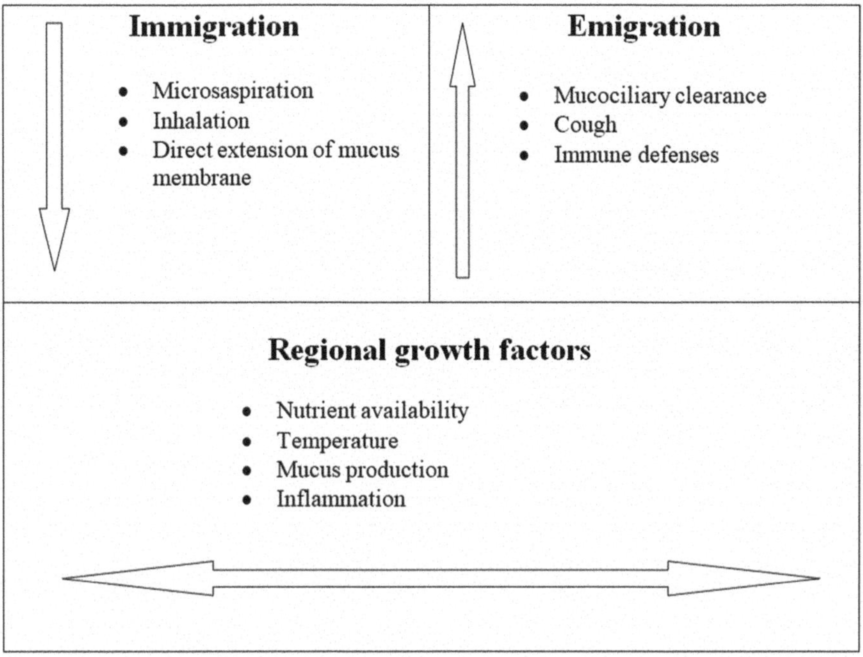

Fig. 7.2 Determinants of respiratory microbiome

(TSLP). These factors could contribute to additional inflammation by releasing IL-5 and IL-13 resulting in Goblet cell hyperplasia, a condition of Goblet cell expansion in allergic states with IL-13 production. This leads to an anaerobic environment in a system with excessive mucus secretion that enhances bacterial colonization by inhibiting phagocytosis. During health, the lung or respiratory environment does not allow robust microbial growth. Inflammation alters the regional factors and further contributes to the imbalanced composition in the microbiome.

7.4 Reactive Nitrogen Species (RNS) in Respiratory Tract Inflammation

Reactive nitrogen species (RNS) are produced in response to inflammatory stimuli. These RNS are synthesized enzymatically by inducible nitric oxide synthase (iNOS) in leukocytes and myelocytes (Bogdan 2015). Nitric oxide (NO) in reaction with free oxygen species O_2^- gives peroxynitrite ($ONOO^-$) which further breaks into nitrites (NO_2^-) and nitrates (NO_3^-). Nitrates (NO_3^-) act as a terminal electron acceptor and support the anaerobic respiration and outgrowth of Gammaproteobacteria via denitrification. Gammaproteobacteria gets benefits from the inflammation in the respiratory tract (Rivera-Chavez et al. 2013; Winter et al. 2013; Winter and Baumler 2014; Lopez et al. 2012, 2015; Bliska and van der Velden 2012; Vazquez-Torres and

Baumler 2015; Spees et al. 2013). RNS has antimicrobial properties that destroy the native microbes in respiratory tract resulting in the alteration of microbiome composition towards Gammaproteobacteria.

$NO + O_2^- \rightarrow ONOO^- \rightarrow NO_3^- + NO_2^-$.

(Nitric (Peroxynitrite) (Nitrate) (Nitrite) Oxide)

7.5 Dynamics of Respiratory Microbiome in Health and Disease

The normal composition of microbial communities gets altered during disease and with the treatment of antimicrobial agents resulting in disruption of native microbiome (Alicia and Allan 2018). The state of imbalance in the microbial communities is called as respiratory dysbiosis. Certain factors contribute to the dysbiosis of respiratory microbiome majorly smoking, antibiotics, chronic respiratory disorders (CRDs), and lung transplantation. Altered respiratory microbiome and byproducts of inflammation initiate pathogenesis in respiratory system, therefore affecting the constancy of determinants of respiratory microbiome, i.e., balance among the three factors, immigration, emigration, and inflammation, is lost (Fig. 7.2).

7.5.1 Smoking

Cigarette smoke contains various harmful water-soluble compounds and toxic gases such as Nicotine, ammonia, benzopyrene, hydro-quinone, CO, CO_2, formaldehyde, cadmium, nitrogen oxide, acetone and acrolein, etc (WHO Global Tuberculosis control 2011). Nicotine is the main pharmacogenic factor that causes addiction to cigarette and it is easily absorbed by mucus membrane in oral and respiratory lining affecting the innate immunity (Bjartveit and Tverdal 2005). Smoking is one of the major factors that affect the lung microbiome by causing inflammation in the respiratory tract. In a study conducted by Zhang et al., 40 numbers of eight-week-old male mice was grouped into smoking and non-smoking mice (20 each), and exposed smoking group mice to smoke for 2 h per day for 90 days. Concentrations of IL-6 and C-reactive protein (CRP) were determined by ELISA and microbial communities were identified by 16S rRNA profiling. From this study, it was concluded that weight of smoking mice increased very slowly compared to non-smoking group. Although there is no change in IL-6 and CRP concentration between the two groups, there is congestion and inflammation densely in the smoking group than in non-smoking mice. Also, there is a change in microbial composition in both the groups at the phylum and genus level. Proteobacteria and Firmicutes are the dominant phyla in both smoking and non-smoking groups. *Trichococcus*, *Escherichia*, *Shigella,* and Oxalobacteriaceae are the unique taxa identified in the smoking mice group. And in non-smoking mice, *Oceanospirillales*, *Lactobacillus,* and Lactobacteriaceae members are identified to be unique (Zhang

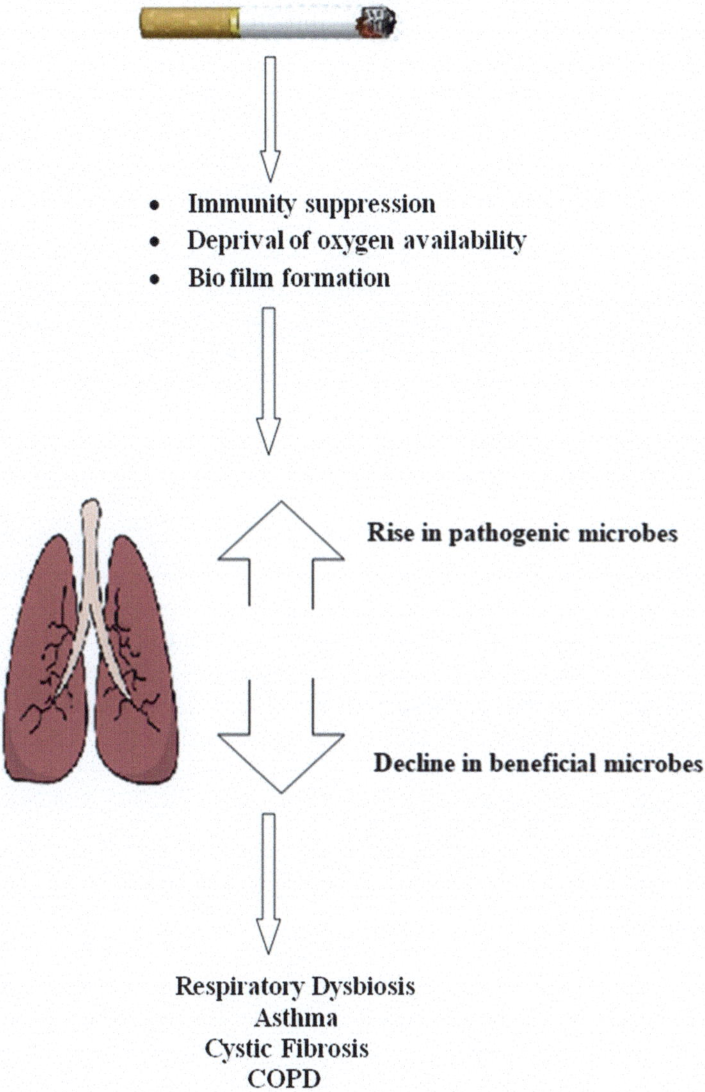

Fig. 7.3 Effect of smoking on respiratory microbiome

et al. 2018). Apart from disturbing the microbiome in respiratory tract, smoking exasperates lung inflammation by overexpressing TNF-α (tumor necrosis factor-α), IL-6, and monocyte chemoattractant protein-1 (MCP-1) (Gualano et al. 2008; Vlahos et al. 2006; Guerassimov et al. 2004). And also smoking can increase the incidence of pulmonary tuberculosis (Fig. 7.3). This was well documented by the Department of Pulmonary Medicine in their research in the year 2014, where patients are categorized into group I (control) and group II ($n = 60$). From the common radiological findings, it was identified that the incidence of pulmonary tuberculosis increased with the duration of smoking time among the patients (Mishra and Srivatsava 2017).

Some observations concluded that smoking can disturb only oral microbiome but not lung microbiome. This observation was found out by Alison et al. in lung HIV microbiome project (LHMP), 2013 promoted by National Heart, Lung and Blood Institute in characterizing the lung microbiota. Oral washes and Bronchoscopic Alveolar Lavages (BAL) are collected from sixty-four enrolled volunteers (45 non-smokers and 19 smokers) (Fig. 7.3). Microbial communities were characterized by culture-independent methods by 16S rRNA sequence. Most of the oral native bacteria are also present in lung. Specific bacteria such as Enterobacteriaceae, *Haemophilus*, *Ralstonia,* and *Methylobacterium* are in the outer portion of lungs. *Tropheryma* was present only in lungs (not a resident of oral microbiome). The lung microbiome did not witness any alteration in communities due to smoking. Mouth microbes differed between smokers and non-smokers especially in *Gemella*, *Neisseria,* and *Porphyromonas* (Morris et al. 2013).

"Vicious Circle" concept proposed by Mammen et al. explains that the harmful compounds in cigarette smoke can impair the innate immune defense and disturb the microbiome of respiratory tract resulting in disproportion and disturbance in taxonomic composition and relations among the microbial communities (Mammen and Sethi 2016). Direct exposure and contact of cigarette smoke through the respiratory air tract can lead to respiratory diseases such as asthma, COPD, cystic fibrosis, and lung cancer. The effect of smoking in causing diseases and dysbiosis is represented in Table 7.1. The impact of smoking on the microbiome can be direct or indirect. Smoking suppresses immunity, deprives oxygen availability, and promotes biofilm formation (Huang and Shi 2019) (Fig. 7.3).

7.5.2 Antibiotics

Antibiotics are being widely used to fight against infections. But tremendous and overuse of them is a menace to mankind. Antibiotics inhibit and kill the microbes. Due to the bactericidal property of antibiotics, the beneficial microbes in the microbiome may get altered resulting in change in composition of microbiome (dysbiosis). Even the results from the reports regarding the effect of antimicrobial agents on the microbiome are unclear; several research works are currently focusing on the mechanisms underlying antibiotic-mediated dysbiosis. Many antimicrobial agents are the derivatives of toxic and harmful elements such as lead, silver, arsenic, and mercury. For example, the first industrially synthesized antibiotic Salvarsan is a derivative of arsenic (Waksman and Woodruff 1940). The main problem associated with antibiotic usage is the evolution of antibiotic-resistant microbes. Firstly, the microbes in the community that are highly susceptible to antibiotics are eliminated and only the resistant species remains alive (Table 7.2). These resistant microbes now adopt some of the potential mechanisms such as drug degradation, antibiotic

Table 7.1 Influence of smoking on respiratory microbiome

Disease	References	Sampling	Rich microbes	Depleted microbes
Asthma	Simpson et al. (2016)	Subgingival plaque sample of human	• *Fusobacterium.* • *Prevotella.* • *Selenomonas.*	Not identified
Airway	Charlson (2010)	Nasopharyngeal and oropharyngeal swabbings of human	Oropharynx • *Megasphaera.* • *Veillonella* spp. Nasopharynx- • *Eggerthella.* • *Erysipelotrichacea.* • *Anaerovorax.* • *Eubacterium* spp.	Oropharynx • *Capnocytophaga.* • *Fusobacterium.* • *Neisseria* spp. Nasopharynx • *Shigella* spp.
Airway	Zhang et al. (2018)	Lung sample of mice	• *Trichococcus*, • *Escherichia.* • *Shigella*, • Oxalobacteraceae.	• *Oceanospirillales.* • *Lactobacillus.* • Lactobacillaceae. • *Enterobacter.* • *Acidimicrobiales.* • Caulobacteraceae. • Phyllobacteriaceae. • *Raoultella.* • Caulobacteraceae.

Table adapted and modified from Huang and Shi (2019)

efflux from membrane pumps, changing the membrane domains making antibiotics impermeable into a cell, etc., to defeat the functionality of antibiotics. After a period, these resistant strains multiply in respiratory tract. The composition of microbiota before and after antibiotic usage may be the same, but the genomic shape of microbial communities differs (Looft and Allen 2012; Jernberg et al. 2010; Raymond et al. 2016). Antibiotics at higher concentrations conk outs the respiratory system badly. Sometimes antibiotics trigger the expression of pathogenic virulence factors and lead to biofilm formation. Biofilms are a defense property of microbes towards antibiotics. These physiological alterations in the microbial ecosystem can affect the health in causing surface-associated infections in the respiratory tract (Martínez 2017) (Fig. 7.4).

7.6 Chronic Respiratory Disorders

Chronic respiratory disorders (CRDs) such as chronic obstructive pulmonary disease (COPD), asthma, cystic fibrosis (CF), and idiopathic pulmonary fibrosis (IPF) alter the dynamics of microbiome by affecting the immigration, emigration, and regional conditions of respiratory ecosystem resulting in twisting out the shape of respiratory microbial ecosystem. CRDs are commonly due to gastroesophageal reflux, elevating the volumes of microaspiration. Mucociliary clearance and cough are the factors responsible for the elimination of microbes from the ecosystem. Chronic lung

Table 7.2 Antibiotic-resistant bacteria in respiratory microbiome

Antibiotic	Resistant microbe	Resistance mechanism	References
Ampicillin	*Haemophilus influenza*, ampicillin-susceptible *Staphylococcus aureus*	Penicillinases	Maddocks and May (1969), Connell et al. (2013)
Penicillin	Streptococci	Penicillinases	Kundsin and Miller (1964)
B-lactam antibiotics	*Haemophilus influenza*, *Streptococcus pneumonia*	β-Lactamases	Armbruster et al. (2010), Budhani and Struthers (1998)
Amoxicillin	*Streptococcus pneumonia*	β-Lactamases	Weimer et al. (2011)
Chloramphenicol	*Streptococcus pneumonia*	Chloramphenicol acetyltransferase	Sorg et al. (2016)
Ceftazidime	*Pseudomonas aeruginosa*	β-Lactamases	Sherrard et al. (2016)
Gentamicin	*Staphylococcus aureus*	Aminoglycoside-modifying enzymes	De Leon et al. (2014)
Imipenem and ceftazidime	*Pseudomonas aeruginosa*	β-Lactamases	Kataoka and Tanaka (2003)

Adapted and modified from Vandeplassche et al. (2019)

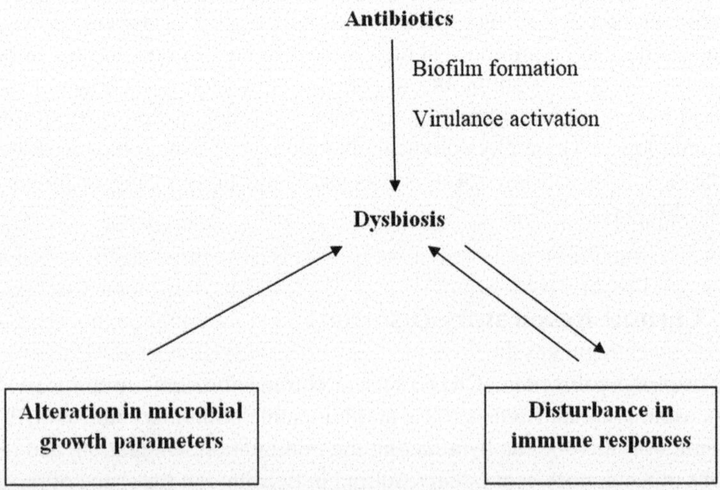

Fig. 7.4 Antibiotics on normal respiratory microbiome

diseases modulate the inflammatory response in the respiratory system. This was demonstrated by Segal et al. by comparing the levels of microbiome detected in BAL. Levels of anaerobic oral commensals like *Prevotella* and *Veillonella* rise abundantly during the CRDs associated with inflammation and rise in lymphocyte levels (Segal et al. 2013). Generally, CRDs are characterized by exacerbation period. Pulmonary function declines due to exacerbations with a significant rise in mortality and morbidity rates.

7.6.1 COPD

Chronic obstructive pulmonary disease (COPD) is a chronic respiratory disorder that is characterized by airflow blockage in the respiratory tract (Fig. 7.5). This is caused due to emphysema or chronic bronchitis. Smoking is the major risk factor in the development of COPD in humans. Factors that contribute to airflow obstruction include alveolar wall destruction, enlargement of mucus glands in bronchi, and hyperresponsiveness in the airway (Barnes 2000; Tashkin et al. 1992). Many culture-independent methods have determined a diverse pulmonary microbiome in COPD exacerbations from BAL and sputum cultured samples. These exacerbations are due to inflammation and dynamic changes in respiratory microbiome (SZE et al. 2012; Millares et al. 2014; Huang et al. 2014; Pragman et al. 2012). *Haemophilus, Pseudomonas,* and *Moraxella* communities are abundant during COPD. Sometimes, the microbiome proportion can alter towards the *Proteobacterium* in COPD exacerbation. COPD exacerbations may also be triggered by viral infections (Hewitt et al. 2015). The relationship between viral infections and respiratory microbiome was poorly understood. When patients are infected with rhinovirus experimentally, they developed features of COPD (Papi et al. 2006; Rohde et al. 2003; Mallia et al. 2011; Seemungal et al. 2001). It was known clearly that microbial communities shift towards *Proteobacteria* phylum and *Pseudomonas* species increase during COPD exacerbations (Molyneaux et al. 2013; Millares et al. 2014). GOLD stage 4 is the severe stage in COPD. Sequence studies from COPD patients demonstrated that

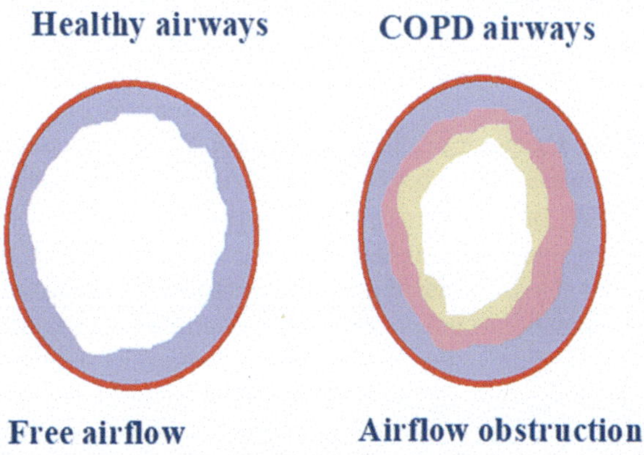

Fig. 7.5 Schematic representation of normal and COPD airways. Airflow obstruction in COPD air tract is due to mucus glands enlargement and accumulation of mucus narrowing the airway and blocking the passage of air

levels of *Firmicutes* increased severely in GOLD stage 4. An increase in *Firmicutes* contributes to an increase in *Lactobacillus* genus. Alveoli and airways in COPD patients contain a distinct microbiome (Dickson et al. 2013). Introduction of a new pathogen into the respiratory tract can lead to inflammation (Mammen and Sethi 2016). In a study conducted by Pragman et al. it was observed that Firmicutes dominated both Proteobacteria and Actinobacteria (Pragman et al. 2012). In another cohort study, Proteobacteria, Firmicutes, and Actinobacteria were found to be the most prevalent phyla in both normal and exacerbated patients. *Streptococcus* and *Haemophilus* are identified at higher levels in both groups (Millares et al. 2015).

COPD is majorly caused by smoking. Smoking alters the dynamics of host–microbe interactions and results in respiratory abnormalities (Garmendia et al. 2012). Cigarette smoke is a risk factor to many of the beneficial microbes in respiratory tract (Arcavi and Benowitz 2004). Somehow antibiotic treatment also disturbs the microbial diversity and distribution during COPD (Flagnan et al. 2007). Dynamic changes in respiratory virome mediated by smoking could lead to COPD. Lung contains some virus (virome). Virus-like phages cause opportunistic infections by mediating the immune microbial competition (Klainer and Beisel 1969). Gregory et al. first conducted a study and demonstrated that smoking lowers the virome community by changing the levels of IL-8 and arachidonic acid. Both IL-8 and arachidonic acid are the factors that are responsible for COPD (Gregory et al. 2018). Several factors that are responsible for COPD and their effects on the microbiome and clinical outcomes of dysbiosis need to be investigated further.

7.6.2 Asthma

Microbiome composition slightly alters towards the *Proteobacteria*. This was identified by Hitly et al. in asthma patients. *Proteobacteria*, especially *Haempohilus* is the dominant microbial population in the bronchial trees of asthmatic patients to controls (Hilty et al. 2010; Marri et al. 2013; Huang et al. 2011). Majority of the asthma exacerbation cases are due to viral infections (Nicholson et al. 1993; Johnston et al. 1995). Some of the fungal species can also contribute to uneasiness during asthma. From sputum culturing, *Aspergillus* and *Penicillium* were found to be the dominant fungal species in asthmatics (Agbetile et al. 2012) (Fig. 7.6).

Several studies have strongly focused on the food habits in studying asthma exacerbations. Some works strongly suggest that foods with high fiber content reduce the risk of asthma. High fiber helps in the normal functioning of a healthy microbiota. High fiber foods improved and raised the levels of short chain fatty acids (SCFA) and ensured protection against asthma inflammation in murine models (Trompette et al. 2014). Furthermore, many research works on ovalbumin-induced asthma have discussed that native microbiota promotes Th 17-dependent neutrophil inflammation (Lemaire et al. 2011). Similarly, antibiotics administration can lead to inflammation in the airways and reduces microbial diversity (Russell et al. 2013). Interactions among the host-microbial communities in the respiratory tract are very much essential in understanding the mechanisms underlying the dysbiosis during

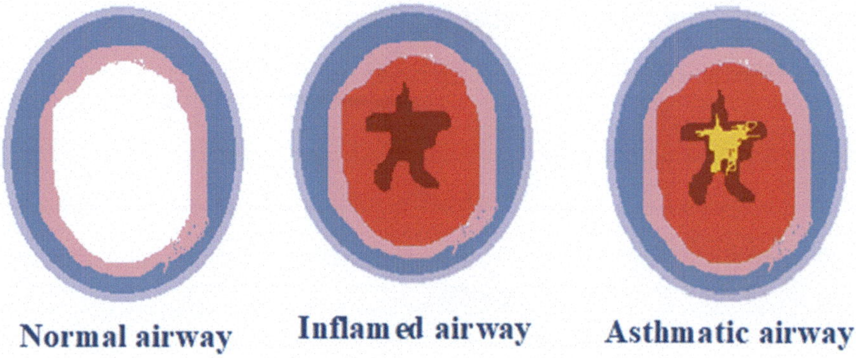

Fig. 7.6 Normal airway, inflamed airway, and asthmatic airway

asthma (Fig. 7.6). As like COPD, normal airway is obstructed in asthma. According to "microbiota hypothesis" (framed by correlating the antibiotics, allergies, and gut microbiota) deviation in GI microbiota due to antibiotic usage and poor food diets in western areas leads to disruption of immunologic tolerance (Huffnagle 2010) (Fig. 7.5). Smoking is another risk factor in shaping the microbiome of lower airway. Based on 16SrRNA sequencing, the bacterial species that are dominant in ex-smokers were found to be *Fusobacteria*, *Firmicutes,* and *Bacteroidetes*. *Proteobacteria* is lower in non-smokers (Colak et al. 2015; Simpson et al. 2016).

7.6.3 Cystic Fibrosis

Cystic fibrosis (CF) is caused due to mutation in the CFTR gene (cystic fibrosis transmembrane conductance regulator), where several organs such as lungs and GI tract are affected. Water and ion concentration in epithelial tissues are maintained in equilibrium by cAMP-regulated chloride and bicarbonate transport channel. This equilibrium in water to ion concentration is disturbed during CF disease (Saint-Criq and Gray 2017). During CF, the mucus-producing cells are mainly affected by respiratory dysbiosis leading to mucus accumulation in the air tract (Fig. 7.7), thereby disturbing the normal environment in the airway (Willger et al. 2012). *Pseudomonas aeruginosa*, a gram-negative bacillus is relatively linked to CF clinically (Harris and De 2007; Tunney et al. 2008). Apart from Bacteriome, Virome (Virus) and mycobiome (Fungi) were also characterized from the respiratory tract of the CF patients. Species of *Candida*, *Aspergillus*, *Scedosporium,* and *Malassezia* are detected in CF (Delhaes et al. 2012; Willger et al. 2014; Mounier et al. 2014; Willner et al. 2009; Kramer et al. 2015). An unknown set of viral and bacteriophage groups may also play a major role in disease progression (Billard et al. 2017; Fancello et al. 2011; Willner et al. 2009). CF is associated with chronic infections and bronchiectasis (enlargement of lung airway). Some other bacteria such as methicillin-resistant *Staphylococcus aureus* (MRSA), *Burkholderia cepacia*, and some non-tuberculosis

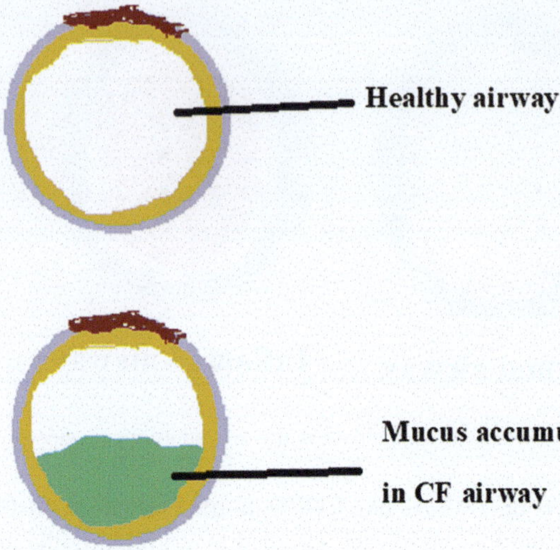

Fig. 7.7 Healthy airway and CF airway with mucus accumulation

bacteria like *Mycobacterium abscessus* and *Mycobacterium avium* are also involved in contributing to morbidity in pulmonary CF patients (Gilligan 2014; Dasenbrook et al. 2010; Martiniano and Nick 2015). It is very difficult to diagnose and treat some pathogens like *Achromobacter spp* (*ruhlandii* and *xylosoxidans*), *Stenotrophomonas maltophilia*, etc. during CF. Anaerobic bacteria in respiratory airways play a key role in downregulating the expression of the NOS (nitric oxide synthase) gene and release IL-6, IL-8, CSF (colony stimulating factor), and granulocyte macrophages in excess amounts. *Germella*, *Actinomyces*, *Neisseria*, *Granulicatella*, *Rothia* are the genera that are found frequently in the airway of CF patients (Fig. 7.7) (Acosta et al. 2017; Rogers et al. 2003; Coburn et al. 2015; Surette 2014). Obligate anaerobes like *Veillonella*, *Prevotella*, *Porphyromonas,* and *Fusobacterium,* etc. are also seen in CF airways (Sherrard et al. 2016; Huang and LiPuma 2016; Mirkovic et al. 2015). In more than 30% of children (<5 years) cases and 10% of adults with CF, *Haemophilus influenza* was identified as the more frequent and potential colonizing pathogen in the respiratory tract (Lipuma 2010). The proportion of lung microbiome during CF may not be constant. It changes over time. For example, microbial diversity is more in childhood than in adult age. A decline in microbial diversity shifts the composition of lung microbiome towards Pseudomonadaceae (mostly *P. aeruginosa* as the dominating one). This is because of over antibiotic usage and the selection of drug-resistant microbes in the community (McGuigan and Callaghan 2015; Cox et al. 2010; Zhao et al. 2012).

7.6.4 Idiopathic Pulmonary Fibrosis

IPF is a fatal chronic respiratory disease that occurs in the lung parenchyma (Raghu et al. 2015). Unlike COPD and asthma, several active pathogenic microbes are responsible for exacerbation in IPF (Collard et al. 2007) (Fig. 7.5). *Streptococcus* and *Staphylococcus* are the pathogens that are majorly involved in causing IPF (Han et al. 2014). In a study conducted by Molyneaux et al, *Haemophilus*, *Streptococcus*, *Neisseria,* and *Veillonella* were identified as enriching pathogens in IPF patients by analyzing the BAL fluid (Molyneaux et al. 2014). Signaling of defective toll-like receptor TLR3 initiates disease in IPF. rs3775291(L412F) is an SNP in TLR3 that leads to a change in leucine to phenylalanine encoding a different protein. L412F is a potential marker associated with defective TLR3. L412F polymorphism results in inflammation and desensitized interferon response towards TLR3 activation. This is due to defective pulmonary fibroblasts in IPF. This L412F polymorphism is the major confirmation for mortality and morbidity in IPF patients (O'Dwyer et al. 2013). Apart from the alterations in the microbiome or pathogenic entry, presence of a certain microbe in the lung microbiota can drive towards IPF progression (Knippenberg et al. 2015). For example, pneumolysin is a Pneumococcus-associated toxin that causes fibrosis in animal alveoli.

7.6.5 Lung Transplantation

In many of the final stages of respiratory disorders, lung transplantation is the main and only option left (Christie et al. 2012). Infections are associated with mortality and morbidity in patients who have undergone lung transplantation during pneumonia and bronchiolitis obliterans syndrome (BOS). BOS is a chronic infection that damages the air passages in lung (Botha et al. 2008; Husain and Singh 2002; Khalifah et al. 2004). Lung transplantation may alter the host defense in patients disturbing the regional microbiota (Duncan and Wilkes 2005; Kotloff and Thabut 2011). Through the culture-independent methods and metagenomic approaches, *Pseudomonas aeruginosa* and *fluorescens* were identified as the dominant species in patients who have undergone lung transplantation. Out of the two, *P. aeruginosa* is the most predominant pathogen with a high risk of BOS. Whereas *P. fluorescens* confers little acute infection only. The two species *aeruginosa* and *fluorescens* of *Pseudomonas* have unique microbial and genomic characteristics and are more divergent than the other microbes in the microbiome (Dickson et al. 2014b).

7.7 Conclusions and Future Perspectives

With advances in metagenomics and next-generation sequencing methods, it became easy in sequencing the genome of microbes and assembling the genomic data. Culture-independent methods have potentially identified the components of respiratory microbiome in different regions of respiratory tract such as alveoli, upper respiratory tract, etc. Understanding the microbiome of respiratory tract in normal homeostatic health conditions and during disease exacerbations helps in studying the microbiome signatures and dynamics and also it is very useful in therapeutic approaches during any imbalance in the microbiome. With advanced instrumentation facility and emerging novel technologies in the fields of microbiology, genomics, and molecular biology, it became very easy in studying the dynamics of respiratory microbiome and its interaction with the other microbiomes in the body such as oral microbiome, skin, vaginal, and gut microbiome.

Acknowledgments The authors gratefully acknowledge Vignan's Foundation for Science, Technology and Research (deemed to be university) and Krishna University, Machilipatnam for the support extended.

Conflict of Interest The authors declare that they have no competing interests.

References

Acosta N, Whelan FJ, Somayaji R (2017) The evolving cystic fibrosis microbiome: a comparative cohort study spanning 16 years. Ann Am Thorac Soc 14:1288–1297

Agbetile J, Fairs A, Desai D, Hargadon B, Bourne M, Mutalithas K, Edwards R, Morley JP, Monteiro WR, Kulkarni NS (2012) Isolation of filamentous fungi from sputum in asthma is associated with reduced post-bronchodilator FEV1. Clin Exp Allergy 42:782–791

Alicia BM, Allan RG (2018) The human respiratory microbiome: implications and impact. Semin Respir Crit Care Med 39:199–212

Amberson JB (1954) A clinical consideration of abscesses and cavities of the lung. Bull Johns Hopkins Hosp 94:227–237

Arcavi L, Benowitz NL (2004) Cigarette smoking and infection. Arch Intern Med 164:2206–2216

Armbruster CE, Hong W, Pang B (2010) Indirect pathogenicity of *Haemophilus influenzae* and *Moraxella catarrhalis* in polymicrobial otitis media occurs via interspecies quorum signaling. mBio 1:e00102

Barnes PJ (2000) Chronic obstructive pulmonary disease. N Engl J Med 343:269–280

Billard L, Le Berre R, Pilorge L (2017) Viruses in cystic fibrosis patients' airways. Crit Rev Microbiol 43:690–708

Bisgaard H, Hermansen MN, Buchvald F (2008) Childhood asthma after bacterial colonization of the airway in neonates. *N Engl J Med* 357(15):1487–1149

Bjartveit K, Tverdal A (2005) Health consequences of smoking 1–4 cigarettes per day. Tob Control 14:315–320

Bliska JB, van der Velden AW (2012) *Salmonella* "sops" up a preferred electron receptor in the inflamed intestine. MBio 3:e00226

Bogdan C (2015) Nitric oxide synthase in innate and adaptive immunity: an update. Trends Immunol 36:161–178

Botha P, Archer L, Anderson RL, Lordan J, Dark JH (2008) *Pseudomonas aeruginosa* colonization of the allograft after lung transplantation and the risk of bronchiolitis obliterans syndrome. Transplantation 85:771–774

Budhani RK, Struthers JK (1998) Interaction of *Streptococcus pneumoniae* and *Moraxella catarrhalis*: investigation of the indirect pathogenic role of beta-lactamase-producing moraxellae by use of a continuous-culture biofilm system. Antimicrob Agents Chemother 42:2521–2526

Charlson ES (2010) Disordered microbial communities in the upper respiratory tract of cigarette smokers. PLoS One 5(12):e15216

Christie JD, Edwards LB, Kucheryavaya AY, Benden C, Dipchand AI (2012) The registry of the international society for heart and lung transplantation: 29th adult lung and heart-lung transplant report. J Heart Lung Transplant 31:1073–1086

Coburn B, Wang PW, Diaz Caballero J (2015) Lung microbiota across age and disease stage in cystic fibrosis. Sci Rep 5:10241

Colak Y, Afzal S, Nordestgaard BG, Lange P (2015) Characteristics and prognosis of never-smokers and smokers with asthma in the Copenhagen general population study. A prospective cohort study. Am J Respir Crit Care Med 192:172–181

Collard HR et al (2007) Idiopathic pulmonary fibrosis clinical research network. Acute exacerbations of idiopathic pulmonary fibrosis. Am J Respir Crit Care Med 176:636–643

Connell JL, Ritschdorff ET, Whiteley M (2013) 3D printing of microscopic bacterial communities. Proc Natl Acad Sci USA 110:18380–18385

Cox MJ, Allgaier M, Taylor B (2010) Airway microbiota and pathogen abundance in age-stratified cystic fibrosis patients. PLoS One 5:e11044

Dasenbrook EC, Checkley W, Merlo CA, Konstan MW, Lechtzin N, Boyle MP (2010) Association between respiratory tract methicillin-resistant S*taphylococcus aureus* and survival in cystic fibrosis. *JAMA* 303(23):2386–2392

DeLeon S, Clinton A, Fowler H (2014) Synergistic interactions of *Pseudomonas aeruginosa* and *Staphylococcus aureus* in an *in vitro* wound model. Infect Immun 82:4718–4728

Delhaes L, Monchy S, Fréalle E (2012) The airway microbiota in cystic fibrosis: a complex fungal and bacterial community–implications for therapeutic management. PLoS One 7:e36313

Dickson RP, Erb-Downward JR, Freeman CM, Mc Closkey L, Beck JM, Huffnagle GB (2015) Spatial variation in the healthy human lung microbiome and the adapted island model of lung biogeography. Ann Am Thorac Soc 12(6):821–830

Dickson RP, Erb-Downward JR, Huffnagle GB (2013) The role of the bacterial microbiome in lung disease. Expert Rev Respir Med 7(03):245–257

Dickson RP, Erb-Downward JR, Huffnagle GB (2014a) Towards an ecology of the lung: new conceptual models of pulmonary microbiology and pneumonia pathogenesis. Lancet Resp Med 2:238–246

Dickson RP, Huffnagle GB (2015) The lung microbiome: new principles for respiratory bacteriology in health and disease. PLoS Pathog 11:e1004923

Dickson RP, Martinez FJ, Huffnagle GB (2014b) The role of the microbiome in exacerbations of chronic lung diseases. Lancet 384(9944):691–702

Dickson RP, Erb-Downward JR, Freeman CM, Walker N, Scales BS, Beck JM, Martinez FJ, Curtis JL, Lama VN, Huffnagle GB (2014c) Changes in the lung microbiome following lung transplantation include the emergence of two distinct pseudomonas species with distinct clinical associations. PLoS One 9(5):e97214

Duncan MD, Wilkes DS (2005) Transplant-related immunosuppression: a review of immunosuppression and pulmonary infections. Proc Am Thorac Soc 2:449–455

Fancello L, Desnues C, Raoult D (2011) Bacteriophages and diffusion of genes encoding antimicrobial resistance in cystic fibrosis sputum microbiota. J Antimicrob Chemother 66:2448–2454

Flagnan JL, Brodie EL, Weng L (2007) Loss of bacterial diversity during antibiotic treatment of incubated patients colonized with Pseudomonas aeroginosa. J Clin Microbiol 45:1954–1962

Garmendia J, Morey P, Bengoechea JA (2012) Impact of cigarette smoke exposure on host–bacterial pathogen interactions. Eur Respir J 39:467–477

Gilligan PH (2014) Infections in patients with cystic fibrosis: diagnostic microbiology update. Clin Lab Med 34:197–192

Gleeson K, Eggli DF, Maxwell SL (1997) Quantitative aspiration during sleep in normal subjects. Chest 111:1266–1272

Gregory AC, Sullivan MB, Segal LN, Keller BC (2018) Smoking is associated with quantifiable differences in the human lung DNA virome and metabolome. Respir Res 19:174

Gualano RC, Hansen MJ, Vlahos R, Jones JE, Park-Jones RA, Deliyannis G (2008) Cigarette smoke worsens lung inflammation and impairs resolution of influenza infection in mice. Respir Res 9:53–69

Guerassimov A, Hoshino Y, Takubo Y, Turcotte A, Yamamoto M, Ghezzo H (2004) The development of emphysema in cigarette smoke exposed mice is strain dependent. Am J Respir Crit Care Med 170(9):974–980

Han MK et al (2014) Lung microbiome and disease progression in idiopathic pulmonary fibrosis: an analysis of the COMET study. Lancet Respir Med 2:548–556

Harris JK, De Groote MA, Sagel SD (2007) Molecular identification of bacteria in bronchoalveolar lavage fluid from children with cystic fibrosis. Proc Natl Acad Sci U S A 104:20529–20533

Hewitt R, Farne H, Ritchie A, Luke E, Johnston SL, Mallia P (2015) The role of viral infections in exacerbations of chronic obstructive pulmonary disease and asthma. Ther Adv Respir Dis 10 (2):158–174

Hilty M, Burke C, Pedro H, Cardenas P, Bush A, Bossley C, Davies J, Ervine A, Poulter L, Pachter L, Moffatt MF, Cookson WO (2010) Disordered microbial communities in asthmatic airways. PLoS One 5:e8578

Huang C, Shi G (2019) Smoking and microbiome in oral, airway, gut and some systemic diseases. J Transl Med 17:225

Huang YJ, LiPuma JJ (2016) The microbiome in cystic fibrosis. Clin Chest Med 37:59–67

Huang YJ, Nelson CE, Brodie EL, DeSantis TZ, Baek MS, Liu J, Woyke T, Allgaier M, Bristow J, Weiner-Kronish JP, Sutherland ER et al (2011) Airway microbiota and bronchial hyperresponsiveness in patients with suboptimally controlled asthma. J Allergy Clin Immunol 127(2):372–381

Huang YJ, Sethi S, Murphy T, Nariya S, Boushey HA, Lynch SV (2014) Airway microbiome dynamics in exacerbations of chronic obstructive pulmonary disease. J Clin Microbiol 52:2813–2823

Huffnagle GB (2010) The microbiota and allergies/asthma. PLoS Pathog 6(5):e1000549

Huffnagle GB, Dickson RP, Lukacs NW (2016) The respiratory tract microbiome and lung inflammation: a two-way street. Mucosal Immunol 10(2):299–306

Human Microbiome Project Consortium (2012) Structure, function and diversity of the healthy human microbiome. Nature 486:207–221

Husain S, Singh N (2002) Bronchiolitis obliterans and lung transplantation: evidence for an infectious etiology. Semin Respir Infect 17:310–314

Huxley EJ, Viroslav J, Gray WR, Pierce AK (1978) Pharyngeal aspiration in normal adults and patients with depressed consciousness. Am J Med 64(4):564–568

Jernberg C, Lofmark S, Edlund C (2010) Long-term impacts of antibiotic exposure on the human intestinal microbiota. Microbiology 156(1):3216–3223

Johnston SL, Pattemore PK, Sanderson G, Smith S, Lampe F, Josephs L, Symington P, O'Toole S, Myint SH, Tyrrell DA (1995) Community study of role of viral infections in exacerbations of asthma in 9–11 year old children. BMJ 310:1225–1229

Kataoka D, Tanaka Y (2003) The clinical aspects of beta-lactam-resistant *Stenotrophomonas maltophilia*. Yonago Acta Med 46:91–102

Khalifah AP, Hachem RR, Chakinala MM, Schechtman KB, Patterson GA (2004) Respiratory viral infections are a distinct risk for bronchiolitis obliterans syndrome and death. Am J Respir Crit Care Med 170:181–187

Klainer AS, Beisel WR (1969) Opportunistic infection: a review. Am J Med Sci 258:431–456

Knippenberg S, Ueberberg B, Maus R (2015) *Streptococcus pneumoniae* triggers progression of pulmonary fibrosis through pneumolysin. Thorax 70(7):636–646

Kotloff RM, Thabut G (2011) Lung transplantation. Am J Respir Crit Care Med 184:159–171

Kramer R, Sauer-Heilborn A, Welte T (2015) Cohort study of airway mycobiome in adult cystic fibrosis patients: differences in community structure between fungi and bacteria reveal predominance of transient fungal elements. J Clin Microbiol 53:2900–2907

Kundsin RB, Miller JM (1964) Significance of the *Staphylococcus aureus* carrier state in the treatment of disease due to group a streptococci. N Engl J Med 271:1395–1397

Lemaire MM, Dumoutier L, Warnier G, Uyttenhove C, Van Snick J, de Heusch M, Stevens M, Renauld JC (2011) Dual TCR expression biases lung inflammation in DO11.10 transgenic mice and promotes neutrophilia via microbiota-induced Th17 differentiation. J Immunol 187:3530–3537

Lighthart B (2000) Mini review of the concentration variations found in the alfresco atmospheric bacterial populations. Aerobiologia 16:7–16

Linh Nguyen DN, Viscogliosi E, Delhaes L (2015) The lung mycobiome: an emerging field of the human respiratory microbiome. Front Microbiol 6:89

Lipuma JJ (2010) The changing microbial epidemiology in cystic fibrosis. Clin Microbiol Rev 23:299–323

Looft T, Allen HK (2012) Collateral effects of antibiotics on mammalian gut microbiomes. Gut Microbes 3(5):463–467

Lopez CA, Rivera-Chavez F, Byndloss MX, Baumler AJ (2015) The periplasmic nitrate reductase NapABC supports luminal growth of *Salmonella enterica* serovar *typhimurium* during colitis. Infect Immun 83:3470–3478

Lopez CA, Winter SE, Rivera-Chávez F, Xavier MN, Poon V, Nuccio SP, Tsolis RM, Bäumler AJ (2012) Phage-mediated acquisition of a type III secreted effector protein boosts growth of *Salmonella* by nitrate respiration. mBio 3(3):e00143–e00112

Maddocks JL, May JR (1969) "Indirect pathogenicity" of penicillinase-producing enterobacteria in chronic bronchial infections. Lancet 1:793–795

Mallia P, Message SD, Gielen V, Contoli M, Gray K, Kebadze T, Aniscenko J, Laza-Stanca V, Edwards MR, Slater L, Papi A, Stanciu LA, Kon OM, Johnson M, Johnston SL (2011) Experimental rhinovirus infection as a human model of chronic obstructive pulmonary disease exacerbation. Am J Respir Crit Care Med 183:734–742

Mammen MJ, Sethi S (2016) COPD and the microbiome. Respirology 21:590–599

Marri PR, Stern DA, Wright AL, Billheimer D, Martinez FD (2013) Asthma-associated differences in microbial composition of induced sputum. J Allergy Clin Immunol 131:346–352

Martínez JL (2017) Effect of antibiotics on bacterial populations: a multi-hierarchical selection process. F1000Research 6:51

Martiniano SL, Nick JA (2015) Nontuberculous mycobacterial infections in cystic fibrosis. Clin Chest Med 36:101–115

Mc Guigan L, Callaghan M (2015) The evolving dynamics of the microbial community in the cystic fibrosis lung. Environ Microbiol 17:16–28

Millares L, Ferrari R, Gallego M, Garcia-Nunez M, Perez-Brocal V, Espasa M, Pomares X, Monton C, Moya A, Monso E (2014) Bronchial microbiome of severe COPD patients colonised by Pseudomonas aeruginosa. Eur J Clin Microbiol Infect Dis 33:1101–1111

Millares L, Perez-Brocal V, Ferrari R, Gallego M, Pomares X, Garcia-Nunez M, Monton C, Capilla S, Monso E, Moya A (2015) Functional metagenomics of the bronchial microbiome in COPD. PLoS One 10(12):e0144448

Mirkovic B, Murray MA, Lavelle GM (2015) The role of short-chain fatty acids, produced by anaerobic bacteria, in the cystic fibrosis airway. Am J Respir Crit Care Med 192:1314–1324

Mishra DP, Srivastava A (2017) Effect of smoking in development of pulmonary tuberculosis. J Adv Med Dent Scie Res 5(10):99–102

Molyneaux PL, Cox MJ, Willis-Owen SA, Mallia P, Russell KE, Russell AM, Murphy E, Johnston SL, Schwartz DA, Wells AU, Cookson WO, Maher TM, Moffatt MF (2014) The role of bacteria in the pathogenesis and progression of idiopathic pulmonary fibrosis. Am J Respir Crit Care Med 190:906–913

Molyneaux PL, Mallia P, Cox MJ, Footitt J, Willis-Owen SA, Homola D, Trujillo-Torralbo MB, Elkin S, Kon OM, Cookson WO, Moffatt MF, Johnston SL (2013) Outgrowth of the bacterial airway microbiome after rhinovirus exacerbation of chronic obstructive pulmonary disease. Am J Respir Crit Care Med 188:1224–1231

Morris A, Beck JM, Schloss PD, Campbell TB, Crothers K, Curtis JL, Flores SC, Fontenot AP, Ghedin E, Huang L, Jablonski K (2013) Comparison of the respiratory microbiome in healthy nonsmokers and smokers. Am J Respir Crit Care Med 187(10):1067–1075

Mounier J, Gouëllo A, Keravec M (2014) Use of denaturing high-performance liquid chromatography (DHPLC) to characterize the bacterial and fungal airway microbiota of cystic fibrosis patients. J Microbiol 52:307–314

Nicholson KG, Kent J, Ireland DC (1993) Respiratory viruses and exacerbations of asthma in adults. BMJ 307:982–986

NIH HMP Working Group (2009) The NIH human microbiome project. Genome Res 19:2317–2323

O'Dwyer DN, Armstrong ME, Trujillo G, Cooke G, Keane MP, Fallon PG, Simpson AJ, Millar AB, McGrath EE, Whyte MK, Hirani N, Hogaboam CM, Donnelly SC (2013) The toll-like receptor 3 L412F polymorphism and disease progression in idiopathic pulmonary fibrosis. Am J Respir Crit Care Med 188:1442–1450

Pace NR (1997) A molecular view of microbial diversity and the biosphere. Science 276:734–740

Papi A, Bellettato CM, Braccioni F, Romagnoli M, Casolari P, Caramori G, Fabbri LM (2006) Infections and airway inflammation in chronic obstructive pulmonary disease severe exacerbations. Am J Respir Crit Care Med 173:1114–1121

Pragman AA, Kim HB, Reilly CS, Wendt C, Isaacson RE (2012) The lung microbiome in moderate and severe chronic obstructive pulmonary disease. PLoS One 7:e47305

Quinn LH, Meyer O (1929) The relationship of sinusitis and bronchiectasis. Arch Otolaryngol Head Neck Surg 10:152

Raghu G et al (2015) An official ATS/ERS/JRS/ALAT clinical practice guideline: treatment of idiopathic pulmonary fibrosis. An update of the 2011 clinical practice guideline. Am J Respir Crit Care Med 192:e3

Raymond F, Deraspe M, Boissinot M (2016) Partial recovery of microbiomes after antibiotic treatment. Gut Microbes 7(5):428–434

Relman DA, Falkow S (2001) The meaning and impact of the human genome sequence for microbiology. Trends Microbiol 5:206–208

Rivera-Chávez F, Winter SE, Lopez CA, Xavier MN, Winter MG, Nuccio SP, Russell JM, Laughlin RC, Lawhon SD, Sterzenbach T, Bevins CL, Tsolis RM, Harshey R, Adams LG, Bäumler AJ (2013) *Salmonella* uses energy taxis to benefit from intestinal inflammation. PLoS Pathog 9(4):e1003267

Rogers GB, Hart CA, Mason JR (2003) Bacterial diversity in cases of lung infection in cystic fibrosis patients: 16S ribosomal DNA (rDNA) length heterogeneity PCR and 16S rDNA terminal restriction fragment length polymorphism profiling. J Clin Microbiol 41:3548–3558

Rohde G, Wiethege A, Borg I, Kauth M, Bauer TT, Gillissen A, Bufe A, Schultze-Werninghaus G (2003) Respiratory viruses in exacerbations of chronic obstructive pulmonary disease requiring hospitalisation: a case-control study. Thorax 58:37–42

Russell SL, Gold MJ, Willing BP, Thorson L, McNagny KM, Finlay BB (2013) Perinatal antibiotic treatment affects murine microbiota, immune responses and allergic asthma. Gut Microbes 4:158–164

Saint-Criq V, Gray MA (2017) Role of CFTR in epithelial physiology. Cell Mol Life Sci 74:93–115

Seemungal T, Harper-Owen R, Bhowmik A, Moric I, Sanderson G, Message S, Maccallum P, Meade TW, Jeffries DJ, Johnston SL, Wedzicha JA (2001) Respiratory viruses, symptoms, and inflammatory markers in acute exacerbations and stable chronic obstructive pulmonary disease. Am J Respir Crit Care Med 164:1618–1623

Segal LN, Alekseyenko AV, Clemente JC, Kulkarni R, Wu B, Gao Z, Chen H, Berger KI, Goldring RM, Rom WN, Blaser MJ, Weiden MD (2013) Enrichment of lung microbiome with supraglottic taxa is associated with increased pulmonary inflammation. Microbiome 1(1):19

Sherrard LJ, Bell SC, Tunney MM (2016) The role of anaerobic bacteria in the cystic fibrosis airway. Curr Opin Pulm Med 22:637–643

Simpson JL, Daly J, Baines KJ, Yang IA, Upham JW, Reynolds PN, Hodge S, James AL, Hugenholtz P, Willner D, Gibson PG (2016) Airway dysbiosis: *Haemophilus influenzae* and Tropheryma in poorly controlled asthma. Eur Respir J 47:792–800

Sorg RA, Lin L, van Doorn GS (2016) Collective resistance in microbial communities by intracellular antibiotic deactivation. PLoS Biol 14:e2000631

Spees AM, Wangdi T, Lopez CA, Kingsbury DD, Xavier MN, Winter SE, Tsolis RM, Bäumler AJ (2013) Streptomycin-induced inflammation enhances *Escherichia coli* gut colonization through nitrate respiration. mBio 4(4):e00430–e00413

Stahl PD, Wileman TE, Diment S, Shepherd VL (1984) Mannose-specific oligosaccharide recognition by mononuclear phagocytes. Biol Cell 51:215–218

Surette MG (2014) The cystic fibrosis lung microbiome. Ann Am Thorac Soc 11(Suppl. 1):S61–S65

Sze MA, Dimitriu PA, Hayashi S, Elliott WM, McDonough JE, Gosselink JV, Cooper J, Sin DD, Mohn WW, Hogg JC (2012) The lung tissue microbiome in chronic obstructive pulmonary disease. Am J Respir Crit Care Med 185:1073–1080

Tashkin DP, Altose MD, Bleecker ER (1992) The lung health study: airway responsiveness to inhaled methacholine in smokers with mild to moderate airflow limitation. The lung health study research group. Am Rev Respir Dis 145:301–310

Trompette A, Gollwitzer ES, Yadava K, Sichelstiel AK, Sprenger N, Ngom-Bru C, Blanchard C, Junt T, Nicod LP, Harris NL, Marsland BJ (2014) Gut microbiota metabolism of dietary fiber influences allergic airway disease and hematopoiesis. Nat Med 20:159–166

Tunney MM, Field TR, Moriarty TF (2008) Detection of anaerobic bacteria in high numbers in sputum from patients with cystic fibrosis. Am J Respir Crit Care Med 177:995–1001

Vandeplassche E, Tavernier S, Coenye T (2019) Influence of the lung microbiome on antibiotic susceptibility of cystic fibrosis pathogens. Eur Respir Rev 28:190041

Vazquez-Torres A, Baumler AJ (2015) Nitrate, nitrite and nitric oxide reductases: from the last universal common ancestor to modern bacterial pathogens. Curr Opin Microbiol 29:1–8

Venkataraman A, Bassis CM, Beck JM, Young VB, Curtis JL, Huffnagle GB (2015) Application of a neutral community model to assess structuring of the human lung microbiome. M Bio 6 (1):02284–02214

Vlahos R, Bozinovski S, Jones JE, Powell J, Gras J, Lilja A (2006) Differential protease, innate immunity, and NF-kappa B induction profiles during lung inflammation induced by subchronic cigarette smoke exposure in mice. Am J Physiol Lung Cell Mol Physiol 290:L931–L945

Waksman SA, Woodruff HB (1940) The soil as a source of microorganisms antagonistic to disease-producing Bacteria. J Bacteriol 40(4):581–600

Weimer KE, Juneau RA, Murrah KA, Pang B, Armbruster CE, Richardson SH, Swords WE (2011) Divergent mechanisms for passive pneumococcal resistance to β-lactam antibiotics in the presence of *Haemophilus influenzae*. J Infect Dis 203(4):549–555

Willger SD, Grim SL, Dolben (2012) Spatial distribution of microbial communities in the cystic fibrosis lung. ISME J 6:471–474

Willger SD, Grim SL, Dolben EL (2014) Characterization and quantification of the fungal microbiome in serial samples from individuals with cystic fibrosis. Microbiome 2:40

Willner D, Furlan M, Haynes M (2009) Metagenomic analysis of respiratory tract DNA viral communities in cystic fibrosis and non-cystic fibrosis individuals. PLoS One 4:e7370

Winter SE, Baumler AJ (2014) Dysbiosis in the inflamed intestine: chance favors the prepared microbe. Gut Microbes 5:71–73

Winter SE, Winter MG, Xavier MN, Thiennimitr P, Poon V, Keestra AM, Laughlin RC, Gomez G, Wu J, Lawhon SD, Popova IE, Parikh SJ, Adams LG, Tsolis RM, Stewart VJ, Bäumler AJ (2013) Host-derived nitrate boosts growth of *E. coli* in the inflamed gut. Science 339 (6120):708–711

Woese CR, Fox GE (1977) Phylogenetic structure of the prokaryotic domain: the primary kingdoms. PNAS 74(11):5088–5090

Zhang R, Chen L, Cao L, Li KJ, Huang Y, Luan XQ, Li G (2018) Effects of smoking on the lower respiratory tract microbiome in mice. Respir Res 19:253

Zhao J, Schloss PD, Kalikin LM (2012) Decade-long bacterial community dynamics in cystic fibrosis airways. Proc Natl Acad Sci U S A 109:5809–5814

Understanding the Interplay Between the Host Immune–Microbiome Interactions: A State of the Art Review

8

Kuraganti Gunaswetha, Edla Sujatha, and
Pallaval Veera Bramhachari

Abstract

The microbiome and immune system are effectively impacting each other to endure, characterizing the healthy individual's dysfunctional equilibrium. The human gut has the most complex microbiota of the multitude of non-sterile cavities, with a solid effect on host homeostasis and immunostasis, making it fundamental for looking after health. Simultaneously, mammalian immunity is formed by resident bacteria. At the point when this immune system–microbiota partnership is working appropriately, it takes into account the acceptance of defensive reactions to pathogen just as the upkeep of regulatory pathways engaged with the support of resilience to harmless antigens. The investigation of the microbiome–immune system crosstalk has shown a solid association between microbial communities and the advancement of hypersensitive infections and asthma. Interruption of the microbiome affects the host's safe reaction and can prompt infection pathogenesis. Disease and remedial medicines, then again, affect microbial populaces. A preview of the present status of the microbiome–immune system in host weakness to pathogens, extreme hypersensitivity responses, autoimmunity, chronic inflammation, and cancer research is emphasized in this review. The turn of events and use of next-generation DNA sequencing strategies have changed gut microecology, considering new experiences into the synthesis of the intestinal microbiota and it connects to an

Kuraganti Gunaswetha, Edla Sujatha and Pallaval Veera Bramhachari contributed equally with all other contributors.

K. Gunaswetha · E. Sujatha (✉)
Department of Microbiology, Kakatiya University, Warangal, Telangana, India
e-mail: sujathaedla_1973@kakatiya.ac.in

P. V. Bramhachari
Department of Biotechnology, Krishna University, Machilipatnam, Andhra Pradesh, India

P. V. Bramhachari (ed.), *Microbiome in Human Health and Disease*,
https://doi.org/10.1007/978-981-16-3156-6_8

assortment of diseases. We additionally examine how we have gotten familiar with the associations between resident microbes and the immune system, just as the outcomes of these outcomes for human health.

Keywords

Microbiota · Immune system · Epithelial barrier · Host–microbial relationship · Gut microbiota

8.1 Introduction

The alliance between the microbiome and immune system has proven to be the most significant regarding human health and disease. The historical outlook of the immune system and microbiome interaction came into understanding in 1798 by the outbreak of smallpox and its vaccination study by Edward Jenner. According to Edward Jenner, the exposure of viruses from cowpox scabs can eventually protect the individual from the smallpox virus (Yoshikawa et al. 2014). Later Louis Pasteur proved this technique of vaccination against fowl cholera and anthrax disease. He demonstrated that the old culture of *Pasteurella multocida* bacterium and heated culture of *Bacillus anthracis* of fowl cholera and anthrax disease have not caused any disease, respectively (Berche 2012). Thus he showed that general principles of vaccination could be applied to other diseases other than smallpox to protect humans against other infections. Later in the 1890s the protection caused by vaccination as stated by Emilvon Behring and Shibasaburo Kitasato that the presence of some antibodies in blood are associated with protection against disease (Hajj Hussein et al. 2015). Paul Ehrlich later proved that antibodies not only protect animals against bacteria but also against some foreign toxins too. The idea of the presence of antibodies as protective agents in blood was proved and accepted by many scientists. The relation between microbes and the immune system was studied and enlighten by introducing the techniques and discoveries like passive immunization for tetanus toxin, phagocytes, tuberculin reaction by Robert Koch, anaphylaxis by Charles Richet, clonal selection theory by Medawar, radioimmunoassay by Rosalyn Yalow, the principle of the production of monoclonal antibodies by Kohler and Milstein. The study of the microbiome and immune system gained a huge spectacular growth in research which continues to the present.

Antibodies are protein in nature and their appearance in blood serum is triggered when the animal is exposed to foreign substances that are infectious. The foreign bacterium as toxin substances that induce the production of antibodies is called antigens. The immune response against antigens can be categorized into antibody and cell-mediated immune responses. The cell-mediated response functions through detection and elimination of viral infected abnormal cells and chemically modified cells, where on the other hand antibody-mediated response protects against bacteria and parasites that act as non-cell-associated invaders. In contrast, cell-mediated immunity provides resistance to many bacteria and parasite infections, while antibodies provide immunity against viruses and contribute towards graft rejection.

The study of the immune module enables us in understanding the different physiopathological behavior of the human microbiome which is the complex mixture of bacteria, fungi, protozoa, viruses, and archaea (Geva Zatorsky et al. 2017). The genetic immunological factors and notably diet and environmental biodiversity are the two core drivers of microbiome composition in humans. The genetic immunological factors are more localized to the urinogenital tract microbiome (Gupta et al. 2017). Hence the composition of the human microbiome has its importance according to its site of accuracy (Mezouar et al. 2018). In mammals, the host differential organismal lineage through time and type of diet from evolution are the two factors that altered the microbiome composition that acts on different subsets of bacteria at different temporal scales (Groussin et al. 2017). The human microbiome composition is different or can be said smaller in size compared to that of their ancestor gut lineage. This substantial change in the microbiome is due to environmental changes and diet respective of industrial and western world environment (Clayton et al. 2016).

Mycobiome also a composite microbiome in the human lineage induces the human immune system. Till now more than 400 heterogenous fungal species have been isolated from the human gut and skin region (Halwachs et al. 2017; Belkaid and Naik 2013). The host physiology is usually influenced by bioactive molecules secreted by fungal species, especially the gut fungal microbiome through Dectine-1 receptor which regulates the host immune system (IS) (Iliev et al. 2012). Any kind of changes in the composition of the bacterial microbiome and mycobiome lead to chronic diseases that may include gut inflammation, colitis exacerbation in the host.

8.2 Microbiome–Immune System Interaction

The human immune system and diverse gut flora have co-evolved to form a defense barrier against pathogens along with developing tolerance towards beneficial microorganisms. This mutualistic relationship of the host immune system and gut microbiome enables the researchers to study the gut microflora as 70–80% of immune cells are found in the gut itself (Ley et al. 2006; Kumar et al. 2017). Host-microbiome mutualism to a great extent surpasses the sole metabolic and nutritional aspects and includes the interaction between the microbiome and host immune system. Co-evolution, however, has produced unavoidable interconnections between the physiologies of microbial communities and their hosts that reach out to past metabolic functions. In the connection between the microbiota and the immune system, the interconnections are especially evident (Hooper et al. 2012).

8.3 Experimental Tools for Analyzing the Microbiota Immune System Interactions

Quite a bit of our present comprehension of microbiota–immune system interaction has been procured from "gnotobiotics." Gnotobiotic animals also are known as germ-free animals, especially rodents, have become basic experimental tools for figuring out which host immune capacities are hereditarily encoded and which requires microbial interactions (Hooper et al. 2012). Germ-free animals are removed from the mother's womb surgically without exposing it to bacteria, fungi, viruses, and any kind of eukaryotic parasites and raised in sterile isolators. The germ-free animals are observed to be having altered immune systems lymph nodes, hearts, metabolisms, lungs, and even reproductive capacities (Heijtz et al. 2011). Germ-free animals can fill in as living test tubes for the inception of a single microbial species or a particular defined mixture of species; these can also be a tool for studying their microbiologically sterile state (Luczynski et al. 2016).

The advancement of gnotobiotic experiments are of firstly, the germ-free mouse strains can be both genetically targeted and wild-type inbred isogenic strains, where the different constituents of the immune system contribute towards the host-microbial mutualism would thus be able to be ascertained by looking at the impacts of microbial colonization in genetically modified and wild-type mice (Geuking et al. 2011; Vaishnava et al. 2011). Secondly, the composition of human microbiota can be defined from the polymorphism of the bacterial gene using the next genome sequence by encoding the 16S ribosomal RNA sequence. These analyses enable us to characterize the distinct effects on the host immune system by constructing defined microbiota (Talham et al. 1999). The species contrasts can be shut utilizing mice with a characterized adapted microbiota, with this developing technology. Forthcoming, there is even the possibility of humanized isobiotic mice that additionally have a refined immune system (Goodman et al. 2011). A third development has been the improvement of trial systems that permit the dissecting of the mutualistic relationship between the host immune system and microbial colonization. This cannot be accomplished by antimicrobial treatment alone since small portions of the bacterial colony persist. The explore of the deletion of bacterial strains enables the study of mucosal immune induction, functional protection, and memory of the host immune system since after deletion of bacteria the animal becomes germ-free. The deletion of bacteria can only be grown in culture but does not persist in vivo (Hapfelmeier et al. 2010). By comparing germ-free and colonized mice using metabolomic and transcriptomic tools the effect of inhabitant microbiota on host physiology can be obtained. DNA microarrays, nuclear magnetic resonance spectroscopy, and mass spectroscopy are some of the tools which have prompted a point by point comprehension of how microbiota shape numerous aspects of development, host physiology, and immunity (Peterson et al. 2007) as well as by giving significant bits of knowledge into how microbiota impact metabolic signaling in mammalian hosts (Holmes et al. 2011).

8.4 Interaction of Microbiota–Immune Homeostasis

The microbial community in the human gut plays a crucial part in the development of the immune system and protection against pathogens in much diversified beneficial functions. The gut microbiota of the host is ascertained by mainly host genetics and environmental exposure. Intestinal bacteria regulate the immune homeostasis by sending the signals for regulating anti- and pro-inflammatory host immune responses and priming the systemic immune responses by T regulatory cells (Tregs) and T helper 17 (Th17) cells (Korn et al. 2007). Inflammation and sepsis are common issues that are caused by the disruption of the interplay of host intestinal tissue and microbiota. Differentiation of T regulatory cells is generally facilitated by gut commensal microbiota, where induction of Th17 cells is caused by segmented filamentous bacteria (SFB) (Chaudhry and Rudensky 2013). Changes in the configuration of Th17 cells may lead to disease susceptibility as they are important in maintaining a defense mechanism against invading bacterial pathogens (Sheridan and Lefrançois 2010).

8.5 Innate Immunity with Epithelial Barrier Defense

To decrease the invasion and tissue damage caused by bacteria it is at most important to keep the distance of intestinal microbiota with intestinal epithelial cells (IECs) (Brown et al. 2013). Gnotobiotic mice and antibiotic treated mice show lower expression of bacterial pattern recognition receptors due to the absence of bacteria in IECs which leads to the increase in disease susceptibility. The innate immune epithelial barrier includes the presence of a mucus layer also called mucin glycoprotein secreted by goblet cells, which is a vital component in segregating the microbiota from the intestinal epithelium (Faderl et al. 2015). The large intestine is covered with both outer and inner mucus layers, whereas the small intestine is covered with an inner discontinuous mucus layer secreted along the apical surface. Susceptibility towards pathogen and commensal bacteria-induced inflammation is proportional to the thickness of the inner mucus layer by limiting the direct contact of bacteria with epithelial cells, where the outer mucus layer provides glycans to the mucosa-associated microbiota (Taherali et al. 2018). Along with the mucus layer, innate epithelial barrier defense includes antimicrobial peptides (AMPs) secreted by paneth cells. Extensively secreted AMPs play a crucial role in segregating the microbiota from the epithelium; therefore, a decrease in the number of paneth cells from epithelium results in an invasion of the epithelial barrier by symbiotic and pathogenic microbes leading to inflammation (Peterson and Artis 2014). The innate epithelium barrier defenses also have innate lymphoid cells (ILCs) which confine much of the microbial community of the lumen of the intestinal tract. The role and secretion of mucus, AMPs in innate epithelial barrier defense are explained in Fig. 8.1.

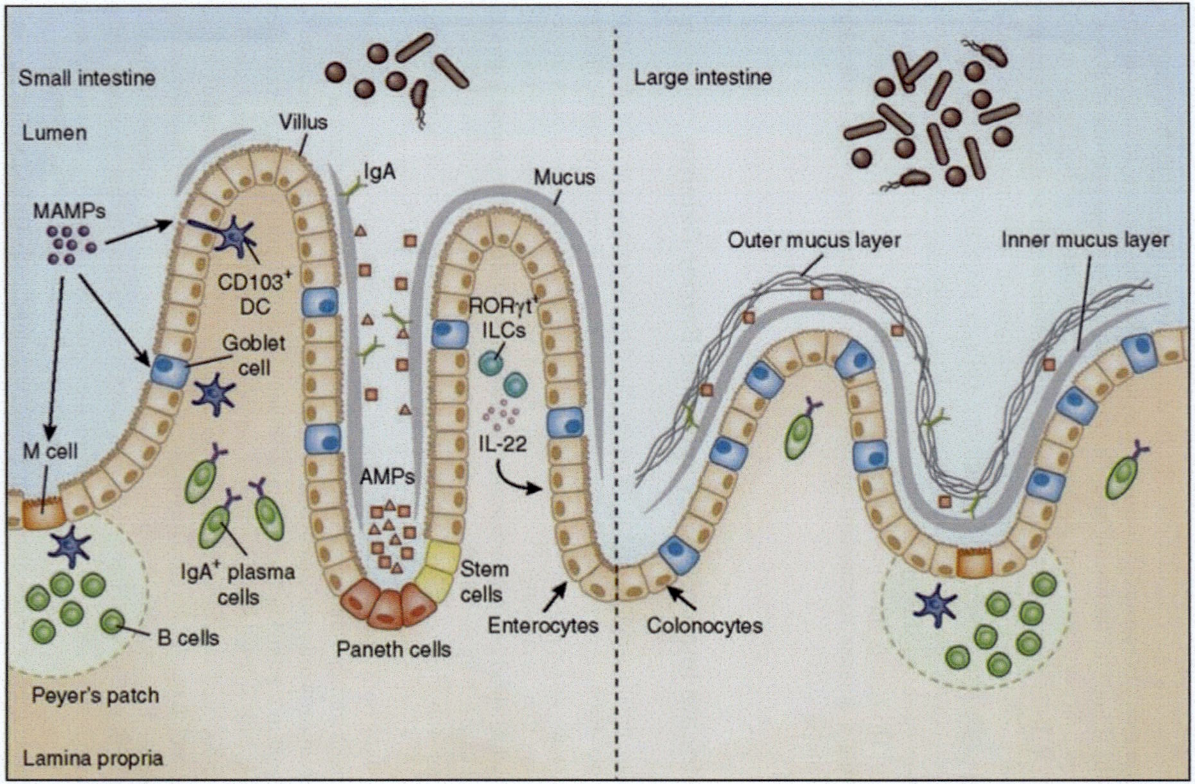

Fig. 8.1 Anatomical containment of the microbiota along the intestine. The intestinal epithelium barrier comprises a single layer of enterocytes or colonocytes, protected by the immune system. In the small intestine, mucus laer is discontinuous, with fewer goblet cells. The enterocytes are enriched with paneth cells that

Fig. 8.1 (continued) secrete AMPs, which can cross-link with the mucus layer. Through this barrier, sampling of MAMPs can be mediated through antigen uptake by M cells and gobet cells to dendritic cells (DCs), along with direct transepithelial luminal sampling from DCs. RORγt ILCs can sense microbial signals and produce IL-22 to aid in IEC barrier function. Commensal-specific IgA is produced by plasma cells in the lamina propria, mediated by DCs in a T cell-independent mechanism. The large intestine uses a thick, continuous mucus layer to compartmentalize the microbiota, with IgA and AMPs having a secondary role. (Brown et al. 2013)

8.6 Innate Immune Cell's Barrier

The epithelial enterocyte barrier cells secreting the usual amount of mucus, AMPs, pattern recognition receptors (PRRs), and secreted IgA molecules promote intestinal homeostasis and tolerant response to the microbiota, where the microbiota is segregated away from the IECs, and the intestinal immune system directs a largely tolerant response to the resident commensal bacteria (Maynard et al. 2012). The tolerogenic immune cell responses to the microbiota are regulated by the secretion of cytokines TGF-β, BAFF, IL-33, TSLP, and APRIL stimulated by the microbe-associated molecular patterns (MAMPs) by binding to the pattern recognition receptors (PRRs) (De Souza and Fiocchi 2016). This cytokine condition enriches CD103+ dendritic cells (DCs), which aid in the improvement of Treg cells secreting IL-10 and TGF-β. Treg cells and CD103+ DCs invigorate the production of commensal-specific IgA. IECs barrier wholeness is enhanced by the secretion of IL-22 by RORγt ILCs in this specific condition. DCs and macrophages secret IL-12 and IL-23 which in turn promote TH1 and TH17 response. These T helper cells secrete high levels of IFN-γ and IL-17A, respectively, and T-bet+ ILCs also accumulate to produce IFN-γ (Brown et al. 2013). Infringement in the epithelial barrier by the microbiota n this circumstance can likewise prompt more significant levels of B cells secreting commensal-specific IgG (Fig. 8.2). In immune deficiency diseases like HIV, IBD infections the innate barrier defect occurs and a potentially harmful pro-inflammatory response stimulates by the intestinal immune system to the microbiota to clear invading bacteria by promoting dysbiosis (Hunt et al. 2014).

$CD4^+$ T cells are commonly divided into regulatory T (Treg) cells and conventional T helper (Th) cells. Regulatory T cells (Tregs) are important for the induction and maintenance of peripheral tolerance (Yong et al. 2007). It is observed that in the large intestine the proportion of Treg cell proportions is significantly larger in the $CD4^+$ T cell population, whereas in the small intestine there are only fewer $CD4^+$ T cells and Th17 cells. Nevertheless, the gnotobiotic mice or antibiotic-treated mice have depicted an excellent reciprocicity of this condition which occurred by the absence of transcription factor Helios expression (Alexander et al. 2014). When these gnotobiotic mice were colonized with commensal microbiota the restoration of Tregs was observed this might due to these cells have a unique TCR repertoire that confers specificity for commensal microbiota, and/or that commensal microbiota products can influence the Th17:Treg balance, potentially by favoring Treg cell differentiation or expansion (Littman and Pamer 2011).

The innate lymphoid cells (ILCs) of lymphoid lineage, developed from common lymphoid progenitor cells, emerge as the first line of defense at mucosal barriers by lacking rearranged Ag-receptors. Three groups of ILCs regulate cytokine production as encounters with microbial and inflammation. cNK cells and helper ILC1 secrete th1-type cytokine IFN-γ that are included in group 1 ILCs where IL-4, IL-5, and/or IL-13, TH2-type cytokines production is characterized by group 2 ILCs, fetal lymphoid tissue-inducer (LTi) cells, and ILC3s with cytotoxicity receptor (NCR). NKp46 (NCR + ILC3s) are included in group 3. Group 3 ILCs develop and produce

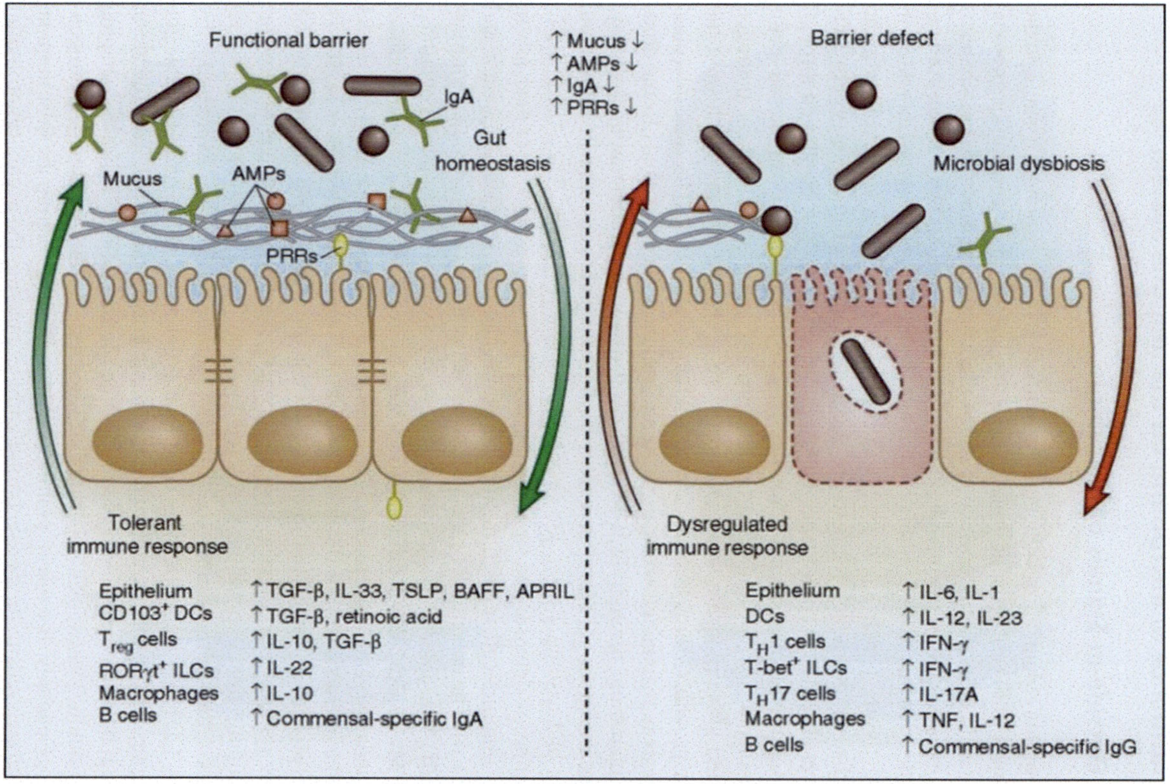

Fig. 8.2 Immune factors that are engaged with a functional barrier in the intestinal immune tract and permitting the concurrence of commensal bacteria, standing out from a barrier defect prompting dysbiosis and dysregulated immune responses (Brown et al. 2013)

IL-22 a Th17-type cytokine through transcription factor RORγt and also protect the mucosal tissues from damage and inflammation.

8.7 Influence of Microbiota on the Immune System

8.7.1 Outside-In Signals of Microbiota

The microbiota colonization influences the formation of the lymphoid tissue, and the development of the subsequent immune system was studied by comparing gnotobiotic mice and colonized mice which revealed that commensal microbiota influences the immune system through outside-in signals. As it is well known that gastrointestinal tissue contains a high number of myeloid and lymphoid cells embedded in lymphoid tissue, which are indeed influenced by microbiota in developing the lymphoid structures which leads to the adequate function of immune system cells (Hooper et al. 2012). The influence on lymphoid structures can be studied through commensal microbiome since it confers resistance through direct commensal–pathogen interaction by consuming particular nutrients that are required for pathogen growth and also produce a small chain of fatty acids like acetate which limit the pathogen growth (Momose et al. 2008; Fukuda et al. 2011). The host-microbial homeostasis is maintained when the hosting immune system responds to the commensal microbiota through TLR-MyD88 (Toll-like receptors) signaling pathway. Toll-like receptors (TLRs) perceive particular pathogen-related sub-atomic examples and assume a basic job in innate immune responses. They take part in the mainline of defense against attacking pathogens and assume a critical job in survival, immune cell regulation, inflammation, and proliferation (Kawasaki and Kawai 2014). The enactment of the TLR signaling pathway starts from the cytoplasmic Toll/IL-1 receptor (TIR) domain that associates with a TIR domain-containing adaptor, MyD88 (Myeloid differentiation primary response 88). Upon incitement with ligands, MyD88 initiates IL-1 receptor-related kinase-4 (IRAK-4) to TLRs through the interaction of the death domains of both molecules (Reuven et al. 2014).

RegIIIγ is an epithelial antimicrobial protein that is induced by the MyD88 depended on bacterial signals, lipopolysaccharide (LPS), and flagellin proteins also inducing this expression (Vaishnava et al. 2011). The flagellin protein stimulates the production of IL-23 through TLR5 expressed by CD103 + CD11b + dendritic cells of lamina propria, which in turn stimulate the IL-22 expression by innate lymphoid cells (Kinnebrew et al. 2012). The outside-in signals can be studied in understanding the microbiome in influencing and shaping the host immune system by comparing germ-free gnotobiotic and specific pathogen-free mice. Generally, the role of commensal microbiota in shaping the host immune system through outside-in signals is studied by comparing the axenic (gnotobiotic) mice and specific pathogen-free mice. A reserch report stated that commensal microbiota when colonized in axenic mice, developed Peyer's patches, mature mesenteric lymph nodes, increased levels of antimicrobial peptides, high production of immunoglobulin A (IgA), dendritic cells (DCS), B and T cells, and also developed goblet cells with a thick layer of mucus, which were absent in early axenic mice and were normalized with the colonization of commensal microbiome (Macpherson et al. 2004; Sommer and Backhed 2013).

8.7.2 Inside-Out Signals of Microbiota

To maintain the integrity of the immune system, the host immune system should have homeostatic relation with microbiota, which is accomplished by maintaining the controlled interactions between microbiota and host tissue along with the limited composition of microbial affiliation. Inside-out signal deals with the internal, physical, and chemical control of the immune system on microbiota localization and composition of the microbiota (Hooper et al. 2012). As the immune system counters microbiota at an enormous rate compared to any other organ, it should control as well as avoid the unnecessary innate immune signals that emerge from variation in microbiota consortia without causing damage to the host metabolic functions. In the intestine, to reduce the pathological reactions caused by microbial interaction with the epithelial surface is achieved by separating the mucosal immune response through stratification and compartmentalization (Mezouar et al. 2018).

Stratification is reducing or minimizing the interaction between the intestinal luminal microbiota with the epithelial cell surface, which is achieved by immune effectors which function in stratifying the microbiota. Mucin glycoproteins form a thick viscous layer at the intestinal epithelial cell surface which is secreted by goblet cells that are embedded in enterocytes of the intestine. The mucus layer is differentiated in the colon, the outer mucus layer is exposed to a high number of bacteria where the inner mucus layer withstands the bacterial penetration due to dense impermeable mucus layer which consists of O-glycosylated MUC2 mucin with ZG16 and β-defensins proteins (Hansson and Johansson 2010; Bergström et al. 2016). On the contrary, small intestine contains a heterogeneous mucus layer that is outer and inner mucus layers which are not distinct (Johansson et al. 2011). The intestinal mucus layer is rich in paneth cells, RegIIIγ C-type lectin is antibacterial and also minimizes the bacterial penetration which is controlled by toll-like receptors (TLRs) and intestinal specific immunoglobulin A (IgA) produced by dendritic cells by interacting with B and T cells in Peyer's patches (Macpherson and Uhr 2004; Cash et al. 2006; Salzman et al. 2007). Where, in gnotobiotes, the commensal bacteria stratification is dependent on IL-22 and IL-23 produced by the ILC3 (Mao et al. 2018).

Mucosal compartmentalization is the function that confines the commensal bacteria from penetrating the luminal epithelial barrier by less exposure to the systemic immune system (Mezouar et al. 2018). Although some of the intestinal bacteria inevitably penetrate the laminal intestinal epithelial barrier causing immune responses like phagocytosis by lamina propria macrophages such as dendritic cells (DCs) that are embedded along lamina propria of the epithelial lining and then migrate to the mesenteric lymph node (Iliev et al. 2012). These DCs in the mesenteric lymph node induce the secretion of protective IgA antibodies and get distributed along the surface of the lamina by activating the B and T cells through interacting with the Peyer's patch (Macpherson and Uhr 2004). Later these DC's cells after activating B and T cells leave through the thoracic duct and reach systemic circulation and induce the mucosal response throughout the mucosal surface of the host (Macpherson and Uhr 2004). Defective in the compartmentalization was observed in immune-deficient mice, i.e., in engineered mice, the priming of serum IgG antibodies response toward commensal bacteria is due to lack of IgA antibodies, indicating the bacteria exposed to the systemic immune system. This is because the mice do not have TLR-mediated MyD88 signals for recognizing bacteria, which results in the successful penetration of commensal bacteria by crossing epithelial barrier and phagocytic cells (Slack et al. 2009).

8.8 Microbiota Protection Against Autoimmune Diseases

The interaction between the host immune system and microbiota favors the autoimmune disease by segmented filamentous bacteria in Type1 diabetes (T1D), where the damage of pancreatic islets of Langerhans leads to the imbalance of insulin secretion. With some genetic predispositions and diabetogenic T cell populations with defined CD_4 and CD_8, non-obese diabetic (NOD) mouse acts as a prompt model of T1D (Schmidt et al. 1999). In the isogenic NOD colony, the prevalence of T1D depends upon the presence of both pathogenic and microbiota diversity since they act as determining factors (Hooper et al. 2012). Where, on the other hand, congenic NOD mice with single-locus MyD88 deficiency show the same incidence of T1D as the parent NOD strain (Kendall et al. 2009). The over access of intestinal microbiota penetration with an epithelial barrier, interaction of commensal microbiota with the systemic immune system are the major effects of host-microbiome mutualistic relation which occurs due to the deficiency of MyD88 (Kinnebrew et al. 2010; Carvalho et al. 2012). Since commensal bacteria help in the development of the host immune system in immune-deficient patients, a little change in the commensal bacteria community affects and triggers the inflammation responses. For example, lesions in salivary glands, lungs, and feet are caused due to defect in tyrosine phosphate SHP-I signal that leads to an autoinflammatory syndrome which is microbiota dependent (Croker et al. 2008).

8.9 Microbiota in Modulating Immune Responses in Cancer

The effect of microbial composition on carcinogenesis and inflammation is tissue-specific. Generally, the tumor-promoting effects of microbiota on the host immune system are spontaneous, genetically driven, and cancer-induced through carcinogen (Yu et al. 2010; Lofgren et al. 2011; Li et al. 2012; Grivennikov et al. 2012). It was observed that the microbiota shows both tumor-promoting and also anti-tumor effects. For example, the bacterial extract mixture of *Bacillus Calmette-Guerin* (BCG) is used in the treatment of bladder cancer (Dias et al. 2018). The antitumor effects are due to the innate immunity by converting tumor tolerance to the anti-tumor response by TLR and NOD-like receptors of bacterial components (Fukata and Abreu 2007; Garaude et al. 2012). Pattern recognition receptors (PRR) generally initiate regulatory responses by monitoring microbial status and barrier integrity may also promote resistance to cell death and by triggering cancer-promoting inflammation. Genotoxins and tumor-promoting metabolites are the carcinogenic molecules that are released by microbes, which are recognized by the TLRs and microbe-associated molecular patterns (MAMPs) that promote carcinogenesis (Ochi et al. 2012; Sheflin et al. 2014).

The recent method was developed by studying the commensal microbiota of the host in immune therapy by adverting of ICBs. ICBs are the immune checkpoint blockers that exhibit the function of reactivating T cells in an ineffective tumor microenvironment in response to tumor antigens. The efficacy of ICB is greatly influenced by the gut microbiome community type and number (Schwabe and Jobin 2013). The blocked off two checkpoints Cytotoxic T lymphocyte-association protein 4 (CTLA-4) and programmed-cell-death protein 1 (PCD-1) by monoclonal antibodies are well known. Recent studies emphasized that distinct gut bacterial species greatly influenced the immune-stimulatory and anti-tumor effects of the CTLA-4 antibody (Clemente et al. 2012). The composition of gut microbiota in the host shows a high frequency of difference in responses to PD1 blocker (Sivan et al. 2015). For example, the natural gut microbiota *Lactobacillus johnsonii* and *Enterococcus hirae* get altered and translocated by cyclophosphamide an immunostimulatory alkylating agent, by which the bacteria stimulate the production of specific Th17 and Th1 cell subsets (Sharma and Allison 2015).

8.10 Role of Metabolites in Microbiota and Immune System

Microbiota and its metabolites are significant instigators of host physiology and pathophysiology through the control of a huge range of inflammatory, metabolic, and indeed, even behavioral procedures (Hsiao et al. 2013). Distinctive microbiome setups produce, adjust, and corrupt a huge exhibit metabolite, thereby giving utilitarian complementation to the metabolic limits of the host; e.g., complex proteins furthermore, carbohydrates that cannot be corrupted by the host can be processed by the microbial community (Nicholson et al. 2012). Intervened by metabolite signaling through a progression of innate immune receptors of microbial metabolites, the microbiota relates to the intimate communication with its eukaryotic host.

Single chain fatty acids (SCFAs) are one of the most produced molecules by gut bacteria that control multiple aspects of metabolism and immunity. PPARγ intracellular receptors, the surface proteins GPR41 and GPR43, and GPR109a butyrate receptors detect the SCFAs like propionate, acetate, and butyrate which are the products of dietary fiber fermentation (Alex et al. 2013). Administration of SCFAs leads to alterations in hematopoiesis, resulting in an intensified myeloid output due to elevated numbers of myeloid precursors (Khosravi et al. 2014). These precursors promote the clearance of systemic infection and ameliorates the allergic reactions (Balmer et al. 2014; Trompette et al. 2014).

8.11 Advancements to Identify Mechanisms of Immune Interaction

The recent novel mechanisms have been created by the scientist to study the microbial host interactions. By performing integration analysis of microbiota and host data sets complementing them with the results of visual analysis for assessing both the microbiome and host immune system can be successfully done by high dimensional technologies t deciphering the novel pathways (Siebert et al. 2019). The integrated high dimensional "omic" techniques are used to measure a vast number of analytes/metabolites which serve as biomarkers that release in different proportions based on response to treatment of diseases (Matson et al. 2018). The study of these analytical biomarkers leads to the study of immune responses to therapy and hypotheses of mechanical insights into the immune system (Arneson et al. 2017). The high dimensional omic technologies include metatranscriptomic sequencing, shotgun metagenomics, and metabolomics, these advancements profoundly describe the phenotypes of both host and microbiome generating speculations concerning immune modulation of significant disease environment (Lozupone 2018). Due to poor comprehension of what these organisms, transcripts, and genes do, translation of results and potential ramifications of correlations between organisms, transcripts, and qualities turn out to be a very test in utilizing these progressions.

8.12 Conclusion and Future Perspectives

Even though propels in cutting edge (next-generation) sequencing and bioinformatics have been significant drivers of progress in human microbiome research, there have likewise been other key advances that have empowered unthinking bits of knowledge. These remember propels for our capacity to culture diverse intestinal microbes and to genetically manipulate bacteria so that the impacts of addition or loss of specific functions can be assessed, the development of animal models, for example, gnotobiotic mice for building up causality, and mix of other front line innovations, for example, metabolomics. Despite these gains, there are as yet numerous difficulties, for example, the high number of qualities and metabolites of obscure capacity and of bacteria with ineffectively comprehended properties that dodge development and the absence of accessibility of tools to genetically manipulate the majority of those that we can culture. The most energizing and translational work incorporates complex multi-omic and bioinformatics strategies with corroborative test work to build up an unthinking connection between microorganisms and disease.

It is currently certain that the safe framework assumes a focal job informing the structure of the microbiota just as its nearness to have tissues. Simultaneously, inhabitant microorganisms give flags that encourage ordinary insusceptible framework improvement and impact the resulting safe reactions. Interruption of these complex and dynamic connections can have significant ramifications for host health. Notwithstanding, there are as yet significant gaps in our comprehension of how the host immune system regulates the microbiota, and of how the microbiota shape has invulnerability. The inquiries that remain are testing and will require the advancement of new tools and approaches. At last, these endeavors should prompt further understanding to have microbial connections and give compelling new chances to improve human health.

Acknowledgments Kuraganti Gunaswetha, Edla Sujatha, Pallaval Veera Bramhachari are grateful to Kakatiya University, Warangal and Krishna University, Machilipatnam for the support extended.

Conflict of Interest The authors declare that they have no competing interests.

References

Alex S, Lange K, Amolo T, Grinstead JS, Haakonsson AK, Szalowska E, Koppen A, Mudde K, Haenen D, Roelofsen H, Houtman R (2013) Short-chain fatty acids stimulate angiopoietin-like 4 synthesis in human colon adenocarcinoma cells by activating peroxisome proliferator-activated receptor γ. Mol Cell Biol 33(7):1303–1316

Alexander KL, Targan SR, Elson CO III (2014) Microbiota activation and regulation of innate and adaptive immunity. Immunol Rev 260(1):206–220

Arneson D, Shu L, Tsai B, Barrere-Cain R, Sun C, Yang X (2017) Multidimensional integrative genomics approaches to dissecting cardiovascular disease. Front Cardiovasc Med 4:8

Balmer ML, Schürch CM, Saito Y, Geuking MB, Li H, Cuenca M, Kovtonyuk LV, McCoy KD, Hapfelmeier S, Ochsenbein AF, Manz MG (2014) Microbiota-derived compounds drive steady-state granulopoiesis via MyD88/TICAM signaling. J Immunol 193(10):5273–5283

Belkaid Y, Naik S (2013) Compartmentalized and systemic control of tissue immunity by commensals. Nat Immunol 14(7):646–653

Berche P (2012) Louis Pasteur, from crystals of life to vaccination. Clin Microbiol Infect 18:1–6

Bergström JH, Birchenough GM, Katona G, Schroeder BO, Schütte A, Ermund A, Johansson ME, Hansson GC (2016) Gram-positive bacteria are held at a distance in the colon mucus by the lectin-like protein ZG16. Proc Natl Acad Sci 113(48):13833–13838

Brown EM, Sadarangani M, Finlay BB (2013) The role of the immune system in governing host-microbe interactions in the intestine. Nat Immunol 14(7):660–667

Carvalho FA, Aitken JD, Vijay-Kumar M, Gewirtz AT (2012) Toll-like receptor–gut microbiota interactions: perturb at your own risk! Annu Rev Physiol 74:177–198

Cash HL, Whitham CV, Behrendt CL, Hooper LV (2006) Symbiotic bacteria direct expression of an intestinal bactericidal lectin. Science 313(5790):1126–1130

Chaudhry A, Rudensky AY (2013 Mar 1) Control of inflammation by integration of environmental cues by regulatory T cells. J Clin Invest 123(3):939–944

Clayton JB, Vangay P, Huang HU, Ward T, Hillmann BM, Al-Ghalith GA, Travis DA, Long HT, Van Tuan B, Van Minh V, Cabana F (2016) Captivity humanizes the primate microbiome. Proc Natl Acad Sci 113(37):10376–10381

Clemente JC, Ursell LK, Parfrey LW, Knight R (2012) The impact of the gut microbiota on human health: an integrative view. Cell 148(6):1258–1270

Croker BA, Lawson BR, Rutschmann S, Berger M, Eidenschenk C, Blasius AL, Moresco EM, Sovath S, Cengia L, Shultz LD, Theofilopoulos AN (2008) Inflammation and autoimmunity caused by a SHP1 mutation depend on IL-1, MyD88, and a microbial trigger. Proc Natl Acad Sci 105(39):15028–15033

De Souza HS, Fiocchi C (2016) Immunopathogenesis of IBD: current state of the art. Nat Rev Gastroenterol Hepatol 13(1):13

Dias LP, Luzo ÂCM, Volpe BB, Durán M, Galdames SE, Ferreira LA, Durán N, Fávaro WJ (2018) Effects of intravesical therapy with platelet-rich plasma (PRP) and Bacillus Calmette-Guérin (BCG) in non-muscle invasive bladder cancer. Tissue Cell 52:17–27

Faderl M, Noti M, Corazza N, Mueller C (2015) Keeping bugs in check: the mucus layer as a critical component in maintaining intestinal homeostasis. IUBMB Life 67(4):275–285

Fukata M, Abreu MT (2007) TLR4 signalling in the intestine in health and disease. Biochem Soc Trans 35(6):1473–1478

Fukuda S, Toh H, Hase K, Oshima K, Nakanishi Y, Yoshimura K, Tobe T, Clarke JM, Topping DL, Suzuki T, Taylor TD (2011) Bifidobacteria can protect from enteropathogenic infection through production of acetate. Nature 469(7331):543–547

Garaude J, Kent A, van Rooijen N, Blander JM (2012) Simultaneous targeting of toll-and nod-like receptors induces effective tumor-specific immune responses. Sci Transl Med 4 (120):120ra16–120ra16

Geuking MB, Cahenzli J, Lawson MA, Ng DC, Slack E, Hapfelmeier S, McCoy KD, Macpherson AJ (2011) Intestinal bacterial colonization induces mutualistic regulatory T cell responses. Immunity 34(5):794–806

Geva-Zatorsky N, Sefik E, Kua L et al (2017) Mining the human gut microbiota for immunomodulatory organisms. Cell 168(5):928–943

Goodman AL, Kallstrom G, Faith JJ, Reyes A, Moore A, Dantas G, Gordon JI (2011) Extensive personal human gut microbiota culture collections characterized and manipulated in gnotobiotic mice. Proc Natl Acad Sci 108(15):6252–6257

Grivennikov SI, Wang K, Mucida D, Stewart CA, Schnabl B, Jauch D, Taniguchi K, Yu GY, Österreicher CH, Hung KE, Datz C (2012) Adenoma-linked barrier defects and microbial products drive IL-23/IL-17-mediated tumour growth. Nature 491(7423):254–258

Groussin M, Mazel F, Sanders JG, Smillie CS, Lavergne S, Thuiller W, Alm EJ (2017) Unraveling the processes shaping mammalian gut microbiomes over evolutionary time. Nat Commun 8(1): 1–12

Gupta VK, Paul S, Dutta C (2017) Geography, ethnicity or subsistence-specific variations in human microbiome composition and diversity. Front Microbiol 8:1162

Hajj Hussein I, Chams N, Chams S, El Sayegh S, Badran R, Raad M, Gerges-Geagea A, Leone A, Jurjus A (2015) Vaccines through centuries: major cornerstones of global health. Front Public Health 3:269

Halwachs B, Madhusudhan N, Krause R, Nilsson RH, Moissl-Eichinger C, Högenauer C, Thallinger GG, Gorkiewicz G (2017) Critical issues in mycobiota analysis. Front Microbiol 8:180

Hansson GC, Johansson ME (2010) The inner of the two Muc2 mucin-dependent mucus layers in colon is devoid of bacteria. Gut Microbes 1(1):51–54

Hapfelmeier S, Lawson MA, Slack E, Kirundi JK, Stoel M, Heikenwalder M, Cahenzli J, Velykoredko Y, Balmer ML, Endt K, Geuking MB (2010) Reversible microbial colonization of germ-free mice reveals the dynamics of IgA immune responses. Science 328 (5986):1705–1709

Heijtz RD, Wang S, Anuar F, Qian Y, Björkholm B, Samuelsson A, Hibberd ML, Forssberg H, Pettersson S (2011) Normal gut microbiota modulates brain development and behavior. Proc Natl Acad Sci 108(7):3047–3052

Holmes E, Li JV, Athanasiou T, Ashrafian H, Nicholson JK (2011) Understanding the role of gut microbiome–host metabolic signal disruption in health and disease. Trends Microbiol 19 (7):349–359

Hooper LV, Littman DR, Macpherson AJ (2012) Interactions between the microbiota and the immune system. Science 336(6086):1268–1273

Hsiao EY, McBride SW, Hsien S, Sharon G, Hyde ER, McCue T, Codelli JA, Chow J, Reisman SE, Petrosino JF, Patterson PH (2013) Microbiota modulate behavioral and physiological abnormalities associated with neurodevelopmental disorders. Cell 155(7):1451–1463

Hunt PW, Sinclair E, Rodriguez B, Shive C, Clagett B, Funderburg N, Robinson J, Huang Y, Epling L, Martin JN, Deeks SG (2014) Gut epithelial barrier dysfunction and innate immune activation predict mortality in treated HIV infection. J Infect Dis 210(8):1228–1238

Iliev ID, Funari VA, Taylor KD, Nguyen Q, Reyes CN, Strom SP, Brown J, Becker CA, Fleshner PR, Dubinsky M, Rotter JI (2012) Interactions between commensal fungi and the C-type lectin receptor Dectin-1 influence colitis. Science 336(6086):1314–1317

Johansson ME, Larsson JM, Hansson GC (2011) The two mucus layers of colon are organized by the MUC2 mucin, whereas the outer layer is a legislator of host–microbial interactions. Proc Natl Acad Sci 108(Suppl 1):4659–4665

Kawasaki T, Kawai T (2014) Toll-like receptor signaling pathways. Front Immunol 5:461

Kendall PL, Moore DJ, Hulbert C, Hoek KL, Khan WN, Thomas JW (2009) Reduced diabetes in btk-deficient nonobese diabetic mice and restoration of diabetes with provision of an anti-insulin IgH chain transgene. J Immunol 183(10):6403–6412

Khosravi A, Yáñez A, Price JG, Chow A, Merad M, Goodridge HS, Mazmanian SK (2014) Gut microbiota promote hematopoiesis to control bacterial infection. Cell Host Microbe 15 (3):374–381

Kinnebrew MA, Buffie CG, Diehl GE, Zenewicz LA, Leiner I, Hohl TM, Flavell RA, Littman DR, Pamer EG (2012) Interleukin 23 production by intestinal CD103+ CD11b+ dendritic cells in response to bacterial flagellin enhances mucosal innate immune defense. Immunity 36 (2):276–287

Kinnebrew MA, Ubeda C, Zenewicz LA, Smith N, Flavell RA, Pamer EG (2010) Bacterial flagellin stimulates toll-like receptor 5—dependent defense against vancomycin-resistant Enterococcus infection. J Infect Dis 201(4):534–543

Korn T, Oukka M, Kuchroo V, Bettelli E (2007) Th17 cells: effector T cells with inflammatory properties. Semin Immunol 19(6):362–371

Kumar V, Abbas AK, Aster JC (2017) Robbins basic pathology e-book. Elsevier, Amsterdam

Ley RE, Peterson DA, Gordon JI (2006) Ecological and evolutionary forces shaping microbial diversity in the human intestine. Cell 124(4):837–848

Li Y, Kundu P, Seow SW, de Matos CT, Aronsson L, Chin KC, Kärre K, Pettersson S, Greicius G (2012) Gut microbiota accelerate tumor growth via c-Jun and STAT3 phosphorylation in APC min/+ mice. Carcinogenesis 33(6):1231–1238

Littman DR, Pamer EG (2011) Role of the commensal microbiota in normal and pathogenic host immune responses. Cell Host Microbe 10(4):311–323

Lofgren JL, Whary MT, Ge Z, Muthupalani S, Taylor NS, Mobley M, Potter A, Varro A, Eibach D, Suerbaum S, Wang TC (2011) Lack of commensal flora in helicobacter pylori–infected INS-GAS mice reduces gastritis and delays intraepithelial neoplasia. Gastroenterology 140 (1):210–220

Lozupone CA (2018) Unraveling interactions between the microbiome and the host immune system to decipher mechanisms of disease. Msystems 3(2):e00183–e00117

Luczynski P, McVey Neufeld KA, Oriach CS, Clarke G, Dinan TG, Cryan JF (2016) Growing up in a bubble: using germ-free animals to assess the influence of the gut microbiota on brain and behavior. Int J Neuropsychopharmacol 19(8):pyw020

Macpherson AJ, Uhr T (2004) Induction of protective IgA by intestinal dendritic cells carrying commensal bacteria. Science 303(5664):1662–1665

Mao K, Baptista AP, Tamoutounour S, Zhuang L, Bouladoux N, Martins AJ, Huang Y, Gerner MY, Belkaid Y, Germain RN (2018) Innate and adaptive lymphocytes sequentially shape the gut microbiota and lipid metabolism. Nature 554(7691):255–259

Matson V, Fessler J, Bao R, Chongsuwat T, Zha Y, Alegre ML, Luke JJ, Gajewski TF (2018) The commensal microbiome is associated with anti–PD-1 efficacy in metastatic melanoma patients. Science 359(6371):104–108

Maynard CL, Elson CO, Hatton RD, Weaver CT (2012) Reciprocal interactions of the intestinal microbiota and immune system. Nature 489(7415):231–241

Mezouar S, Chantran Y, Michel J, Fabre A, Dubus JC, Leone M, Sereme Y, Mège JL, Ranque S, Desnues B, Chanez P (2018) Microbiome and the immune system: from a healthy steady-state to allergy associated disruption. Hum Microbiom J 10:11–20

Momose Y, Hirayama K, Itoh K (2008) Effect of organic acids on inhibition of Escherichia coli O157: H7 colonization in gnotobiotic mice associated with infant intestinal microbiota. Antonie Van Leeuwenhoek 93(1–2):141–149

Nicholson JK, Holmes E, Kinross J, Burcelin R, Gibson G, Jia W, Pettersson S (2012) Host-gut microbiota metabolic interactions. Science 336(6086):1262–1267

Ochi A, Nguyen AH, Bedrosian AS, Mushlin HM, Zarbakhsh S, Barilla R, Zambirinis CP, Fallon NC, Rehman A, Pylayeva-Gupta Y, Badar S (2012) MyD88 inhibition amplifies dendritic cell capacity to promote pancreatic carcinogenesis via Th2 cells. J Exp Med 209(9):1671–1687

Peterson DA, McNulty NP, Guruge JL, Gordon JI (2007) IgA response to symbiotic bacteria as a mediator of gut homeostasis. Cell Host Microbe 2(5):328–339

Peterson LW, Artis D (2014) Intestinal epithelial cells: regulators of barrier function and immune homeostasis. Nat Rev Immunol 14(3):141–153

Reuven EM, Fink A, Shai Y (2014) Regulation of innate immune responses by transmembrane interactions: lessons from the TLR family. Biochim Biophys Acta 1838(6):1586–1593

Salzman NH, Underwood MA, Bevins CL (2007) Paneth cells, defensins, and the commensal microbiota: a hypothesis on intimate interplay at the intestinal mucosa. Semin Immunol 19 (2):70–83

Schmidt D, Amrani A, Verdaguer J, Bou S, Santamaria P (1999) Autoantigen-independent deletion of diabetogenic CD4+ thymocytes by protective MHC class II molecules. J Immunol 162 (8):4627–4636

Schwabe RF, Jobin C (2013) The microbiome and cancer. Nat Rev Cancer 13(11):800–812

Sharma P, Allison JP (2015) The future of immune checkpoint therapy. Science 348(6230):56–61

Sheflin AM, Whitney AK, Weir TL (2014) Cancer-promoting effects of microbial dysbiosis. Curr Oncol Rep 16(10):406

Sheridan BS, Lefrançois L (2010) Intraepithelial lymphocytes: to serve and protect. Curr Gastroenterol Rep 12(6):513–521

Siebert JC, Görg C, Palmer B, Lozupone C (2019) Visualizing microbiome–immune system interplay. Immunother 11(2):63–67

Sivan A, Corrales L, Hubert N, Williams JB, Aquino-Michaels K, Earley ZM, Benyamin FW, Lei YM, Jabri B, Alegre ML, Chang EB (2015) Commensal Bifidobacterium promotes antitumor immunity and facilitates anti–PD-L1 efficacy. Science 350(6264):1084–1089

Slack E, Hapfelmeier S, Stecher B, Velykoredko Y, Stoel M, Lawson MA, Geuking MB, Beutler B, Tedder TF, Hardt WD, Bercik P (2009) Innate and adaptive immunity cooperate flexibly to maintain host-microbiota mutualism. Science 325(5940):617–620

Sommer F, Bäckhed F (2013) The gut microbiota—masters of host development and physiology. Nat Rev Microbiol 11(4):227–238

Taherali F, Varum F, Basit AW (2018) A slippery slope: on the origin, role and physiology of mucus. Adv Drug Deliv Rev 124:16–33

Talham GL, Jiang HQ, Bos NA, Cebra JJ (1999) Segmented filamentous bacteria are potent stimuli of a physiologically normal state of the murine gut mucosal immune system. Infect Immun 67 (4):1992–2000

Trompette A, Gollwitzer ES, Yadava K, Sichelstiel AK, Sprenger N, Ngom-Bru C, Blanchard C, Junt T, Nicod LP, Harris NL, Marsland BJ (2014) Gut microbiota metabolism of dietary fiber influences allergic airway disease and hematopoiesis. Nat Med 20(2):159–166

Vaishnava S, Yamamoto M, Severson KM, Ruhn KA, Yu X, Koren O, Ley R, Wakeland EK, Hooper LV (2011) The antibacterial lectin RegIIIγ promotes the spatial segregation of microbiota and host in the intestine. Science 334(6053):255–258

Yong Z, Chang L, Mei YX, Yi L (2007) Role and mechanisms of CD4+ CD25+ regulatory T cells in the induction and maintenance of transplantation tolerance. Transpl Immunol 17(2):120–129

Yoshikawa T, Saijo M, Morikawa S (2014) Emergence of zoonotic orthopox virus infections. In: Viral infections and global change, vol 377. Wiley

Yu LX, Yan HX, Liu Q, Yang W, Wu HP, Dong W, Tang L, Lin Y, He YQ, Zou SS, Wang C (2010) Endotoxin accumulation prevents carcinogen-induced apoptosis and promotes liver tumorigenesis in rodents. Hepatology 52(4):1322–1333

Part III

Microbiome for Human Health: Clinical Applications

Intestinal Microbiome–Macromolecule Signaling That Mediates Inflammation and Immune System Interaction

9

S. Anju, Y. Aparna, K. Anuradha, Pallaval Veera Bramhachari, and J. Sarada

Abstract

An established microbiome maintains the homeostasis of the host systems and entails specified tasks in stages of development of the host immune system. Synchrony between the host metabolism and the microbiota is critical for a competent immune system. *Clostridium* strains have been initiating the production of Tregs, which considerably are recognized to curb induced colitis in the experimental conditions. *Faecalibacterium prausnitzii* is a clostridial organism protecting patients from the onset of inflammatory bowel disease (IBD). The signal transduction pathway of *Bacteroides fragilis* is known to regulate the production of factors responsible for the differentiation of IL-10 secreting Tregs. Polysaccharide A secreted by *Bacteroides fragilis* induces Treg cell development utilizing TLR2 signaling pathway. It can also potentiate and trigger the signal transducer of transcription factor 3 –STAT3 and recruit the Th17 cells. The gut biota also is responsible for sustaining the recruitment of Th1/Th17 and protect against pathogens. Studies reveal $CD^{+}8$ cells delegating cell-mediated response get triggered as part of the skin flora *Staphylococcus epidermidis.*

An alteration in the normal flora results in an imbalance of the regulatory network. The loss of equilibrium in the commensal community is termed as 'dysbiosis'. The role of dysbiotic microbiota is indispensable in the creation of an inflammatory environment in the gut. Dysbiosis in the microbial flora can also lead to chronic inflammation as part of colonic carcinogenesis. These microbes secrete the pro-inflammatory MAMPs or metabolites that may exert damage on

S. Anju (✉) · Y. Aparna · K. Anuradha · J. Sarada
Bhavan's Vivekananda College of Science, Humanities, and Commerce, Sainikpuri, Secunderabad, India

P. V. Bramhachari
Department of Biotechnology, Krishna University, Machilipatnam, Andhra Pradesh, India

P. V. Bramhachari (ed.), *Microbiome in Human Health and Disease*,
https://doi.org/10.1007/978-981-16-3156-6_9

the host organs. *Fusobacterium nucleatum* triggers the inflammatory signaling pathways and functions to shield the cells from the immune system. It manages to abolish the NK cell-mediated destruction of the cancerous cells. It also causes enriched myeloid cell infiltration. Even macrophages and dendritic cells that primarily mediate inflammation are induced by the microbiota and their metabolic products.

Keywords

Microbiota · Intestinal microbiome · Inflammation · Immune system

9.1 Introduction

There are about a thousand different types of bacteria in the gut microbiota part of the colon of humans among which the majority of them remain uncultured. The microbiome part of the animal system in the gut has a very specific metabolic profile that is likely to complement that of the host. This results in the unique biochemical profile of any individual and therefore results in the health status as well as the immunity. The human gut microbes have a collection of metabolic enzymes that can complement the host metabolism. Certain enzymes are unique to these microorganisms which help the host in the biosynthesis of vitamins, and breakdown of polysaccharides, polyphenols, etc. Several studies involving the fecal microbiota have established the role of the microbiome in the development of gastrointestinal diseases. There is also evidence indicating the establishment of diabetes and obesity as an influence of gut microbes (Rowland 2018). Above all this, there is known to be a bidirectional communication through the gut–brain axis (Yarandi 2016; Rowland et al. 2018) (Table 9.1).

Several signaling pathways trigger the immune system as an interaction based on the metabolic products and the microbe's part of the gut. In certain cases of obesity and metabolic syndrome-related disorders, there is evidence that the gut microbes interact with the innate immune receptors and lead to disorders like acute inflammation. Contrarily, the right interactions of the probiotic microbes can overcome the negative outcomes that lead to obesity and are known to release the anti-obesity factors. There is an inevitable role of the microbes in the development of the gut-associated lymphoid tissue (GALT) and immune tolerance (Cavalcante et al. 2015). The inflammatory and oxidative stress responses that are supposed to regulate cardiovascular function are modulated by adenosine monophosphate-activated protein kinase (AMPK)/nicotinamide adenine dinucleotide phosphate (NADPH). The concentration of AMPK/NADPH is identified to have been a result of the metabolism of short-chain fatty acids by the gut microbiota. The mucosal immune system is the primary site for many allergens and the development of immunological response. To tackle the allergic responses on this basis is to contemplate the probiotic diet as one of the solutions (Rautava et al. 2005).

Table 9.1 Pathways of carbohydrate metabolism and end products of various gut microbes

Gut microbiota	Pathway of carbohydrate metabolism	Metabolic end product released
Methanobrevibacter smithii	Methanogenesis	Methane
Roseburia sps	Butyryl pathway	Butyrate
Blautia hydrogenotrophica	Wood Ljungdahl pathway	Acetate, formate
Desulphovibrio	Sulphate reduction	Hydrogen sulphide
Ruminococcus bromii	Glycolysis	Ethanol
Eubacteriumrectale	Butyryl pathway	Butyrate
Eubacterium halli	Glycolysis	Lactate
Anaerostipes sps	Butyryl pathway	Butyrate
Coprococcus cactus	Butyryl pathway	Butyrate
Coprococcus eutactus	Butyryl pathway	Butyrate
Faecalibacterium frausnitzii	Butyryl pathway	Butyrate
Coprococcus comes	Butyryl pathway	Butyrate
Megasphaera elsdenii	Acrylate pathway	Prpionate
Bacteroidetes	Succinate pathway	Propionate
Veillonella sps	Succinate pathway	Propionate

There exists a host–microbe cross-talk, an important factor for the reciprocal advantage and undisturbed coexistence. It is also assumed that a greater understanding of probiotic interactions can lead to the elucidation of gut-related or many other disease aberrancies. They are a means to induce the anti-inflammatory effects which are essential for immune regulation. There are diverse immune-regulatory responses that are known to modulate the $T_H{}^1$ and $T_H{}^2$ regulator T-cell subclass. These cells are regulated and suppressed concerning the specificity of the antigen and based on the clonal anergy mechanism. There are gut-derived regulatory cells $T_H{}^3$ and $T_R{}^1$ that show the effects by the production of the cytokines (transforming growth factor) TGF-β and IL-10, respectively. A great deal of control is established in protecting the host from the atopic immune response and autoimmune disorders by generating an adequate immune response (Rescigno 2014). Microbiome and metabolome evaluation has revealed that microbial metabolites like short-chain fatty acids, acetate, and butyrate are important in the immunity, inflammation, regulation of epigenetic mechanisms, and overall intestinal integrity (Neu and Pammi 2018).

9.2 Interactions of Microbiota and Immune System

The intestinal microbiome has an indispensable role in the metabolism of complex macromolecules and simultaneously channelizing them to various physiological functions. The microorganisms in the gut establish in the surface epithelial cells and interact with the mucosal lymphoid tissue, to result in the maturation of the immune system. In the early developmental phases, the immune system educates itself to distinguish between beneficial and harmful microbes. One of the studies has established the role of the zebrafish gut microbiota in maintaining the alkaline phosphatase activity and maintaining the expression of the glycan (Bates et al. 2006).

The microbiota in various parts of the human body tends to establish the homeostasis along with the immunostasis (Veronica Lazar 2018). The gut of human beings is heavily populated with bacterial species which are of numbers nearing $100–400 \times 10^{12}$. The mucosal layer of the GIT (gastrointestinal tract) is a double layer that is formed by the O-glycosylated mucin protein which is encoded by the MUC2 gene of the mucin protein family. The bacterial species mostly adhere to the outer mucus layer; however, the inner mucus layer forms the physical barrier that limits the entry and interaction of bacteria with the epithelium. The majority of the microbial species that get established in the intestinal tract are acquired through the mother's milk in early life. They are generally predominated by *Bifidobacterium* and *Lactobacillus* species. With the advancement of the formative years from childhood to adult life, augmentation of the food sources increases the complexity and diversity of the bacterial population. These bacterial species are now included with Bacteroides, Parabacteroides, and Clostridium (Firmicutes). Bacterial numbers increase progressively toward the large intestine from the stomach or duodenum. The largest diversity of taxa and the numbers of bacterial species are identified in the colon that is represented as $10^9–10^{12}$ c.f.u. (colony forming units) per ml that is about 99% of the total GI population (Arnold et al. 2019).

The development and the mediation of the immune system that is part of the gut-associated lymphoid tissues (GALT), Peyer's patches, isolated lymphoid follicles, and mesenteric lymph nodes have been known to be influenced greatly by the gut microbiome. During the immune system development of an individual, the distinction between the self and nonself by the innate cells is by and large due to the microbes and their products released in the host system. They are responsible for the activation and maintenance of the innate hematolymphoid cells, natural killer cells (NK cells), cytotoxic, non-cytotoxic, and helper lymphoid cells. Also, the NK cells and ILC1 produce huge amounts of IFN-γ, antimicrobial peptides (AMPs), granulysin, defensins, and RegIIIγ that regulate the ecological balance of the population of the microbiota and also bring in the component of the immune surveillance. The microbial products like tryptophan, polysaccharide A, and α-galactosylceramide are known to stimulate the production of IL-22, IL-17, and IgA that play a key role as immune mediators. Thus, the microbes need to be part of the host as they involve as a closed framework along with the immune system to either suppress or stimulate the immune response (Belizário et al. 2018).

A chronic inflammatory condition of the gastrointestinal tract, clinically the condition referred to as inflammatory bowel disease, is a perfect example of the immune response showing exaggeration. Many studies have reflected that this condition is influenced largely by dietary factors. Some of the crucial components in the diet such as the fiber content and vitamin D are known to have importance in the development of the disease. The imbalance in the gut microbiome also greatly alters the nutritional status of the host and that has an impact on the innate and

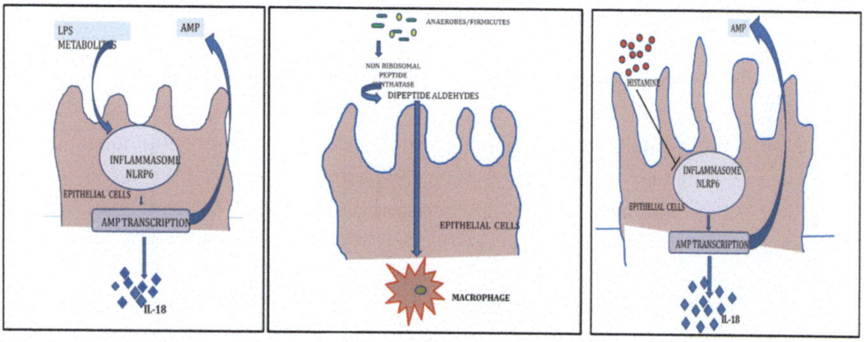

Fig. 9.1 Host–microbiota interactions mediating inflammation

adaptive immune responses. Microbial metabolites derived from their metabolism supplement the host nutrition, and these macromolecules mediate the regulation of the immune responses that can lead to a specific measure of the inflammation expressed (Celiberto et al. 2018).

Filamentous bacteria, Bacteroides, Proteobacteria, and Acinetobacter that are associated with closer to zones of intestinal epithelium raise the strongest immune response. The capsular polysaccharide of *Bacillus fragilis* stimulates anti-inflammatory cytokines like IL-10, leading to colonization and promoting necessary immunosuppression in the intestine. Simultaneously, the outer membrane vesicles of these bacteria activate the autophagy that induces the T regulatory cells resulting in suppression of mucosal inflammation. Enteric *Citrobacter*, *Clostridium ramosum*, and *Mucispirillum* are some of the commensal bacteria that are relevant, and regulate the immune responses that are critical in maintaining the intestinal homeostasis (Blander et al. 2017). (Fig. 9.1).

9.3 Mediators of Inflammation

The gut microbiome and immune system are essential parts of gut–brain communication that involves neuroendocrine and autonomic nervous systems. Enteric neurons communicate the intestinal conditions to intestinal muscularis macrophages via β2-adrenergic receptors and to the vagus nerve. Numerous members of microbiota produce neurotransmitters and neuropeptides like dopamine and acetylcholine. These molecules in a study have been found to induce intestinal epithelial cells of the mouse to release molecules that modulate signaling within the enteric nervous system. Also, it has been concluded that microbes, like *Enterococcus* and *Bacteroidetes*, noticeably increased in numbers in the presence of GABA-supplemented groups. KEGG enrichment analysis revealed that the nitrogen metabolism, sphingolipid signaling pathway, sphingolipid metabolism, and microbial metabolism in different environments were enriched in the GABA1 group (Rees et al. 2018).

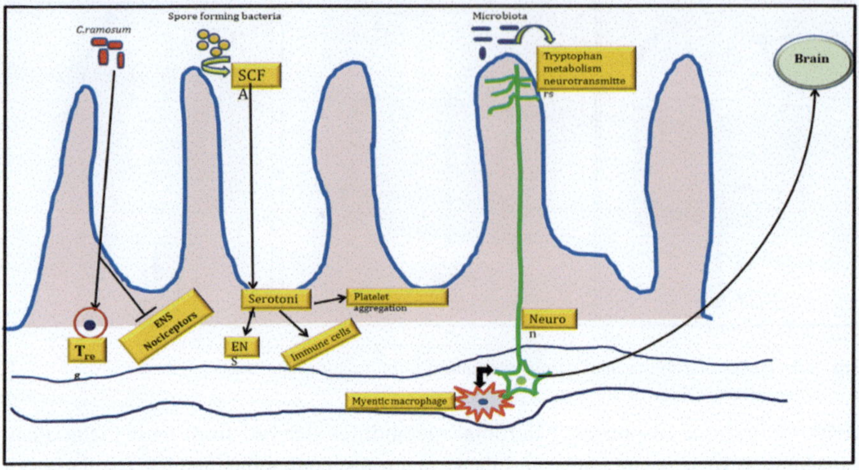

Fig. 9.2 Mediation of neuroinflammation by microbiota

Alterations in microbial metabolism were very much concurrent with changes in the abundances of *Enterococcus* and *Bacteroidetes*. In conclusion, GABA supplementation can improve intestinal mucosal immunity by promoting jejunal SIgA secretion, which might be related to the T-cell-dependent pathway and transformed gut microbiota structure and metabolism (Zhao et al. 2020). Intestinal inflammatory disorders are identified with neurophysiological and behavioral symptoms. Some disorders of the central nervous system (CNS) are found to be accompanied by intestinal complications. Certain observations indicate that intestinal and nervous system physiologies are functionally linked. Multiple pathways that are having bidirectional communication between the intestine and the CNS are collectively referred to as the gut–brain axis. Some of the microbes naturally colonizing the mammalian gastrointestinal (GI) tract play a causative role in regulating CNS function, development, and host behavior. However, members of the gut microbiota are potent modulators of intestinal, systemic, and CNS-resident immune cell function, implying the role of gut–brain interactions that may involve the host immune systemic disorders associated with the gut microbiota including neuroinflammatory, neuropsychiatric, and neurodegenerative disorders that also have significant inflammatory manifestations (Fung 2020) (Fig. 9.2).

Current advances in technology have enabled deep sequencing and analysis of members and signals of microbial communities. In a healthy state, the microbiome is composed of commensals and their genes and phenotypes that may be selected by the immune system to endure in symbiosis. These extremely synchronized signals are modulated by a network of microbial and host metabolites (Kleinstein et al. 2020; Rosen and Palm 2017).

Shreds of evidence in a study indicate that modulation of central nervous system by the microbiome occurs mostly through a neuroimmune and neuroendocrine mechanism that involves the vagus nerve communication which is mediated by several microbially derived molecules like short-chain fatty acid, secondary bile acid, and tryptophan metabolism. These molecules propagate signals by the interaction of endocrine cells and the mucosal immune system. But some cross the intestinal barrier and thereby enter the systemic circulation. Some may cross the blood–brain barrier in addition to generating these metabolites that activate endogenous central nervous signaling mechanisms. The microbiota can independently produce several neuroactive molecules such as gamma-aminobutyric acid, norepinephrine, and dopamine that bind to relevant receptors eliciting the host immune response (Gilbert et al. 2018; Zinöcker and Lindseth 2018).

Very important signaling pathways are carried out in the gut by microbes and their metabolites. These macromolecules communicate with central nervous system that is involved in the communication with the endocrine system. There are at least 12 different types of cells with several subtypes and subgroups which along with the cells of the intestine that get triggered with combinations of molecules are crucial in these signaling pathways. Intestinal cells that are interspersed between epithelial cells throughout containing more than 20 different types of signaling molecules that are released in the response to chemical or mechanical stimuli. These molecules enter the systemic circulation and reach the central nervous system. These molecules regulate the receptors involved in satiety and hunger, and thereby have been identified on the cells which are activated by microbial metabolites including bile acids. Endogenous molecule synthesis like bile acids is heavily influenced by the downstream metabolism by the gut microbiota. Expression of farnesoid X receptor (FXR) that is activated by bile acid leads to the production of fibroblast growth factor 19 (FGF19). Furthermore, this leads to action on peptide or neuropeptide that is supposed to work for improved central regulation of energy and glucose metabolism. Some of the intestinal G protein-coupled bile acid receptors, like TGR5 expressed on the intestinal surface as L cells, are also activated by bile conjugates that result in the glucose homeostasis. Commensal bacteria with the gut-associated immune system have a substantial role in the gut microbial regulation of autoimmunity and inflammation. Microbiota influences the development and function of the central nervous system of resident immune cells like microglia influencing its maturation (Martin et al. 2018).

Products of bacterial metabolism like short-chain fatty acids (SCFAs) and hydrogen sulfide act as messengers to colon epithelial and immune cells. SCFAs are known to be very crucial in maintaining colon homeostasis. Acetate, propionate, and butyrate are the three categories of SCFAs produced in the colon by bacterial fermentation of carbohydrates. These are identified as an important source of energy for epithelial cells in the colon. SCFAs interact with the ligands in the intestine and modulate glucose metabolism. SCFAs regulate (peroxisome proliferator gamma coo activator) PGC1α which is a master regulator in mitochondrial biogenesis. These molecules together coordinate in glucose uptake, oxidative phosphorylation, and fatty acid oxidation. An inflammasome NLRP3 (nucleotide binding oligomerization domain) is activated by the SCFAs that induce the release of IL-18. This event regulates the microbiome consortium and takes over the inflammatory responses. Bacteria like *E.coli* and *Salmonella* in the intestine are known to produce

Table 9.2 Cytokine modulation: gut mucosal innate mechanisms

Lactobacillus sps.	IL-22
Bifidobacterium	IL-10
E. coli	IL-6, IL-8
Microbial DNA	INF -β
Epithelial cells and submucosa cells	IL-10,IL-22,IL-4, IL-6, IL-33,IL-18, IL-13,IL-21,IL-17,IL-1β, TNF, IFN, TGF-β

anti-inflammatory hydrogen sulphide as a by-product of the degradation of sulfur-containing amino acids (Jackson and Theiss 2019; Blander et al. 2017) (Table 9.2).

More intake of calories leads to the accumulation of fat, leading to lipotoxicity that results in the production of effector molecules (cytokines) and cells that are primarily involved in innate immunity. This causes a low-profile inflammatory condition due to the accumulation and activation of mast cells, dendritic cells, and macrophages of several tissues. Several inflammatory pathways are activated during the attenuation of insulin signaling. This in turn leads to several metabolic abnormalities. The incidence of chronic inflammation in obesity-related cases is mainly due to the immune mediators triggered in impaired insulin action (Boulangé et al. 2016.

Nutrients influence the way the immune cells are triggered for any immune response. Most importantly, the macrophages that are known to function differently in various pathological conditions are accordingly stimulated by diversified macromolecules. Macrophages exist in two major phenotypes (M1 & M2) that are stimulated differentially based on the consumption of glucose, oxygen, and glutamine. The signal molecules involved in the activation are bacterial lipopolysaccharide (LPS), pro-inflammatory components TNF, interferon-γ, IL-1, interleukin-4, and interleukin-13. Anti-inflammatory cytokines, TGF, IL-10, and glucocorticoids are crucial in the regulation of the inflammation. The effector activities of the macrophage M1 phenotype are supported by the chemical mediators ROS (reactive oxygen species), ATP, NADPH, nitric oxide, and utilization of the glucose (Belizário et al. 2018).

9.4 Mitochondria and Gut Microbiome Signaling

Epithelial barrier function is essential in intestinal homeostasis. Histone deacetylase (HDAC) inhibitors when investigated have been found to regulate immune responses. In human and murine colonic epithelial cell lines, the presence of the HDAC inhibitors givinostat and vorinostat improved transepithelial electrical resistance in inflammatory conditions. They also silenced the passage of macromolecules across the epithelial barrier. Mechanistically, these regenerative effects could be linked to increased secretion of transforming growth factor beta1 and interleukin-8, paralleled by differential expression of the tight junction proteins claudin-1, claudin-2, and occludin (Friedrich et al. 2019).

The gut microbes are known to be a dynamic key role-player in regulating the mitochondrial function of the intestinal and in turn influence the immune cells as well as the epithelial cells of the gut. The immune system is greatly influenced by the gut microbiome and the mitochondrial reactive oxygen species from the mitochondrial system. Epithelial barrier integrity and intestinal epithelial mitochondria both serve as a site of action as well as facilitators of inflammatory pathways. Studies on colonic epithelial cells targeting the mitochondrial activity showed stimulation of gut microbiome immune pathways related to inflammation by an AMPK-mediated mechanism. Decreased oxidative phosphorylation is identified to trigger intestinal inflammation. A pro-inflammatory cytokine TNF-α, which plays a central role in IBD inflammation, inhibited mitochondrial function, promoted mtROS accumulation, and thereby resulted in impaired barrier function. These mtROS produced in the immune cells play an important role in the eradication of several pathogens by a bactericidal effect. NLRP3 inflammasomes produce pro-inflammatory cytokines that invade bacteria. Fermentation products of the gut microbiome such as short-chain fatty acid-induced mitochondrial reactive oxygen species production and the immune cells and produced during mitochondrial respiration enhance the oxidative phosphorylation activation and induce Toll-like receptors (TLRs) 1, 2 or 4 at the plasma membrane or endoplasmic reticulum membrane. This controls the rate of bacterial or viral infection and subsequently induces the release of accumulation macrophages as well as neutrophils. However bacterial-induced mitochondrial responses are detrimental to gut epithelial tissues. Epithelial cells of the intestine when showed mitochondrial dysfunction the tissues had lost the ability to tolerate the commensal microbiome and were dependent on the IFNα and TNFα. Mitochondria have an inevitable role in maintaining the homeostasis of the intestine. It can initiate the mediation of the innate and adaptive immune responses, in turn promoting inflammation and immune pathways. Functions of mitochondria are crucial to propagate the role of epithelial barrier integrity during inflammation and play an important role in tumorigenesis. The intestinal microbiome modulates mitochondrial signaling in mucosal cells. Studies to understand mechanisms connecting the gut microbiota and mitochondria of the host during the pathogenesis of IBD and CRC will be the turning point in the targeted therapy. Bacterial and mitochondrial signaling is known to inhibit the apoptotic pathway in the intestinal epithelium. A better understanding of whether direct bacterial toxins or bacterial metabolites are involved in the mechanism whereby bacterial-to mitochondrial signaling enhances colon tumorigenesis is a future gap in the field that can be explored and build new avenues in the treatment strategies (Jackson and Theiss 2019).

9.5 Cytokines Influenced by Gut Microbiota

There are a variety of cytokines produced as a result of the interactions generated due to the number of metabolites released by the gut microbiota. Cytokines play a very important role in generating inflammatory responses in various infectious diseases and thereby develop intestinal and extra intestinal inflammations. This results in generating host inflammatory responses which are otherwise a consequence of microbiota cytokine interactions. Cytokine production may be directly influenced by the microbiota, or sometimes, there are indirect means by which they become responsible for it. These interactions lead to cytokine-mediated molecular pathways throughout the body. These innate inflammatory cytokines are responsible for the intestinal homeostasis (Knight et al. 2018).

Pro-inflammatory interleukin IL-1b initiates a series of cascades of other cytokines and substances that promote inflammation. Stimulation of inflammation and protective role of IL-1b is regulated by the influence of the microbiota and therefore maintains the intestinal homeostasis. Colonic bacteria are known to raise the levels of pro-inflammatory cytokines in the gut and promote the formation of the inflammasome complex. IL-18 promotes the inflammatory responses by inducing the INF-α and also downregulates the IL-22. It is observed that during IBD, alterations in gut microbiota and raise in the macrophages and T-helper cells occur. High levels of IL-6 are found during the *E.coli* infection that triggers the pro-inflammatory responses as a protection to the host. A large range of gut microbiota in the epithelial cells as well as submucosal region produce cytokines IL-18, Il-6, TNF, INF, IL-33, IL-17, IL-21, Il-13, IL-1β, and Il-10, IL-22, IL-4, TGF-β, respectively. The gut has a complex interplay of the cytokines, and therefore, there are critical conditions established whenever conditions that disturb homeostasis are established. Interactions among cytokines based on the concentration levels and the types of receptors or cells there may be anti-inflammatory and pro-inflammatory mediators in the gut. Certain pro-inflammatory mediators are known to promote epithelial proliferation essential for wound closure or healing. On the contrary, it may also promote carcinogenesis if the actual function of healing is not taking place and there is a deviation in the mechanism. It is interesting to note that cytokine profile of the gut can encourage the survival and functioning of the microbiota synergistically to the host mechanisms. At the same time, there is also suppression of the microbiota as adverse effects wherein the immune system is activated to eliminate the microbes in the gut and hence may lead to depletion or reduction of certain species of the normal flora in the intestine (Mendes et al. 2019).

9.6 Conclusion

Microbiota host interactions along with many macromolecules generated and introduced through various pathways modulate inflammatory responses establishing critical conditions. Intestinal inflammation may be due to the activity of microbiota, in correlation with the state of host health or disease or maybe certain other components that are yet to be elucidated. Controlled immune responses that generate appropriate amounts of mediators to either promote or suppress inflammation are very crucial in maintaining the host homeostasis. Host inflammatory response and microbial interactions are dynamic and complex that is important in the regulation of the pro-inflammatory and anti-inflammatory pathways. There is a fact about an indispensable link between the human microbiome and diseases. Analysis of the microbiome components by methods like DNA sequencing of the genomes from the human samples and further introspection using transcriptomes, proteomes, etc., can be instrumental in understanding the dynamics of this invisible community.

Acknowledgments We acknowledge principal and management of Bhavan's Vivekananda College of Science, Humanities and Commerce, Sainikpuri, for their support and encouragement in writing this book chapter.

References

Arnold WM, Hill ES, Fei N, Yee AL, Garcia MS, Cralle LE, Gilbert JA (2019) The human microbiome in health and disease. In: Genomic applications in pathology. Springer, Cham, pp 607–618

Bates JM, Mittge E, Kuhlman J, Baden KN, Cheesman SE, Guillemin K (2006) Distinct signals from the microbiota promote different aspects of zebrafish gut differentiation. Dev Biol 297 (2):374–386

Belizario JE, Faintuch J, Garay-Malpartida M (2018) Gut microbiome dysbiosis and immunometabolism: new frontiers for treatment of metabolic diseases. Mediators Inflamm 2018:2037838

Blander JM, Longman RS, Iliev ID, Sonnenberg GF, Artis D (2017) Regulation of inflammation by microbiota interactions with the host. Nat Immunol 18(8):851–860

Boulangé CL, Neves AL, Chilloux J, Nicholson JK, Dumas ME (2016) Impact of the gut microbiota on inflammation, obesity, and metabolic disease. Genome Med 8(1):1–12

Cavalcante-Silva LH, Galvão JG, Silva JSDFD, Sales-Neto JMD, Rodrigues-Mascarenhas S (2015) Obesity-driven gut microbiota inflammatory pathways to metabolic syndrome. Front Physiol 6:341

Celiberto LS, Graef FA, Healey GR, Bosman ES, Jacobson K, Sly LM, Vallance BA (2018) Inflammatory bowel disease and immunonutrition: novel therapeutic approaches through modulation of diet and the gut microbiome. Immunology 155(1):36–52

Friedrich M, Gerbeth L, Gerling M, Rosenthal R, Steiger K, Weidinger C et al (2019) HDAC inhibitors promote intestinal epithelial regeneration via autocrine TGFβ1 signalling in inflammation. Mucosal Immunol 12(3):656–667

Fung TC (2020) The microbiota-immune axis as a central mediator of gut-brain communication. Neurobiol Dis 136:104714

Gilbert JA, Blaser MJ, Caporaso JG, Jansson JK, Lynch SV, Knight R (2018) Current understanding of the human microbiome. Nat Med 24(4):392–400

Jackson DN, Theiss AL (2019) Gut bacteria signaling to mitochondria in intestinal inflammation and cancer. Gut Microbes 11:1–20

Kleinstein SE, Nelson KE, Freire M (2020) Inflammatory networks linking oral microbiome with systemic health and disease. J Dent Res 99:1131–1139

Knight R, Vrbanac A, Taylor BC, Aksenov A, Callewaert C, Debelius J et al (2018) Best practices for analyzing microbiomes. Nat Rev Microbiol 16(7):410–422

Martin CR, Osadchiy V, Kalani A, Mayer EA (2018) The brain-gut-microbiome axis. Cell Mol Gastroenterol Hepatol 6(2):133–148

Mendes V, Galvao I, Vieira AT (2019) Mechanisms by which the gut microbiota influences cytokine production and modulates host inflammatory responses. J Interferon Cytokine Res 39(7):393–409

Neu J, Pammi M (2018) Necrotizing enterocolitis: the intestinal microbiome, metabolome and inflammatory mediators. In: Seminars in fetal and neonatal medicine, vol 23. WB Saunders, London, pp 400–405

Rautava S, Kalliomäki M, Isolauri E (2005) New therapeutic strategy for combating the increasing burden of allergic disease: probiotics—a nutrition, allergy, mucosal immunology and intestinal microbiota (NAMI) research group report. J Allergy Clin Immunol 116(1):31–37

Rees T, Bosch T, Douglas AE (2018) How the microbiome challenges our concept of self. PLoS Biol 16(2):e2005358

Rescigno M (2014) Intestinal microbiota and its effects on the immune system. Cell Microbiol 16:1004–1013

Rosen CE, Palm NW (2017) Functional classification of the gut microbiota: the key to cracking the microbiota composition code: functional classifications of the gut microbiota reveal previously hidden contributions of indigenous gut bacteria to human health and disease. Bioessays 39 (12):1700032

Rowland I, Gibson G, Heinken A, Scott K, Swann J, Thiele I, Tuohy K (2018) Gut microbiota functions: metabolism of nutrients and other food components. Eur J Nutr 57(1):1–24

Rowland IE (2018) Gut microbiota functions: metabolism of nutrients and other food. Eur J Nutr 2018:1–24

Veronica Lazar LM (2018) Aspects of gut microbiota and immune system interactions in infectious diseases, immunopathology, and cancer. Front Immunol 9:1830

Yarandi SS (2016) Modulatory effects of gut microbiota on the central nervous systen. J Neurogastroenterol 22:201–212

Zhao Y, Wang J, Wang H, Huang Y, Qi M, Liao S et al (2020) Effects of GABA supplementation on intestinal SIgA secretion and gut microbiota in the healthy and ETEC-infected weanling piglets. Mediators Inflamm 2020:1–17

Zinöcker MK, Lindseth IA (2018) The Western diet–microbiome-host interaction and its role in metabolic disease. Nutrients 10(3):365

Microbiome Diagnostics and Interventions in Health and Disease

10

Nandini Dasgupta, Alka Srivastava, Amrita Rao, Vrishali Murugkar, Ravi Shroff, and Gautam Das

Abstract

The growing evidence of literature, correlating the importance of balancing noninfectious microbes inhabiting our bodies in disease and health, has given birth to a new field of medicine—microbiome therapeutics. Composition of microbiome evolves with us right from birth, impacted by several factors like an individual's genetic makeup, quantity and quality of the different foods that we consume, and the environment that we interact with. A change in the composition of our microbiome (gut, skin, lung, gastric, vaginal, oral) may trigger or predispose us to a disease condition before clinical manifestation of symptoms. The current trend to better understand these correlations in health and disease is by leveraging metagenomics, metabolomics, data mining, artificial intelligence, and machine learning tools for human microbiome diagnostics. Encouraging results have been obtained with therapeutic strategies using prebiotics, probiotics, signaling molecules, antimicrobial peptides, and microbiome transplant in alleviating disease symptoms and promoting well-being. This is generating increased interest in the medical and scientific community and awareness in the public. The emerging concepts of 'smart sampling' using 3D printed devices, engineering diagnostic and therapeutic bacteria using synthetic biology, and using microbiome engineering to restore niche-specific balance are some additional paths that scientists are pursuing to arrive at a viable solution. In this article, we will address the challenges and potential solutions of microbiome diagnostics and therapeutics.

N. Dasgupta · A. Srivastava · A. Rao · V. Murugkar · R. Shroff · G. Das (✉)
miBiome Therapeutics LLP, Mumbai, Maharashtra, India
e-mail: gautam@mibiome.com

P. V. Bramhachari (ed.), *Microbiome in Human Health and Disease*,
https://doi.org/10.1007/978-981-16-3156-6_10

10.1 Introduction

Hundred trillion invisible living beings (grouped together as the microbiome) live inside humans. A delicate balance between the quality and quantity of the microbes that live inside our body is vital for our well-being. Growing scientific evidence suggests that the microbiome is intricately connected to human health, wellness, and treatment of certain disorders. As the cells of our bodywork tirelessly to perform the necessary functions for us to live, so do the microbial communities within our body right from our birth. They keep performing jobs that benefit us like breaking down foods, aiding nutrient absorption, protecting against pathogens, and aiding in immunity. According to scientific reports, around 1000 different species of bacteria reside in our gut; therefore, any imbalance ('dysbiosis') can result in a variety of problems to human health. An estimate suggests that 85% 'good bacteria' and 15% 'bad bacteria' are tolerated by the human gut; deviations from this ratio cause perturbations in the digestive system and various illnesses, also affecting our own immunity. Many of us would have experienced that when doctors prescribe antibiotics to kill harmful bacteria that invade us occasionally, they also recommend consuming probiotics as supplements. This is for the simple reason that while antibiotics do an excellent job killing 'bad bacteria,' they also eliminate some of the 'good bacteria' from our system, thus creating an imbalance. This results in diarrhea and many other gastrointestinal problems. Supplementing with probotics helps in restoring the loss of the 'good bacteria' and gradual reversal of symptoms associated with gut dysbiosis.

10.2 Need for Human Microbiome-Based Diagnostics

According to the World Health Organization, a state of good health is not actually the absence of a disease; rather, it is a state of complete mental, physical, and social well-being (WHO 1946). Based on patient examination, clinical history, associated symptoms, and diagnostic tests, health professionals diagnose the nature of the illness. They prescribe appropriate therapeutics, life style changes, or a combination of both, to treat the illness and improve the health of the patient. Diagnostic is a symptom or characteristic of value in diagnosis. Microbiome diagnostics, pertaining to the identification of the imbalance (if any) in the human microbiome, has been more of a research endeavor in the past. With the growing evidence in literature, of changing microbiome profiles (both abundance and diversity) correlations in different disease conditions, its cause and consequences on effectiveness of pharmaceuticals (Vieira-Silva et al. 2020); the importance of microbiome diagnostics as a companion in the clinical medical diagnostics tool box is emerging rapidly (Raes 2016). Diagnostic applications of the microbiome in the past were focused on the pathogenic microbes, but we now know the relevance of monitoring nonpathogenic microbial components of commensals associated with many noncommunicable chronic diseases. For example, Hollister et al. (2019), were capable to clearly differentiate children with irritable bowel syndrome, also called as I.B.S., a chronic condition, from the children who were healthy, by profiling the intestinal microbes, their genes/pathways, and metabolites. With the whole world reeling under the COVID-19 pandemic, a contagious disease whose pathophysiology is not yet completely understood, microbiome diagnostics has the power to query the changes happening at the molecular level in niche-specific microbiomes in both cross-sectional and longitudinal studies in patients (He et al. 2020). The knowledge gathered will empower research scientists and health professionals to build an appropriate repertoire to counter the pathology of SARS-CoV2.

Human orifices, and organs like the gut, lung, and skin, are abundant with niche-specific microbial species including bacteria, archaea, fungi, virus, and protozoa (mainly gut). Each one of us acquires a largely distinctive microbiome early in life. The same may persist with us for years or may undergo changes in compositional diversity/abundance. Such changes may correlate with a change in the environment, health status, or lifestyle. The niche-specific microbiome components are known to differ between environments and populations (Integrative HMP (iHMP) Research Network Consortium 2019), but certain indicator species are conserved across human populations studied around the world. This suggests that a symbiotic relationship between these indigenous organisms and human physiology cannot be ignored. However, the 'cause-consequence' problem of perturbations in the microbiome that are associated with a health condition remains to be understood in detail in several diseases. Some questions still remain: Are the molecular components of an individual's microbiome responsible for health outcomes? How do they combine with and maintain critical physiological processes like the immune system and metabolism?

Recently, gathered evidence clearly suggests that the composition of the gut microbiome has a correlative effect via modulating at least the brain, lung, and liver. Many reviews have described these microbiomes with respect to their modulation and interplay with host factors at a greater depth.

Classically, culturing microorganisms and their identification from various clinical samples were the basis of 'germ theory' for any etiological agent and the earliest tool for microbiome diagnosis. But we now know the unculturable plethora of microbes that inhabit our body, and efforts are ongoing in 'culturomics' to narrow this gap by combining extensive laboratory culturing conditions followed by identification using mass spectrometry (Lagier et al. 2012; Lagier et al. 2018). Attempts at using real-time PCR as a diagnostic for microbiome composition (Ott 2004) at a time when Next-Gen sequencing (NGS) was just launched could not give the true picture of the microbiome since it estimated the abundance of only the dominant 20 pathogenic and commensal species in the intestinal bacterial flora. The advent of NGS has revolutionized the approach that researchers have at their disposal today and together with MALDI-TOF mass spectrometry, it is poised to take microbiome diagnosis to the next level with precise study designing, controls, unbiased analysis of data and reporting. A desirable outcome that would immensely help clinicians to serve their patient better would be the availability of NGS backed biomarker diagnostic assays for a disease condition or its prediction. How to identify these biomarkers? Novel noninvasive diagnostic biomarkers for colorectal cancer diagnosis have been successfully identified (Liang et al. 2017) based on metagenome sequencing analysis of fecal bacterial marker candidates and adapted for a qPCR assay.

10.3 Challenges in Design, Analysis, and Interpretation

For niche-specific microbiome diagnostics, it would be ideal to have reference ranges of microbial species or their metabolites (in a healthy or faulty microbiome) that a doctor could use for differential diagnosis. Among human microbiomes studied worldwide, the gut microbiome has probably been researched the most due to the ease of sample collection, abundance of microbes, and apriori knowledge. Unfortunately, due to many unresolved factors, the availability of a dependable diagnostic test, based on microbiome analysis, is still in the developmental stage. Though there is an abundance of microbiome data, the taxonomic changes identified in a disease are not consistent across different studies. The underlying reasons for the incoherence may have arisen due to variations in sample population (including diet, lifestyle) and the different technological approaches used in the diverse studies. In addition, a key problem in the field is to define the 'healthy' microbiome, owing to the large degree of variation in the microbiome composition among healthy individuals. Hence to stay relevant, all efforts toward identifying microbial markers for disease diagnostics must be based on comparisons with parallel control groups of healthy individuals (Versalovic et al. 2017).

An ambitious project launched in 2007 in the USA (Turnbaugh et al. 2007), termed the HMP or the National Institutes of Health's Human Microbiome Project, was a one of its kind, large-scale initiatives to resolve the burning issues mentioned above (Gevers et al. 2012a, b). The first phase of the program involved generating massive amount of data and putting together the different analysis platforms to determine the composition of the 'healthy' microbiome (absence of evident disease). A baseline adult population (Huttenhower et al. 2012; Lloyd-Price et al. 2017) and 'demonstration' populations with specific disease states were studied to determine characteristic ranges (for some populations) of various microbiome-host parameters. The parameters that were included are as follows: (1) combinations of metabolic functions that are either ubiquitous or specific to the strain; (2) enzymatic repertoires; (3) some host factor, such as race or ethnicity. Information generated comprised of nucleotide sequences of microorganisms and human population (http://hmpdacc.org), protocols for body-wide microbiome sampling and data generation (Aagaard et al. 2013), and computational methods for microbiome analysis and epidemiology (Gevers et al. 2012a, b; Markowitz et al. 2012; Faust et al. 2012). This is a rich community resource for the scientific community. A striking revelation from the HMP1 was that the taxonomic classification of the microbiome was unable to explain host health or disease phenotype; molecular functional analysis of the microbial population or understanding personalized strain-specific makeup was a better correlate (Human Microbiome Project Consortium 2012).

Studies of the gut microbiome from healthy cohorts of other countries such as Denmark (Qin et al. 2010), China (Zhang et al. 2019), and India (Dhakan et al. 2019) have corroborated the same based on observations of some degree of functional redundancy in microbiomes in spite of compositional differences. Diet was a significant player in modulating the composition in these studies.

10.4 An Informed Approach to Next-Gen Sequencing-Based Microbiome Diagnostic Design and Evaluation

Metabolite-related profiling studies are beyond the chapter scope; hence, we are limiting ourselves to nucleic acid-based diagnostics here. Given the diversity of the human microbiome, challenge of limited clinical specimen size, and the large number of samples in cohort studies, Next-Gen sequencing-based approaches (Levy and Myers 2016) are the current method of choice for nucleic acid-based microbiome diagnostics (Fig. 10.1). In order to establish baseline ranges of taxonomic diversity in the HMP1 study, encompassing within and between body sites analyses, to decipher functional commonalities and signature strains across various subjects, NGS-based approaches were adopted. Sequencing profiles based on 16S rRNA gene sequences of 5577 samples and 681 shotgun metagenomes spanning up to 18 body sites and three time points each from 242 healthy adults were analyzed (Human Microbiome Project Consortium 2012). The study was extended (HMP1-II) to 2355 total shotgun metagenomes from 265 healthy adults to identify niche-specific and host-associated microbial community functions and to quantify strain personalization and retention dynamics over time (Lloyd-Price et al. 2017). Subsequently, three iHMP clinical studies served as models of microbiome-associated conditions, wherein the biological properties of both the microbiome and host were studied longitudinally. Microbial community compositions, transcriptomes and proteomes of the microbiomes, global metabolome, and immune and clinical markers from the host were analyzed to generate datasets. Among the conditions, vaginal microbiome of the mother associated with preterm birth, gut microbiome of subjects with inflammatory bowel disease, and gut/nasal microbiomes of type 2 diabetics were chosen. (NIH Human Microbiome Portfolio Analysis Team 2019; Integrative HMP (iHMP) Research Network Consortium 2019).

Country-specific microbiome databases are required to construct meaningful correlations in disease and for the comparison of healthy and diseased individuals. The following critical parameters need attention for any microbiome diagnostic:

1. *Design*: The role of the clinician, statistician, epidemiologist, and research scientist is critical in identifying:
   (a) The sample population size, age group, and location consisting of healthy and affected subjects (clinical diagnosis based, patient consent, diet, clinical history of self and family, medications, supplements, lifestyle, ability to comprehend instructions).

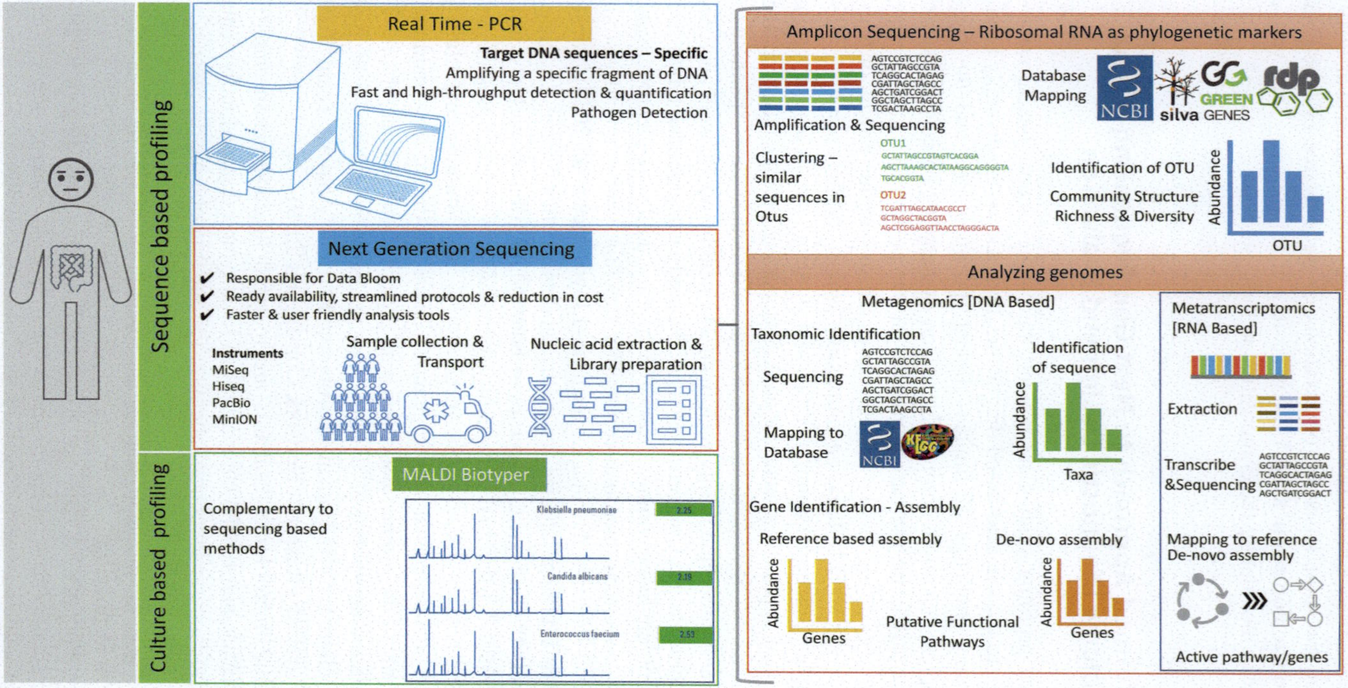

Fig. 10.1 A schematic illustration of few popular methods employed in the microbiome analysis. Sequence-based profiling (top and middle-left panels) has more widespread reach due to the speed, higher throughput, sensitivity, specificity, and the ability to detect smallest signals and even dead bacteria. Culture-based profiling (bottom-left panel) is complementary to the sequence-based profiling; a prototype graph of MALDI-Biotyper is shown, different bacteria have different spectrum and scores. Real-time PCR (top-left panel) is very useful for diagnosis and quantification if the sequence of the organism or a biomarker for a particular disease is known. Next-generation sequencing (middle-left panel) has led to data explosion, due to continuous advancement in the field. It is broadly

Fig. 10.1 (continued) categorized into two types—(**a**) Targeted amplicon sequencing (top-right panel), where 16S/18S ribosomal RNA gene is used as phylogenetic markers for identification of microbes. General flow of the process is shown; (**b**) Holistic approach (bottom-right panel), where the entire genome/metabolites are used for the identification of microbes. It can be further divided into three categories: (1) shotgun based metagenomics, where DNA or genome is the target. General flow of the process is shown, where taxonomic identification is done either by mapping the sequences to the reference genome or assembling the sequences to get gene profile and hence the probable functional pathways; (2) metatranscriptomics, where the transcribed RNA is the target (expression profile) which gives an insight into active pathways/gene; (3) metabolomics, where the metabolites produced by the microbes are the target (not shown here) (Allaband et al. 2019; Schlaberg 2020)

   (b) Which niche to study? (Skin/oral/nasal/gastric/gut/vaginal) Will simultaneous blood tests be needed to record any other parameters?
   (c) Timeline of study: Longitudinal study or cross-sectional study, sampling frequency.
   (d) Appropriate sampling, storage, and transportation: home/clinic collection, preferred time of collection, clear instruction on what and how to collect (stool/urine/sputum); addition of appropriate stabilizer/inhibitor for stabilizing nucleic acids; proper storage at desired temperature (4 °C or − 20 °C).
   (e) Controls: Inclusion of a mock community of several microbes (bacteria/fungi/virus) at varying abundances as a positive control for the process and LoD (limit of detection) determination; a reagent control with sterile water or saline in place of clinical sample to serve as a negative control.
2. *Data generation and analysis*:
   (a) *Isolation of nucleic acid (DNA/RNA/both)*: Commercially available kits or laboratory-developed protocol best suited for the clinical sample type can be evaluated for nucleic acid extraction yield, quality, and removal of inhibitors from mock community to give a statistically sound representation of the richness and abundance of microbial species. Once the isolation protocol meets the quality requirements, the same can be applied for the clinical samples in the study. Efficient conversion of the labile RNA to a more stable cDNA is critical for RNA genomes (RNA viruses) or for transcriptomic analysis of the sample. Methods that are easily adaptable for upscaling and automation are highly desirable to make the process efficiency user agnostic and predictive with turn-around times.
   (b) *Sequencing methodology and platform (targeted amplicon-based/shotgun metagenomics/metatranscriptomics)*: Generally, sequencing library preparation for targeted amplicon-based sequencing includes a polymerase chain reaction step using DNA/cDNA, to generate amplicons of the targeted genetic marker with adapters. The shotgun metagenomics/metatranscriptomics library preparation approach on the other hand is not targeted, but includes a size selection of the double-stranded DNA/cDNA prior to adapter ligation using a series of enzymatic and mechanical manipulations as directed by the manufacturer. Details on the choice of sequencing methodology and platform are out of scope of this chapter. Briefly, NGS platforms can be categorized into two major categories: short-read (e.g., Illumina, Ion Torrent) or long-read (e.g., Pacific Biosciences (PacBio), Oxford Nanopore's MinION) sequencing (Fig. 10.1). For taxonomic profiling, DNA-based targeted amplification of 16S/18S rRNA gene variable region, panel of gene targets, and shotgun metagenomics are commonly used. For molecular function-based querying of the microbiome and to distinguish active from dormant metabolic state, targeted transcript-based or metatranscriptomic methodology is followed. The read depth coverage and sequencing of single reads or paired reads are some other criteria that are taken into consideration.

(c) *Data processing tools and analysis of data*: Upon data acquisition from the sequencers, performing several quality control checks is critical to prepare the data for downstream analytics. Examples are data trimming and the removal of poor-quality reads. Two primary approaches to taxonomic profiling of analysis can be employed. These include de novo assembly-based and 'read alignment to reference-based' methods. Many assembly software such as metaSPADES (Nurk et al. 2017) and MEGAHIT (Li et al. 2015, 2016) can be used to reconstruct genomes from metagenomics sequence data. Once draft genomes are assembled, software such as CONCOCT (Alneberg et al. 2014), or MetaBat (Kang et al. 2015) can perform contig binning and taxonomic profiling. Examples of read-based taxonomic profiling software include Kraken (Wood and Salzberg 2014) and MetaPhlAn2 (Truong et al. 2015). Computational software such as QIIME (Caporaso et al. 2010) and MOTHUR (Schloss et al. 2009) are most commonly employed for targeted amplicon sequence data analysis using operational taxonomic unit (OTU)-based analyses. Aligned read pairs form contigs, followed by clustering of contigs into OTUs based on similarity to reference sequence in a database such as Greengenes (DeSantis et al. 2006) or SILVA (Pruesse et al. 2007) for taxonomic classification. Following classification, community structure as measured via alpha and beta diversity can be examined.

3. *Interpretation of data for diagnostics*:
   (a) *Correlation and association based*: Comparison of the microbial diversity and relative abundance in the healthy group vs. the disease group to derive statistically significant correlations and associations using bioinformatic and statistical tools that take metadata into consideration is critical for unbiased interpretation. This topic will not be covered in detail here.
   (b) *Database richness and accounting for microbiota interactive network*: Ensuring database richness and updating for taxonomy and disease associations for microbiome profiles from various studies are needed to translate the benefit in diagnostic reporting.
   (c) *Predictive medicine*: Since precise microbiome diagnostic needs to be population-specific, every nation may need to determine the microbiome biomarkers for disease diagnosis/prognosis/treatment that is relevant to their population for best patient outcomes. Using Big Data analytics and machine intelligence to discover those correlations and validate them is paving the way for predictive medicine (Fig. 10.2). The following possibilities are very encouraging for the future of medicine:
   - Identify unforeseen mechanistic insights of treatment.
   - Identify associations not yet detected by humans.
   - Identify biomarkers defining a patient's response to treatment.
   - Predict synergism/antagonisms of combination therapies and dosage effects.
   - Potentially minimize side effects and maximize efficacy of treatment.
   - Predictive modeling for the diagnosis and treatment of diseases.

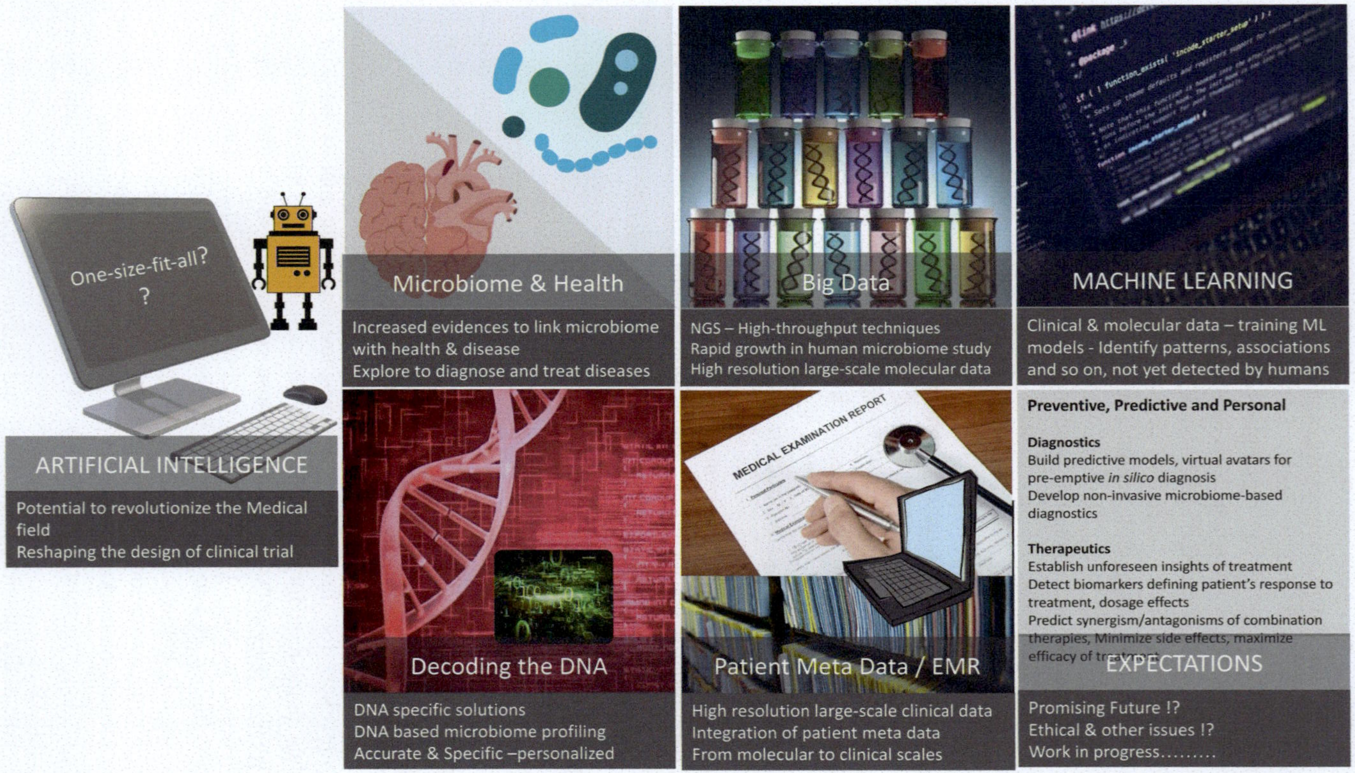

Fig. 10.2 A schematic illustration showing the future of artificial intelligence (AI)/machine learning (ML) in microbiome diagnostics and therapeutics. Artificial intelligence has the potential of removing the 'one-size-fit-all' stigma in the medical field. The progress made in sequencing technology has led to

Fig. 10.2 (continued) an increased interest in microbiome data and can lead to DNA-based specific diagnostics. The integration of high-resolution sequencing data and the patient metadata or clinical data can serve as model for machine learning, which can help in detecting/identifying patterns or biomarkers or insights missed by the human eye. AI/ML is still in nascent phase, and it has huge potential to revolutionize the medical field. However, caution is needed, in terms of training dataset as well as ethics and privacy (Espinoza 2018; Leber et al. 2017; Topçuoğlu et al. 2020)

- Developing noninvasive microbiome-based diagnostics with the help of AI.
- Create virtual preemptive and predictive in silico testing of safer, more effective therapeutics.

Innovations in the field from designing sampling devices, sequencing platforms with better and cheaper technologies, and algorithms for data analysis are forward-facing. Most studies of the gut microbiome study fecal sample, which may not be the best representation of the whole gut microbiota. It is interesting to note that an ingestible, biocompatible, 3D-printed microengineered battery-less pill has shown promise in vitro and in animal models including primates (Rezaei Nejad et al. 2019) to aid in this sampling process.

10.5 The Healthy Gut Microbiome

Among the inhabitant microbes of our gut, bacteria are the most predominant. Though there are ~1000 different bacterial species colonizing our gut, only about 330 of them have been characterized and classified so far. The top inhabiting phyla consists of strict anaerobes (with their relative abundance in parentheses): Firmicutes (64%), Bacteroidetes (23%), Proteobacteria (8%), and Actinobacteria (3%) (Gill et al. 2006; Bäckhed et al. 2012). The constantly changing microbiome responds to our lifestyle modifications, such as diet and exercise, and displays perturbations accordingly (Qin et al. 2010; David et al. 2014; Wu et al. 2011).

Among the benefits we derive from our gut bacteria, commensal species such as *Lactobacillus plantarum* helps in regulating the integrity of intestinal epithelium, which acts as the first physical barrier for enteric pathogens. Short-chain fatty acids (SCFAs) such as acetate, propionate, and butyrate play vital roles in the gut microbiome homeostasis and host immunity. The SCFAs are produced from the breakdown of polysaccharides (mainly dietary fiber) by specific bacteria (El Kaoutari et al. 2013; Canfora et al. 2015; Koh et al. 2016; Miyamoto et al. 2016). It is the tuning of the biochemical pathways in the specific bacteria that result in different by-products though starting with the same dietary fiber source. The major acetate producers belong to the genus Streptococcus, Prevotella, Bifidobacterium, and Clostridium to name a few (Rey et al. 2010). Propionate is produced by Bacteroides spp., Salmonella spp., Dialister spp., Veillonella spp., Roseburia inulinivorans, Coprococcus catus, Blautia obeum, etc., (Louis and Flint 2017). Bacteria belonging to Lachnospiraceae, Ruminococcaceae, and Acidaminococcaceae families are the major butyrate producers in the gut (Duncan et al. 2002). It is reported that the SCFAs along with G-protein coupled receptor 41 (GPR41) and GPR43 present in intestinal epithelial cells can modulate satiety and food cravings (Kim et al. 2013a, b; Ang and Ding 2016). Higher uptake of nutrients is facilitated by suppressing intestinal mobility transit by the secretion of peptide YY (PYY) and glucagon-like peptide-1 (GLP-1) stimulated by SCFAs (Chambers et al. 2018; Ang and Ding 2016).

The major representatives of archaea in the gut microbiome are different species of methanogens and halophiles. Some examples and their relative abundance among gut methanogenic archaea include *Methanobrevibacter smithii* (94%), *M. stadtmanae* (23%), Candidatus *Methanomethylophilus alvus,* and Candidatus *Methanomassiliicoccus intestinalis* (Dridi et al. 2009). Gaseous by-products such as methane and hydrogen are generated by the anaerobes inhabiting the bowel.

The signaling among the gut microbiota, the gut, and the brain by metabolites occurs via neuronal pathways which involves both the central and enteric nervous systems, along with the circulatory system (Cryan and Dinan 2012; Mohajeri et al. 2018). Thus, the significance of the gut microbiome in health and disease is being appreciated like never before by scientists, clinicians, nutritionists, and the informed public.

The healthy adult gut microbiota is highly tolerant in accommodating minor perturbations with respect to its diversity and abundance, due to a temporary change, such as in eating habits, life style, or environment. In a study conducted over a course of 5 years, the individual gut microbiome displayed 60% strain level conservation, where the major contributors were the members of the phyla Bacteroidetes and Actinobacteria (Faith et al. 2013). This conservation or 'longitudinal stability' along with the diversity of the microbiome at individual levels or 'interpersonal diversity' is capable of assigning an unique 'microbial fingerprint' to every individual, based on the identification of >80% of the individuals microbiome composition (Franzosa et al. 2015). Despite its resilience potential, the recent studies indicate 'dysbiosis' of the microbiome to be associated with a major change(s), such as onset of disease, surgery, or antibiotic treatment (Morgan et al. 2012). Some examples of noncommunicable diseases, where correlations between the gut microbiome profile and disease status have been elucidated, will be discussed in subsequent sections of this chapter.

Developing robust microbiome-based therapeutics to restore microbiome balance, maintain the same over a period of time, and prevent relapses of dysbiosis poses a few challenges that need interdisciplinary approaches to offer viable and effective solutions.

10.6 Microbiome Therapeutics

We all consume curd/yogurt in our regular food habits, but it was the curiosity and observation of Elie Metchnikoff, who wondered how a rural Bulgarian community with limited resources for living were able to live longer. He later found out that by manipulating the microbiome, one can increase the life and health spans of humans. He is the father of probiotics.

Traditionally, people in Europe and Japan have relied on fermented food products and the active ingredients that give the health benefits in fermented food products, are the microbes that constitute the food products.

In today's world with high stress, reduced sleep, unbalanced diet, and lack of exercise, it is increasingly important to balance the gut microbiome through supplementation of probiotics. Probiotics not only help in balancing the 'good bacteria' but also keep the 'bad bacteria' away. This in turn helps the digestive system and overall health of an individual.

We also know that imbalance in the microbiome is the underlying cause of many disease conditions. The progressing field of microbiome diagnostics will be best matched, when 'Microbiome Therapeutics' can be customized based on the need of every patient with the right combination of prebiotic, probiotic, and supplements (Fig. 10.3).

10.6.1 Prebiotics Support Probotics

According to the International Scientific Association of Probiotics and Prebiotics, prebiotics are defined as 'a substrate that is selectively utilized by host microorganisms conferring a health benefit' (Gibson et al. 2017). In our microbiome, they promote the absorption of ion and trace element such as that of calcium, iron, and magnesium and modulate cytokine and secretory immunoglobulin A production, via mechanisms involving microbial metabolic products (Holscher 2017). Inulin, fructooligosaccharides (FOS), galactooligosaccharides (GOS), and human milk oligosaccharides (HMOS) are a few prebiotic ingredients in our diet that have a strong correlation in keeping body weight under check (Kim et al. 2019). They are found to stimulate the growth of Bifidobacteria and Lactobacillus species, thus enhancing the availability of SCFAs within the microbiome. Higher SCFA level positively influences satiety and food consumption via improved GLP-1, PYY, and ghrelin production (Cerdó et al. 2019).

10.6.2 Probiotics in Food and as Supplements

Natural probiotics can be obtained from food sources such as curd/yoghurt and fermented foods. Food products that contain probiotics are yoghurt, kefir, cheese, tempeh, kimchi, miso, sauerkraut, and some soy beverages. Freeze-dried bacteria in the form of tablets, capsules, powders and sachets, and ampoules containing bacterial spores, are available commercially from cultured organisms at a defined composition and abundance (Table 10.1, Alfano et al. 2020). In addition, probiotic-fortified foods are also available like juices, chocolates, flour, and cereal. Food and Drug Administration, USA (FDA), regulations allow probiotics to be sold as supplements and not like drugs, for healthy people. For people with illnesses such as irritable bowel syndrome, inflammatory bowel disease, diarrhea (both infectious and antibiotic-induced), urinary tract infections, and eczema, doctor-prescribed probiotics may be given. Pregnant women, infants, young children, and immunocompromised patients should be given probiotics with caution.

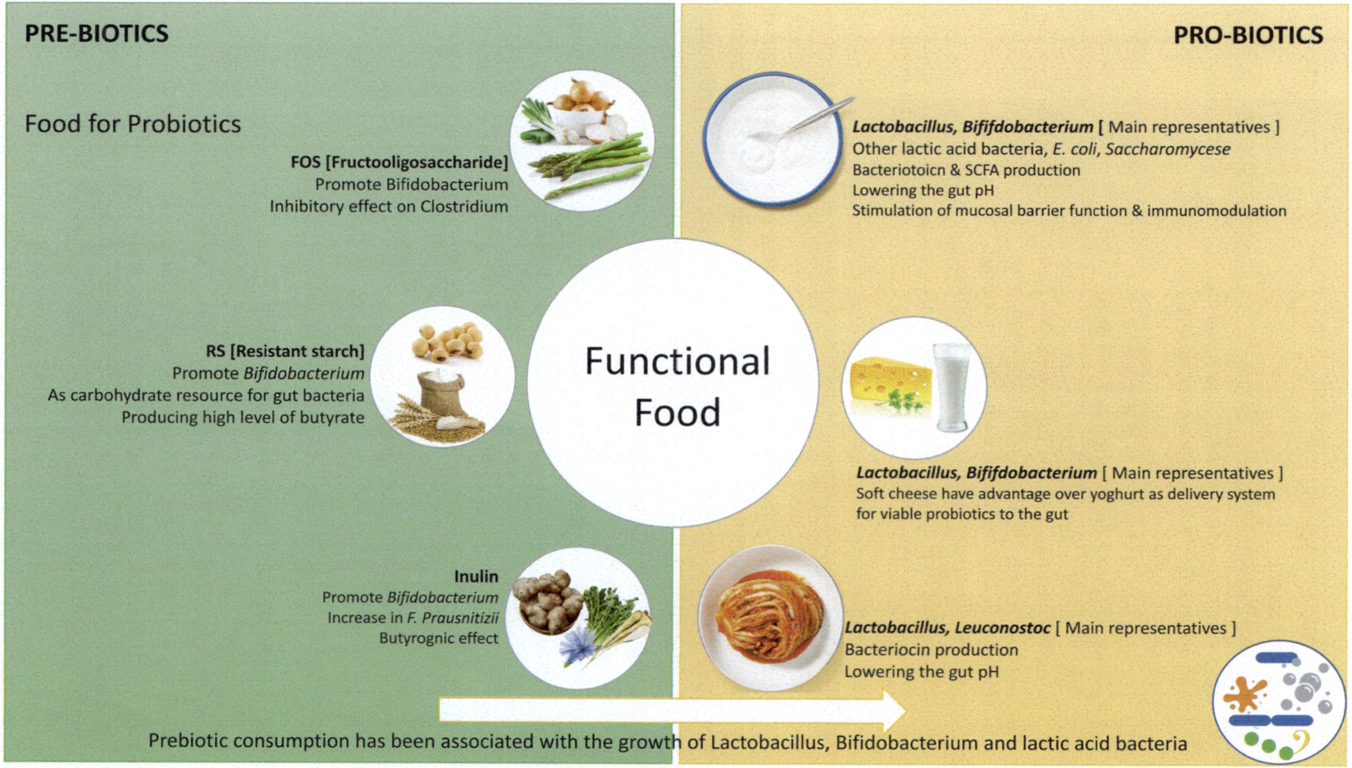

Fig. 10.3 Functional foods have health benefits above and beyond their nutritional value, e.g., providing prebiotics and probiotics which help in balancing the microbiome. Probiotics (right panel) are the live microbes which are beneficial for health and can be obtained from various foods which act as live culture for these beneficial microbes. Prebiotics (left panel) are the food for these live microorganism (probiotics) present in different food and are associated with the growth of beneficial bacteria. (Green et al. 2020; Markowiak and Ślizewska 2017; McBurney et al. 2019; Pandey et al. 2015; Rezac et al. 2018; Terpou et al. 2019)

Table 10.1 List of a few commercially available probiotic preparations

Brand name	Strain	Producer
Dicoflor	*Lactobacillus rhamnosus GG*	AGPHARMA
Enterogermina	*Bacillus clausii*	SANOFI
Enterolactis	*Lactobacillus casei*	SOFAR
Nutriflor	*Lactobacillus acidophilus DDS-1, Lactobacillus bulgaricus DDS-14 Bifidobacterium bifidum, Lactobacillus rhamnosus*	NUTRIGEA
Probactiol duo	*Lactobacillus acidophilus NCFM, Lactobacillus paracasei Lpc-37 Bifidobacterium lactis Bi-07, Bifidobacterium lactis Bi-04*	METAGENETICS
VSL#3	*Streptococcus thermophilus, Bifidobacterium breve, Bifidobacterium longum, Bifidobacterium infantis lactobacillus acidophilus, Lactobacillus plantarum, Lactobacillus paracasei, Lactobacillus delbrueckii subsp. bulgaricus*	FERRING FARMACEUTICI
Yakult	*Lactobacillus casei Shirota*	YAKULT (Tokyo)

10.6.3 Postbiotics

Postbiotics is a therapeutically attractive emerging field that deals with the use of nonviable by-products of microbial growth (metabolites, cell lysis components, enzymes) that have an added health benefit when consumed. Unlike probiotics, which are based on the administration of live organisms, postbiotics are administered in alignment with pharmacokinetic and pharmacodynamic properties. They are being explored since they are abundant at most body sites, are suitable for different routes of administration, have low toxicity potential, are stable in the systemic circulation, and scale up friendly. Bacterial exopolysaccharides (EPS) from *Bifidobacterium* and *Lactobacilli* and extracellular vesicles (EVs) from *Akkermansia muciniphila* and commensal *Escherichia coli* are some such postbiotic examples (Wegh et al. 2019). There are several downsides such as pleiotropic effects, shorter half lives, and high cell-type specificity, which need further exploration and well-designed studies to evaluate them for therapeutics.

10.6.4 Fecal Microbiota Transplantation (FMT)

Fecal microbiota transplantation (FMT) is a procedure of administering donor fecal suspension into the colon of a diseased recipient, thus aiming to restore the disturbed gut microbiota and the associated disease. The first application of FMT in modern medicine was reported in 1965 for *Clostridioides difficile* colitis. Though FMT has benefited many to cure chronic conditions, it is not advisable for all conditions of gut dysbiosis and for all categories of patients. Selection of patients for whom FMT is effective is an important concern since long-term safety of the procedure and outcome for the patient needs careful consideration. Nevertheless, more data from well-designed studies will help in determining the efficacy and safety of the procedure.

10.6.5 Research-Driven Probiotics: The Future

The rationale that the naturally occurring human-associated microorganisms offer myriad of health benefits, is the basis of probiotic therapies. A systematic and well-designed research on specific probiotics designed for an individual is needed (Mimee et al. 2016). This would be possible through the studies on microbiome and microbiome diagnostics. A precise quantization of the microbiome components, the dosage, and regimens all require advanced science. The need of the hour is to develop microbiome-based recommendations for probiotics and probiotics prescribed for specific diseases classified as 'prescription probiotics. Preparing them will involve large-scale culturing and identification of various bacterial species, their long-term storage, and potency testing of the probiotic cocktails tailored to an individual's need. A futuristic probiotic application wherein the probiotic strain can be used for directly delivering anti-inflammatory and intestinal epithelial repair factors to the intestinal tract will allow correction of multiple aberrations in a unified manner.

Genetically engineered probiotics: The use of genetically modified organisms (GMO) in human microbiome therapeutics is still farfetched due to the associated regulatory clearances needed for their safety. Nevertheless, researchers have tested the concept in animal models and have met with success. The probiotic *E. coli* Nissle 1917 was altered to be used as a prophylactic, in order to inhibit virulence of *Vibrio cholerae.* (Hamady et al. 2010). A genetically modified derivative of a vaginal commensal *Lactobacillus jensenii* able to prevent transmission of chimeric simian/human immunodeficiency virus (SHIV) in a rhesus macaque model, when administered is another encouraging study (Motta et al. 2012) demonstrating the utility of engineered probiotic strains. A common side effect of chemotherapy 'oral mucositis,' a condition involving ulcerative lesions, is shown to be benefited from the topical application of an altered *L. lactis* engineered to secrete trefoil factor-1. Data from an early clinical trial for the treatment of the condition displayed good tolerance among patients and could be effective at reducing prevalence (Limaye et al. 2013).

Engineered 'designer' consortia: This concept is based on building a collection of well-characterized probiotic strains that can be custom combined in the laboratory based on the attributes desired for the therapeutic consortia. One such example is elaborated here. Bacteria in the gut generate urease which convert the urea produced by the liver to ammonia and carbon dioxide. Patients with liver deficiency, neurotoxicity and encephalopathy are found to be associated with accumulation of systematic ammonia. In a study, mouse models were treated with antibiotic and polyethylene glycol, resulting in reduction of endogenous microbiota. This was followed by transplantation with a defined microbial community with low urease activity. The microbiota reconstitution was successful in altering community-wide metabolic activity of urea that remained stable for months (Shen et al. 2015). Such designer consortia for human use may be seen in trials soon in the near future.

10.7 The Role of Microbiome Diagnostics and Therapeutics in a Few Disease Scenarios

In this section, select disease conditions primarily influenced by the gut microbiome dysbiosis are presented. A brief pathophysiology of the condition, it's associated microbiome dysbiosis profile and suggested therapeutics for correcting the dysbiosis is summarized in the following sections (Fig. 10.4). Oral, vaginal, skin, and respiratory microbiome modulations are out of scope of this chapter.

10.7.1 Atherosclerosis

Atherosclerosis, a chronic inflammatory disease, and major contributor in CVDs (cardiovascular diseases), is associated with plaque formation consisting of accumulated modified lipids, calcified regions, neurotic cores, inflamed smooth muscle cells, endothelial cells, leukocytes, foam cells and impaired lipid metabolism and endothelial functions (Frostegård 2013). CVD is one of the leading causes of disease and death globally (Benjamin et al. 2018).

Atherosclerosis has been linked to intestinal microbes due to a substance called trimethylamine oxide in recent studies (Chen et al. 2016). In 2015, Cleveland Clinic researchers observed that lecithin and L-carnitine which is present in red meat, egg yolk, etc., can be converted to TMAO wherein intestinal microbes play an important part thus promoting atherosclerosis and speeding up the pathological process of cerebrovascular diseases (Wang et al. 2015a, b).

What's the connect? Intestinal microbes can absorb foods rich in lecithin, choline, and carnitine to produce trimethylamine (TMA, a colorless gas of a foul odor), which is oxidized by flavin monooxygenase (FMO, FMO3 with highest activity) to TMAO in the liver. Atherosclerosis is linked with increased TMAO in the blood (Koeth et al. 2013), which in turn is linked to the diet, intestinal microbes, FMO3 activity, gender, and heredity of host (Seldin et al. 2016). Some mechanisms which TMAO uses to develop atherosclerosis are hampering cholesterol reverse transportation (Koeth et al. 2013), up-regulating the expression of macrophages CD36 and scavenger receptor A1 (SR-A1), encouraging foam cell formation, down-regulating the expression of cholesterol absorption targets ABCG5/8 and NPC1L1 affecting cholesterol metabolism, reducing the expression of cytochrome P450 (CYP) 7A1 and 27A1 in the liver, which in turn reduces the transport of bile acid and clearance of cholesterol and activates monocytes via mitogen-activated kinase and nucleic acid factor-κB signaling pathway by developing vascular inflammation (Seldin et al. 2016). Thus, if the density, richness, and diversity of gut microbiota are improved, it can help in prevention and treatment of atherosclerosis. People having lower species richness and diversity in the gut are prone to develop atherosclerosis (Menni et al. 2018).

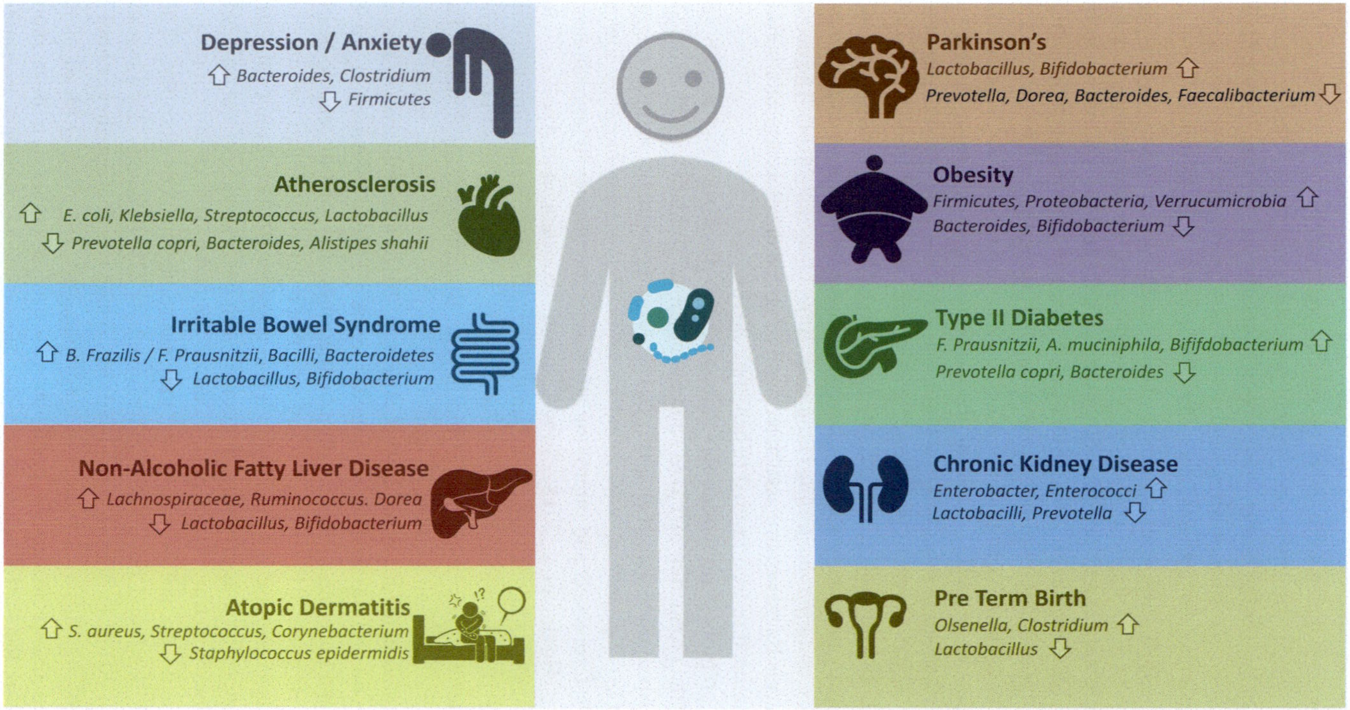

Fig. 10.4 Depiction of various diseases associated with different microbiome (gut, oral, skin, vaginal) and the major upregulated and downregulated microbes involved in said diseases (Álvarez-Mercado et al. 2019; Green et al. 2020; Kho and Lal 2018; Kosti et al. 2020; Liu et al. 2019)

Atherosclerotic patients are found to have a gut microbiota that is less fermentative and more inflammatory (Jie et al. 2017). *Firmicutes* and *Bacteroides* are the major taxa present which seem to be remarkably constant (Huttenhower et al. 2012; Faith et al. 2013). However, in patients with atherosclerosis, *Escherichia coli*, *Klebsiella spp.*, *Enterobacter aerogenes*, *Streptococcus spp.*, *Lactobacillus salivarius*, *Solobacterium moorei*, and *Atopobium parvulum* were found to be increased, whereas *Bacteroides spp.*, *Prevotella copri*, and *Alistipes shahii* were found to be depleted (Jie et al. 2017).

Therapeutics that help: The three 'p' known for the well-being of gut are probiotics, prebiotics, and polyphenols, as these help regulate the gut microbiota composition (Marchesi et al. 2016). Since small intestine cannot absorb polyphenols directly, the bioavailability of polyphenols depends on the gut microbiota and their ability to convert it into components which can be absorbed by the small intestine (Duda-Chodak et al. 2015). Few of the polyphenols which have shown some potential mechanism in atherosclerosis are protocatechuic acid (PCA), quercetin-3-glucuronide, 2,4,5-trimethoxycinnamic acid, gallic acid, and equol (Pieczynska et al. 2020). Resveratrol, a natural phenolic phytochemical, works by reducing TMAO levels by promoting the growth of commensal bacteria such as *Bacteroides*, *Lactobacillus,* and *Bifidobacterium* (Jung et al. 2009; Qiao et al. 2014). The decrease in TMAO levels by resveratrol associated with inhibited development of atherosclerosis has also been proven in vivo (Chen et al. 2016).

As with prebiotics, probiotics like *Lactobacillus plantarum* and *E. aerogenes* could lower the production of TMAO and attenuate the formation of atherosclerotic abrasion in ApoE/mice (Qiu et al. 2017, 2018). Another prebiotic, mannan oligosaccharide (MOS) supplement, was found to regulate gut microbiota by lowering plasma cholesterol levels and improving atherosclerotic plaques in high cholesterol diet-fed mice (Hoving et al. 2018). Stimulation of *Akkermansia* in ApoE –/– mice was linked with berberine, which is found to be effective against atherosclerosis (Zhu et al. 2018). The endothelial function in ApoE –/– mice seems to be improved by administering ITFs, a prebiotic (Inulin-type fructans), as supplement. These inulin-type fructans enhance the formation of butyrate and protect against atherosclerosis formation (Watzl et al. 2005; Catry et al. 2018) as per the recent studies.

Both fish oil and flaxseed oil are found to reduce TMAO by enhancing SCFAs production and lowering LPS generation by microbes, and fish oil seems to be more productive (He et al. 2019). Research suggests that calorie-controlled diet integrated with supervised exercise lowers TMAO levels considerably (Erickson et al. 2019).

FMT is another therapy; however, it comes with risk as well. For example, while beneficial flora is getting transferred, so could the endotoxins or infectious agents present in the donor, and this could start new gastrointestinal complications. This is the reason it has found limited use as treatment for CVD patients (De Leon et al. 2013; Brandt 2013). Further research is needed to take a look at whether or not FMT probably prolongs different aspects of cardiometabolic disorders. Instead of fecal contents, the transplantation of particular group of microbes can be a rational opportunity to FMT. To better define the optimal fecal microbial preparation, dosing, and method of delivery, further research needs to be conducted (Sanchez-Rodriguez et al. 2020).

How the gut dysbiosis and TMAO derived from the microbiota participate in atherosclerosis is yet to be cleared (Zhu et al. 2020a, b). New strategies to prevent or treat the disease can be developed by better understanding of gut microbiota composition, to the development of atherosclerosis (Pieczynska et al. 2020).

10.7.2 Hypertension

Hypertension is among the chief causes of cardiovascular disease and is responsible for global deaths (Go et al. 2014). Although the procedure of how gut microbiota is involved with hypertension is not quite clear, SCFAs and oxidized low-density lipoprotein (ox-LDL) are believed to take some part in it (Ma et al. 2018).

What's the connect? Obese pregnant women with lower blood pressure have shown increase in butyrate-producing bacteria (Gomez-Arango et al. 2016). Gut dysbiosis was improved by fiber and acetate supplementation in a study on hypertensive mice. It led to a surge in *Bacteroides acidifaciens,* which seems to have a defensive role in hypertension/heart failure (Marques et al. 2017). GPR41, GPR43, and GPR109AA are the three G-protein-coupled receptors (GPCRs), regulated by SCFAs. Another type expressed in the kidney is olfactory receptor 78 (Olfr78) regulated by acetate and propionate (Tan et al. 2017). The GPCRs regulated pathways of host can be stimulated by SCFAs which effects the secretion of renin and in turn effects the blood pressure (Furusawa et al. 2013; Pluznick et al. 2013).

Oxidation of LDL causes vasoconstriction leading to hypertension through gut dysbiosis (Packer et al. 2014), by boosting the expression of pro-inflammatory cytokines which induces oxidative stress triggering the Ox-LDL stimulation (Chawla et al. 2011; Peluso et al. 2012). Oxidation of L-arginine by nitric oxide synthase produces nitric oxide (NO) (Ma et al. 2006). Production of NO and endothelin-1 which maintains basic vascular tension and cardiovascular system homeostasis (Boulanger and Lüscher 1990) is hampered due to higher levels of ox-LDL which causes hypertension (Subah Packer 2007). Another cause of hypertension is chronic low-grade inflammation (Schiffrin 2014), which occurs due to depletion in microbial gene richness (Cotillard et al. 2013). Chronic probiotic intake decreases preeclampsia associated with hypertension (Brantsaeter et al. 2011).

Therapeutics that help: The composition of gut microbiota is altered by consuming β-glucan in such a way that it reduces the risk markers associated with CVD as per the single-blind randomized trial (Hoving et al. 2018). When *Lactobacilli* fermented milk was consumed by hypertensive humans, it lowered their blood pressure (Seppo et al. 2003). It is observed in human trials that consumption of at least 10^{11} colony-forming units along with multiple species of probiotics for 8 weeks decreases both systolic and diastolic blood pressures (Khalesi et al. 2014). Long-term administration of probiotics of various *Lactobacillus* bacteria such as *Lactobacillus fermentum* CECT5716 (LC40), *Lactobacillus coryniformis* CECT5711 (K8), and *Lactobacillus gasseri* CECT5714 (LC9) could decrease systolic blood pressure in hypertensive rats (Gómez-Guzmán et al. 2015). Phenylacetyl glutamine, a gut microbiota-derived metabolite, is negatively linked with pulse wave velocity and systolic blood pressure (Menni et al. 2015).

10.7.3 Obesity

Obesity is the manifestation of accumulated fat and is correlated with the progression of many diseases of metabolic origin like cardiovascular disease, type 2 diabetes mellitus, cancer, and nonalcoholic fatty liver disease (Kim et al. 2019). The fact that obese people live 7 years shorter than nonobese people is quite alarming (Van Hul et al. 2018).

What's the connect? Studies from humans and animals have clearly shown a correlation between obesity and gut microbiome. Some examples are as follows: (1) decreased gut diversity (Baothman et al. 2016); (2) increased *Firmicutes* and decreased *Bacteroidetes* (Koliada et al. 2017; Mariat et al. 2009; Greenhill 2015). This increased *Firmicutes/Bacteroidetes* (F/B) ratio facilitates the energy extraction, and this in turn effects the energy storage in the adipose tissue (Mariat et al. 2009; Bell 2015). A significant increase in *Enterobacteriaceae* was observed in obesity (Balamurugan et al. 2010).

There exist a few other mechanisms for the functioning of gut microbiome in influencing obesity. Examples include bile acids that actively help in resolving fat uptake from diet in the small intestine, but they also hamper the growth of different commensal bacteria such as *Lactobacilli and Bifidobacteria* by disorganizing their membrane permeability (Ridlon et al. 2006; Kurdi et al. 2006). Acetate, propionate, and butyrate (SCFAs) which are eventually consumed by various organisms are estimated to have a production rate of 80–200 kcal/day (Riley et al. 2013). Decrease in butyrate-producing bacteria along with reduced intake of dietary carbohydrates such as polysaccharides, vegetable oligosaccharides, and resistant starch was observed in obese patients (Canfora et al. 2019). Another study gave similar results where the levels of fecal butyrate, SCFAs, and *Bifidobacterium* were found to be reduced considerably in obese patients who consumed less fiber (Brinkworth et al. 2009). Metabolic endotoxemia is defined as a chronically high plasma LPS disorder at 10–50 times less than the septic conditions of LPS and is a high-fat dietary elevation of plasma lipopolysaccharide (LPS) as termed by Cani et al. (2007). Dietary increases of endotoxin were linked to enhanced fat deposit, systemic and tissue-specific inflammation, and resistance to insulin (Cani et al. 2007; Amar et al. 2008).

Therapeutics that help: Prebiotics regulate the gut microbiome composition by improving lipid metabolism which is also seen in short-chain FOS treatment in diet-induced obese mice (Cluny et al. 2015) and are known to have antiobesity effects (Barengolts 2016; Nicolucci and Reimer 2017; Delzenne et al. 2011). Animals treated with oligofructose displayed reduction in both triglyceride level and adipose tissue mass (Cluny et al. 2015). It is also reported that α-cyclodextrins supplementation in obese mice (diet-induced) resulted in inflection of gut microbiota and SCFA production (Nihei et al. 2018). Human breast milk is enriched with milk oligosaccharides and serves as wonderful prebiotics candidate. As a prebiotic, it promotes the growth of beneficial bacteria such as *Bacteroides and Bifidobacterium* and hinders the pathogens such as *Campylobacter jejuni*, *Helicobacter pylori,* and *E. coli* (Newburg 2000).

Recent studies have shown that when compared with placebo-treated control animals, supplementation of *Bifidobacterium* species such as *B. breve* B3, *B. infantis*, and *B. longum* and *Lactobacillus species such as L. rhamnosus, L. casei* strain *Shirota* [LAB13], *L. gasseri*, and *L. plantarum* has shown obliteration of weight gain, fat deposits, and white adipose tissue (Barengolts 2016; Kim et al. 2019).

In another study, treatment of obese adults with *L. gasseri* (SBT2055 and BNR17) exhibited reduction in visceral adipose tissue as well as waist size (Kadooka et al. 2010; Kim et al. 2018). Similar study was reported by Pedret et al, where intervention with *Bifidobacterium animalis* subspecies. *Lactis* CECT 8145 reduced waist size, waist circumference/height ratio, and BMI considerably (Pedret et al. 2019). *L. rhamnosus* CGMCC1.3724 therapy displayed weight loss in obese women but nothing significant in obese men (Sanchez-Rodriguez et al. 2020).

Diet is a major player in obesity and has associations with gut microbiota (Brahe et al. 2016). It has been shown that the various diet styles such as western, vegetarian, gluten-free, and the Mediterranean diet disturb gut diversity (Lazar et al. 2019). The Western diet which comprises of high amount of sugar, salt, saturated fats, refined grains, and high fructose corn syrup with lesser amount of fiber is responsible for decrease in total gut microbiota amount along with reduction in beneficial bacteria such as *Lactobacillus sp.* and *Bifidobacterium sp.*, thereby promoting inflammation and changing gut microbiota to obese pattern (Bell 2015; Statovci et al. 2017). One of the mechanisms can be improving energy harvesting by increase in *Firmicutes* for promoting better caloric absorption leading to weight gain (King et al. 2012).

Plant-based diet such as vegetarian and vegan diets is rich in dietary fiber and entails plant-derived products which is known to trigger an increase in the abundance of protective microbiota. This diet promotes an increase in (1) *Bifidobacteria and Lactobacillus*; known intestinal barrier protectors, (2) *Faecalibacterium prausnitzii* and *Roseburia*; butyrate producers, and a decrease in *Escherichia coli* and *Enterobacter cloacae*; inflammation-inducing lipopolysaccharide-producing bacteria, thus ultimately preventing obesity (Tomova et al. 2019; Glick-Bauer and Yeh 2014). Mediterranean diet is majorly comprised of vegetables, olive oil, fruits, a modest amount of poultry, with limited consumption of red meat and dairy products. This dietary habit correlates with higher abundance of *Lactobacillus*, *Bifidobacterium*, and *Prevotella* in the gut, which helps in preventing obesity by improving lipid and cholesterol profiles (Garcia-Mantrana et al. 2018; Coelho and Cândido 2019). Korean traditional diet consists of high amounts of vegetables, fermented foods with modest consumption of legumes and fish. Such a diet helps to prevent obesity by increasing abundance of *Bacteroides* (*Bacteroidaceae*) and *Bifidobacterium* (*Bifidobacteriaceae-Actinobacteria*) while decreasing *Prevotella* (*Prevotellaceae*) (Baik 2018; Lim et al. 2015). With the help of high-fiber diet, obesity can be managed through intestinal SCFA dependent modulation of downstream pathways (Barathikannan et al. 2019). In high-fat diet (HFD) fed mice, when HFD is replaced by treatment with *L. rhamnosus* GG, it reduces adiposity via the heightened production of adiponectin, thereby protecting the animal from insulin resistance as well as helping in diminishing liver adiposity (Kim et al. 2013a, b). Pasteurized nonviable *Akkermansia muciniphila* as a prebiotic treatment showed an increased ability to reduce the development of fat mass along with insulin resistance and dyslipidemia in mice (Plovier et al. 2017; Depommier et al. 2019).

10.7.4 Nonalcoholic Fatty Liver Disease

Nonalcoholic fatty liver disease (NAFLD) is marked by hepatic steatosis and may advance to an inflammatory condition called nonalcoholic steatohepatitis (NASH), liver cirrhosis, and hepatocellular carcinoma. Gut microbiome and certain host factors have been linked to this condition (Grabherr et al. 2019).

What's the connect: A large variation in terms of phylum, family, and genus was observed between healthy controls and NASH patients in several studies. NAFLD patients display upregulation and downregulation of a large array of organisms; the organisms which are enriched are *Bacteroides*, *Ruminococcus*, *Lactobacillus* (Genus), *E. coli* (Species), *Lactobacillaceae* (Family), and *Proteobacteria* (Phylum), while the downregulated organisms are *Oscillibacter, Prevotella, Ruminococcus, Coprococcus* (Genus), *Faecalibacterium prausnitzii* (Species), *Actinobacteria*, *Bacteroidetes*, and *Firmicutes* (Phylum) (Loomba et al. 2017; Del Chierico et al. 2017; Boursier et al. 2016; Da Silva et al. 2018).

NAFLD may be associated with low abundance of *Faecalibacterium prausnitzii*, a butyrate-producing bacterium from *Firmicutes* phylum, which was associated with >5% fat hepatic content and increased adipose tissue inflammation (Munukka et al. 2014, 2017). A gram-negative Proteobacterium *Bilophilia wadsworthia,* has been shown to aggravate high fat diet induced metabolic dysfunctions in mice. The mechanism followed is that it lowers the butyrate metabolism, which leads to disruption of the gut barrier (interrupted tight junctions), thus allowing the LPS circulation from the gut lumen into the incoming portal vein of the liver. Once there, it releases a pro-inflammatory cytokine by acting on the hepatic macrophages and promotes a reduction of bile acids production. All of these leads to a disrupted microbiota and hence the heightened release of LPSs (Feng et al. 2017; Natividad et al. 2018).

Helicobacter pylori is a gram-negative *Proteobacterium* which is responsible for immune resistance contributing to NAFLD. *H. pylori* infection can increase the chances of NAFLD development. However, to comprehend the association between *H. pylori* and NAFLD progression, more clinical studies are needed (Wijarnpreecha et al. 2018; Ning et al. 2019).

In a healthy state, ethyl alcohol is continuously being produced in the gut. In the healthy individual, it gets metabolized in the liver by alcohol dehydrogenase (ADH)/ other hepatic enzymes. However, when alcohol-producing bacteria like *Klebsiella pneumoniae* increase in the gut, they produce reactive oxygen species (ROS) constantly due to exceeding of liver detoxification capacity which promotes hepatic inflammation, often ending in steatohepatitis (Yuan et al. 2019).

Obese and NAFLD animals have shown lesser abundance of *Akkermansia muciniphila,* a gram-negative bacterium from the phylum *Verrucomicrobia* with mucin degrading capacity than in their healthy counterparts (Everard et al. 2013; Zhao et al. 2017).

Therapeutics that help: The consumption of processed foods and beverages containing fructose was seen higher in NAFLD patients (Chen et al. 2017). NAFLD progression is also associated with lower fiber, polyphenols, vitamins (Vitamin D), and mineral nutrients (calcium) intake (Van Herck et al. 2017; Wehmeyer et al. 2016). Polyphenols like quercetin, epigallocatechin gallate, anthocyanins, and resveratrol have also been found to be protective (Wrzosek et al. 2013).

It has been reported that administration of a cocktail of *Lactobacillus acidophilus* ATCC B3208, *Bifidobacterium lactis* DSMZ 32,269, *Bifidobacterium bifidum* ATCC SD6576 and *Lactobacillus rhamnosus* DSMZ 21,690 to adolescents for 12 weeks in the form of probiotic capsules, resulted in substantial decrease in ALT (Alanine aminotransferase), lipid profile and intrahepatic fat content compared to placebo group. In another study, efficacy of 'Symbiter,' containing 14 alive probiotic strains of *Lactobacillus* + *Lactococcus, Bifidobacterium, Propionibacterium,* and *Acetobacter,* is assessed in NAFLD patients and has shown to improve hepatic steatosis, aminotransferase activity, TNF-α, and IL6 levels (Kobyliak et al. 2018). Another multistrain probiotic VSL#3 has been found to protect the integrity of intestinal barrier and diminish endotoxemia and oxidative/ nitrosative stress, thus improving liver pathology in patients suffering from various chronic liver diseases (Loguercio et al. 2005). VSL#3 contains *Bifidobacterium longum,* and it adjusts gut microbiota in such a way that increases the production of conjugated linoleic acid (CLA); this further impacts fatty acid composition in the liver and in a way plays a significant role in therapeutic interventions (Meroni et al. 2019). When *Bifidobacterium longum* is administered in combination with prebiotic fructo-oligosaccharides (FOS), it considerably improves the metabolic and inflammatory markers and fibrosis scores in NASH patients (Malaguarnera et al. 2012).

10.7.5 Type 2 Diabetes

Diabetes mellitus (DM) is marked by chronic hyperglycaemia, as a result of deficits either in insulin secretion, or in insulin action, or both. 90–95% of diabetes cases are of type 2 (T2D) (Woldeamlak et al. 2019).

What's the connect? Bifidobacterium, Bacteroides, Faecalibacterium, Akkermansia, and *Roseburia* were found to be negatively related with T2D, while Ruminococcus, *Fusobacterium*, and *Blautia* were connected positively with T2D (Gurung et al. 2020). Patients treated with metformin or after undergoing gastric bypass surgery have found to be negatively associated with *B. adolescentis, B. bifidum, B. pseudocatenulatum, B. longum, and B. dentium* (Wu et al. 2017; Murphy et al. 2017). *Bacteroides intestinalis, Bacteroides* 203, *and Bacteroides vulgatus* were decreased in T2D patients and *Bacteroides stercoris* were enriched after sleeve gastrectomy (SG) surgery in T2D patients with diabetes remission (Wu et al. 2011; Murphy et al. 2017; Zhang et al. 2013; Karlsson et al. 2013).

Investigations have reported a negative association of *Roseburia inulinivorans, Roseburia_272*, and one unclassified OTU from this genus, with disease (Murphy et al. 2017; Zhang et al. 2013; Karlsson et al. 2013). Lower frequencies of *Faecalibacterium* were reported in patients in two case–control studies (Gao et al. 2018; Salamon et al. 2018). Also, after different types of antidiabetic treatments ranging from metformin and herbal medicine to bariatric surgery, decreased their abundance (Tong et al. 2018; Murphy et al. 2017). Half of the T2D studies showed that out of these five genera, Bacteroides, *Bifidobacterium, Roseburia, Faecalibacterium*, and *Akkermansia*, at least one is reduced, suggesting that they have a role which goes beyond serving as a biomarker (Gurung et al. 2020). There is an increase in organisms like *L. acidophilus*, *L. gasseri*, and *L. salivarius,* and reduction in *L. amylovorus* in T2D patients which signify species specificity (Karlsson et al. 2013; Graessler et al. 2013; Forslund et al. 2015). *L. acidophilus, L. plantarum*, and *L. reuteri* were found to have lower frequencies when compared with controls in this disease (Suceveanu et al. 2018).

Patients with T2D show raised levels of pro-inflammatory cytokines, chemokines, and inflammatory proteins. Also, increased gut permeability allows for passage of gut microbe-derived products into the blood. This causes metabolic endotoxemia, effects glucose homeostasis and insulin resistance in liver, muscle, and fat, and in addition affects the digestion of sugars and production of gut hormones that regulate glucose metabolism (Gurung et al. 2020). The microbes, like *Fusobacterium nucleatum* and *Ruminococcus gnavus* which are potentially harmful in T2D, induces various inflammatory cytokines (Yang et al. 2017a, b; Hall et al. 2017), in other inflammatory diseases.

The *Firmicutes* to *Bacteroidetes* ratio along with proportions of phylum *Firmicutes* and class *Clostridia* was reduced, while class *Betaproteobacteria* was increased in T2DM patients as per one of the case–control study (Larsen et al. 2010). Also, *Faecalibacterium prausnitzii* and genus *Blautia* were diminished in T2DM patients (Navab-Moghadam et al. 2017; Inoue et al. 2017). Increase in serum fructosamine is associated with the decrease in *Prevotellaceae* and increase in *Enterobacteriaceae* (Li et al. 2019a, b).

Therapeutics that help: There have been many animal studies conducted for diabetes. An improved glucose tolerance is seen in several *Bifidobacterium* spps, i.e., *B. bifidum, B. longum, B. infantis, B. animalis, B. pseudocatenulatum, B. Breve* (Le et al. 2015; Moya-Pérez et al. 2015; Kikuchi et al. 2018; Aoki et al. 2017; Wang et al. 2015a, b). In another study, an improvement in glucose tolerance and insulin resistance was seen in diabetic mice by providing *Bacteroides acidifaciens* and *Bacteroides uniformis* (Yang et al. 2017a, b; GauffinCano 2012). In human studies, progress in type 2 diabetes-related symptoms is seen by administration of *L. sporogenes, L. casei Shirota, and L. reuteri* used as monoprobiotics (Hulston et al. 2015; Simon et al. 2015; Asemi et al. 2014, 2016). *Lactobacilli* works more effectively as a part of probiotic cocktail than given individually in majority of cases (Ejtahed et al. 2012; Asemi et al. 2013; Tajabadi-Ebrahimi et al. 2016; Mohamadshahi et al. 2014). Among the several *Lactobacillus* species that have been tested as probiotics, *L. plantarum, L. reuteri, L. casei, L. curvatus, L. gasseri, L. paracasei, L. rhamnosus, and L. Sakei* have shown favorable effects on T2D in mice models (Gurung et al. 2020).

There have been different mechanisms found to reduce gut permeability like (1) Administering *Bacteroides vulgatus and B. Dorei* species (Yoshida et al. 2018); (2) Butyrate, produced by *Faecalibacterium*, via serotonin transporters and PPAR-g pathways (Kinoshita et al. 2002); (3) *Akkermansia muciniphila*, probiotic bacterium using extracellular vesicles which improve intestinal tight junctions via AMPK activation in epithelium (Chelakkot et al. 2018).

There has been an emerging focus on interactions between microbiota and antidiabetic drugs in microbiome research (Gurung et al. 2020). Different combinations have been used and been effective than when administered alone in reducing/improving T2D parameters, like reduced hyperglycemia, adiposity, improved fasting blood glucose, glucose tolerance, and insulin resistance in different studies. For example, (1) probiotic *Bifidobacterium animalis ssp. lactis* 420, prebiotic polydextrose in amalgamation with sitagliptin in diabetic mice (Salamon et al. 2018); (2) prebiotic polysaccharide in combination with metformin and sitagliptin in Zucker diabetic rats' study (Reimer et al. 2014); (3) combination of a prebiotic and metformin in streptozotocin-induced diabetic mice (Zheng et al. 2018).

There has been inconsistency in identification of T2D-associated microbiota in humans because of several elements such as geographic location, race, culture, health status, and drug use. Stool samples are the preferred choice for microbiota analysis in most of the studies due to difficulties in sampling from human intestine, but, however, the stool sample does not represent entire gut microbiome profile (Gurung et al. 2020).

It has been observed that vegetables, fruits, dietary fibers, and medicinal plants reduce or prevent T2DM by raising the level of SCFAs and fine-tuning gut microbiota. Capsaicin can increase the *Firmicutes* to *Bacteroidetes ratio,* Roseburia number, and at the genus level reduce the quantities of *Bacteroides* and *Parabacteroides* which could reduce pro-inflammatory cytokines, such as TNF-α and IL-6 in obese diabetic mice (Song et al. 2017). Pumpkin polysaccharide can exert its antidiabetic effect by increase of selective bacteria, like *Bacteroidetes*, *Prevotella*, and *Deltaproteobacteria* in mice (Liu et al. 2018a, b). The *Lactobacillus rhamnosus* GG fermented carrot juice can improve T2DM in rats by increasing *Oscillibacter* and *Akkermansia* (Hu et al. 2019). The antidiabetic activity of *Momordica charantia* (bitter melon) is improved by *Lactobacillus fermentation*, which in turn can increase *Bacteroides caecigallinarwn*, *Bacteroides thetaiotaomicron*, *Prevotella loescheii*, *Prevotella oralis*, *and Prevotella melaninogenica* (Gao et al. 2018). The phlorizin in many fruits could regulate the blood glucose level by reducing serum LPS and insulin resistance, increasing butyric acid as well as increasing *Akkermansia muciniphila* and *Prevotella* (Mei et al. 2016). It has been reported that the extracts from cinnamon bark can improve glucose tolerance and insulin resistance by reducing *Peptococcus*, and the extracts from grape pomace in diabetic mice could decrease *Desulfovibrio* and *Lactococcus*, and increase *Allobaculum* and *Roseburia* (Van Hul et al. 2018). The inulin reduces the effects of T2DM in diabetic mice by increasing *Cyanobacteria* and *Bacteroides* and decreasing *Deferribacteres* and *Tenericutes* (Li et al. 2019a, b).

10.7.6 Inflammatory Bowel Disease (IBD)

IBD is a long-lasting heterogeneous, GI tract associated disorder, characterized by an inflammatory process and materializing in the form of Crohn's disease (CD) or ulcerative colitis (UC) in different patients and differentially diagnosed by clinicians. Factors that could influence host–microbiome homeostasis such as host genetics, host immune system, shifts in diet, exposure to antimicrobials, urbanization, westernization (Statovci et al. 2017) of life style, and many more could predispose and trigger IBD.

What's the connect? IBD was one of the three conditions that was explored in a longitudinal study of a cohort of 132 individuals over a period of 1 year in the Human Microbiome Program (iHMP) and is the most comprehensive study till date (Lloyd-Price et al. 2017). Interestingly, the dysbiosis consisted of changes in transcription profile of several microbial species and associated changes in host biochemical parameters such as considerable transition in acylcarnitine pools and bile acids and heightened levels of serum antibody. IBD patients go through a cycle of dysbiosis and non-dysbiosis phases. Taxonomic perturbations during dysbiosis included reduction in obligate anaerobes such as *Faecalibacterium prausnitzii* and *Roseburia hominis* and the increase in facultative anaerobes such as *E. coli* in Crohn's disease.

Therapeutics that help: Probiotics, especially Bifidobacteria, are linked to suppression of mucosal inflammation in animal models of IBD. In humans, however, the evidence for probiotic efficacy is less positive for maintenance of remission of CD, based on meta-analysis of eight clinical trials, mainly due to variance in study design. Inclusion of probiotics with conventional treatment for UC did not improve the overall remission rates in mild-to-moderate UC, but it was possible to generate a slight decrease in disease activity. The mixture VSL#3 (composed of four lactobacillus strains; three from bifidobacteria and *S. thermophilus*) has shown positive effects in UC treatment, whereas indications for probiotic efficacy in CD are low. At the current stage, a conclusion has been reached that use of probiotics in IBD cannot be recommended. Studies have been conducted using a diversity of bacterial strains in different clinical situations, but only a few patients have been enrolled in these studies. Hence, more randomized trials with statistically sound study design may give us better clarity (Jadhav et al. 2020).

Among microbially derived products that have been shown to be protective in colitis, polysaccharide A from *Bacteroides fragilis* and butyrate from Clostridial spp. induce peripheral regulatory T cells. Favorable effects of FMT on UC (compared to control treatment) have been published. However, there are many parameters such as donor selection, administration routes, frequency of FMT and the development of easy-to-administer formulation, that need to be optimized and validated before FMT is ready to be offered as a standard therapy for UC.

10.7.7 Irritable Bowel Syndrome (IBS)

IBS is a disorder associated with the gastrointestinal tract and is marked by a long-lasting, persistent discomfort and pain in abdomen and, with altered bowel behaviors. Mainly based on the stool pattern, IBS patients can be categorized into (1) IBS with constipation (IBS-C), (2) IBS with diarrhea (IBS-D), (3) IBS with mixed bowel habits (IBS-M), and (4) unclassified IBS. The other associated comorbidities linked with IBS are profoundly psychiatric in nature such as depression, anxiety, and somatoform disorders (Chong et al. 2019).

What's the connect? In a healthy gut, the mucus epithelium and homeostatic immune responses limit microbes both symbiotic and commensal from breaching the epithelial barrier surface. Nevertheless, when the influx of any agent breaches the barrier, provoking intense immune reactions, it leads to severe inflammation. This in turn impacts the intestinal environment including the alteration in the composition of gut microbiota (Pédron et al. 2016). Such underlying conditions with wider consequences on the gut neuromuscular junction and gut–brain axis could contribute to IBS pathogenesis. Gut dysbiosis accompanying the changes from beneficial bacteria to pathogens in the human gut has been reported (Carroll et al. 2011, 2012). It is not surprising that IBS patients are highly deficient in the beneficial activities of *Lactobacilli* and *Bifidobacteria* (Bellini et al. 2014). Interestingly, there seems to be a direct correlation between the severity of IBS with low levels of exhaled methane, decreased microbial richness, absence of *Methanobacteriales,* and enrichment with *Bacteroides* enterotypes compared to that in controls (Tap et al. 2017).

Disease conditions with altered gut mycobiome profile are attracting the attention of researchers, especially in immunocompromised patients. IBS is another condition where the connection of microbiome alterations and the occurrence of visceral hypersensitivity indicate the role of gut microbiome in the disease pathogenesis (Botschuijver et al. 2017).

Therapeutics that help: Probiotics specially *Lactobacillus plantarum* is found to be most beneficial in IBS by improving visceral sensitivity, intestinal permeability, and inflammation (Ohman and Simrén 2013). Maintaining a food and symptom diary can help patients determine which foods trigger symptoms. Specialized diets such as where fermentable oligo-, di- and monosaccharide, and polyol (FODMAPs) are less may improve symptoms in individual IBS patients, though the safety of this diet such as the possibility of malnutrition needs to be monitored in the long term (Ferreira et al. 2020).

10.7.8 Fecal Microbial Transplantation in *Clostridium difficile* Infections

The commensal microbes of the gut have the capacity to defend against pathogenic invasion by either competing for resources or waging microbial warfare. Consumption of prescribed broad-spectrum antibiotics while controlling the 'bad infection' eliminates these beneficial commensals from the gut. Some individuals are then predisposed to opportunistic pathogens such as *Clostridium difficile* resulting in diarrhea and associated symptoms. Van Nood in 2014 reviewed the efficacy of FMT in treating a variety of dysbiotic states (van Nood et al. 2014). It is encouraging that FMT treatment showed around 90% success rate in treating *C. difficile*-induced severe GI dysbiosis. In a patient with persistent *C. difficile*, associated with diarrhea, the fecal microbiota both 2 weeks and 1-month post-FMT consisted mostly of the bacteria from the healthy donor and normalized the patient's bowel function, thus demonstrating the success of the treatment. Until systematic trials evaluate both safety and effectiveness of FMT, its efficacy is currently on a case-by-case basis. In order to enhance treatment reliability and mitigate safety concerns, it is important to identify the minimal subset of microbes needed to achieve therapeutic efficacy and formulate the treatment using appropriate guidelines (Petrof et al. 2013).

10.7.9 Chronic Kidney Disease

Chronic kidney disease (CKD) is emerging health problem, and its occurrence is due to important risk factors like diabetes, hypertension, and obesity affecting about 10% of the population worldwide (Hall et al. 2014). The microbiome has recently gained lot of importance and known to impart an important contribution in health and disease (Hobby et al. 2019).

What's the connect: Renal failure may occur due to high levels of urea in blood, and in the presence of intestinal bacteria, it is converted to ammonia by urease produced by intestinal bacteria leading to overgrowth of bacterial families. Patients with end-stage renal disease (ESRD) have shown to have an expansion of uricase and indole-forming and p-cresyl-forming enzyme-producing bacteria compared to healthy controls (Wong et al. 2014). The most extensively studied gut-derived uremic toxins, i.e., p-cresyl sulfate and indoxyl sulfate, arise from the colon with increased concentration and decline in renal function (Aronov et al. 2011; Ramezani and Raj 2014). In one study, the patients with ESRD have shown largest increase in *Actinobacteria, Firmicutes*, and *Proteobacteria* compared to healthy controls (Vaziri et al. 2013). *Bifidobacterium catenulatum, Bifidobacterium longum, Bifidobacterium bifidum, Lactobacillus plantarum, Lactobacillus paracasei,* and *Klebsiella pneumoniae* (Wang et al. 2012) are hardly seen in patients on peritoneal dialysis.

Research from different groups have shown that patients undergoing hemodialysis have higher number of *Enterobacteria* and *Enterococci,* facultative anaerobic bacteria, and lower abundance particularly in the genera *Bifidobacteria* with a related higher abundance of *Clostridium perfringens* than healthy controls (Kieffer et al. 2016).

In CKD and ESRD, patients who have reduced intake of fiber and decreased colonic time have encountered comorbidities such as diabetes (Yasuda et al. 2002). The disintegration of the epithelial barrier of the gut is mainly due to ammonia and ammonium hydroxide derived from urea which allows absorption of the toxins produced by microbes, thus leading to systemic inflammation which lead to anemia, protein wasting, and cardiovascular disease (Wong et al. 2014).

Therapeutics that help: In rats with reduced renal function, AST-120, an oral adsorbent, helps in restoring the epithelial tight junction proteins by absorbing uremic toxins produced in gut and thus reducing the levels of endotoxin and markers of oxidative stress and inflammation (Redman et al. 2014). There may be increase in SCFA-producing bacteria due to increased intake of resistant starches reducing loss of renal function, interstitial fibrosis, renal tubular damage, and activation of pro-inflammatory molecules (Vaziri et al. 2014).

The serum levels of indoxyl sulfate have been reduced due to administration of *B. Longum orally* (Takayama et al. 2003). In hemodialysis patients, plasma levels of indoxyl sulfate have been reduced due to increase in dietary fiber (Sirich et al. 2014). In vitro studies and in clinical studies, indoxyl sulfate has been reduced due to *Streptococcus thermophilus* (Vitetta et al. 2019).

The metabolism of gut microbiota of CKD patients can be modulated by dietary fiber which helps in improving CKD by reducing uremic toxins and also improves uremic symptoms and comorbidities in dialysis patients. However, more clinical studies are needed to be carried out in order to evaluate the multidimensional benefits of a fiber-rich diet in CKD and to determine the effect of different kinds of fiber in terms of quality as well as quantity (Camerotto et al. 2019).

10.7.10 Cancer

What's the connect? Gut microbiome has played a significant role in tumor development and effective anticancer therapies (Heshiki et al. 2020). Case in point is the gut flora metabolizing bile acids which then redistribute and regulate the employment of natural killer T cells to tumorous liver cells (Ma et al. 2018). Another example is those gut microbes which metabolize estrogen and hence potentially altering the risk of postmenopausal estrogen receptor-positive breast cancer (Kwa et al. 2016). Another case would be the microbes which take glucocorticoids, present in the gut and urinary tract of men (leading to prostate cancer), and metabolize it to produce 11-oxyandrogens (Devendran et al. 2017, 2018; Zimmermann et al. 2019).

Gut microbiota can act as a tumor promoter (Vivarelli et al. 2019). *Helicobacter pylori* accounts for 90% of gastric cancers and is also considered to be a component of stomach microbiome (Xavier et al. 2020). CagA protein (cytotoxin associated gene A) from *Helicobacter pylori* was the first bacteria-derived protein to have a role in human cancer (Hatakeyama 2017). It assists in increase of gastric cancer by inducing the breakdown of p53 in gastric epithelial cells, by interfering with the host's AKT pathway (Buti et al. 2011). Beside this CagA protein, there are others as well such as *Fusobacterium nucleatum*-derived effector adhesin A (FadA) and *Bacteroides fragilis*-derived metalloproteinase toxin (MP toxin) which promote cell proliferation and cancerogenic transformation of affected host's cells by interacting with epithelial E-cadherin of host, directly or indirectly hampering the intercellular junctions and triggering β-catenin signaling (Murata-Kamiya et al. 2007; Rubinstein et al. 2013; Wu et al. 2007).

Similarly, Salmonella enterica effector avirulence protein A (AvrA) with its intrinsic deubiquitinase activity activates β-catenin thereby enhancing cell proliferation and promoting colonic tumorigenesis (Lu et al. 2014). *Escherichia coli*-derived colibactin and cytolethal distending toxin (CDT), when released near gastrointestinal epithelium, release toxins. This creates break in double-stranded DNA within the host's epithelial cells, thus inducing a temporary cell cycle arrest, contributing to genomic instability which ultimately leads to tumor initiation and progression in those predisposed cells (Lara-Tejero 2000). This is generally encountered during the dysbiosis of gut microbiome led by pathogenic infections (Halazonetis 2004; Frisan 2016).

H. pylori colonization may harm human health by causing gastroesophageal reflux disease and its sequelae, Barrett's esophagus, and adenocarcinoma of the esophagus (Blaser and Atherton 2004) and benefit human health (by protecting against asthma, multiple sclerosis, and IBD (Chen et al. 2017; Kira and Isobe 2019; Piovani et al. 2019; Gravina et al. 2018; Hosseininasab Nodoushan and Nabavi 2019).

There are bacteria which block tumorigenesis inhibiting immune effectors and thus induce cancer formation. *Fusobacterium nucleatum* indirectly helps in onset of cancer by inhibiting its own hosts natural killer (NK) cells (Gur et al. 2015). Few microbiota may obstruct the host hormones metabolism. *Clostridium leptum* and *Clostridium coccoides* secrete β-glucuronidase and increased level of this enzyme when coupled with gut dysbiosis results in deconjugating liver-catabolized and plant-based estrogens. This in turn allows to bind and trigger the estrogen receptors leading to cell proliferation in estrogen related tissues, i.e., as breast and endometrium. There is a heightened risk of breast cancer due to intake of estrogen hormones (Fernández et al. 2018).

Some gut microbiota act as a tumor suppressor. Microbial-derived SCFAs like butyrate and propionate are able to exhibit an anticancer effect by inhibiting host's tumor cells histone deacetylases (Vivarelli et al. 2019). Case in point is the extensively studied LPS (bacterial lipopolysaccharide), which triggers the surface toll-like receptor 4 (TLR4) on the host's cell. This starts a immune response (T-cell mediated) against cancer cells (Paulos et al. 2007). Similarly, the *Salmonella enterica*-derived monophosphoryl lipid A (MPL) is presently used in the vaccine as adjuvant, against anticervical carcinoma (Paavonen et al. 2009). Furthermore, pyridoxine, a group B vitamin from bacteria, is capable of stimulating antitumoral immunosurveillance in host (Aranda et al. 2015). The chapter dedicated to the role of the microbiome in immunotherapy will shed more light on that aspect.

Western diet can cause increase in intestinal tumors due to high animal fat and protein which increases bile secretion into the GI tract (Goncalves et al. 2019). *Bilophila wadsworthia* metabolizes taurine into acetate and ammonia, along with release of hydrogen sulfide, a carcinogenic gas (Ridlon et al. 2006). In colorectal cancer, *Fusobacterium nucleatum* is one of the most predominant bacteria (Bullman et al. 2017).

Therapeutics that help: The use of probiotics synergistically with pain killers such as opioids is being studied (Rousseaux et al. 2007). Many probiotics have shown a potential antineoplastic activity. For example, ferrichrome, a *Lactobacillus casei*-derived metabolite, is able to trigger apoptosis in tumor cells via direct activation of JNK pathway (Konishi et al. 2016). It has been appeared in several studies that *Lactobacilli* may eliminate cancerous or precancerous cells by stimulation of immune cells like NK cells or dendritic cells (DC) or TH1 response in the host (Lenoir et al. 2016; Lee et al. 2004; Baldwin et al. 2010; Takagi et al. 2008). *Lactobacillus casei* when orally administered reduces the reappearance of superficial bladder cancer (Aso and Akazan 1992). *Lactobacillus rhamnosus* GG (LGG) is a gut-resident bacterium which modulates several proliferation pathways of host, for example, mTOR or WNT (Taherian-Esfahani et al. 2016), and thus wield either antiproliferative effects or antimetastatic effects (Orlando et al. 2016; Nouri et al. 2016; Zhao et al. 2017; Behzadi et al. 2017; Chen et al. 2017). This can influence immune system of host, consequently helping in eliminating newly developing cancer cells (Gamallat et al. 2016). LGG can exert an anti-inflammatory profile by triggering an immune response within the normal untransformed gut epithelium (Suzuki et al. 2017) as well as changing gene expression in intestinal porcine epithelial cells and intestine myofibroblasts (Taranu et al. 2018; Uribe et al. 2018). LGG can serve as an apt candidate as a possible adjuvant in integrated anticancer therapies (Vivarelli et al. 2019).

The main purpose of giving probiotics especially *Lactobacilli*, to cancer patients, is to reestablish the levels and function of the beneficial bacteria which gets exhausted after the treatments, thus restoring the microbiota balance in the patients' gut (Zitvogel et al. 2018). At the same time, probiotics may possess a likely risk of development of opportunistic bacteria/infection and exhibit antibiotic resistance when administered to immunocompromised cancer patients (Vanderhoof and Young 2008; Redman et al. 2014).

In 2014, it was observed in a double-blind control trial that Bifilact (*Lactobacillus acidophilus* LAC361 and *Bifidobacterium longum* BB536) administration could considerably reduce both moderate and severe diarrhea induced by the pelvic radiation treatment (Demers et al. 2014). On similar front, in 2015, a probiotic formulation *Colon Dophilus* (a cocktail of ten different strains) was evaluated in a clinical trial in metastatic CRC patients treated with irinotecan-based chemotherapy. Prevention of diarrhea in such patients suggested that the administration of such probiotics facilitates in the reduction of the occurrence as well as severity of diarrhea and gastrointestinal toxicity induced by the chemotherapy and is considered to be safe (Mego et al. 2015). In 2016, it was observed that probiotic *Saccharomyces boulardii* has the ability to downregulate pro-inflammatory cytokines in treated patients, with no beneficial effects on the postoperative infection rates (Consoli et al. 2016). In 2017, it was observed that the epigenetic patterns of tumor tissue from its baseline can be altered, with potential therapeutic benefits by the administration of *Bifidobacterium lactis* and *Lactobacillus acidophilus* to CRC patients (Hibberd et al. 2017).

10.7.11 Mental Disorders

Mood disorders, such as depression and bipolar disorder (BD), have known to cause significant individual and socioeconomic burdens affecting around 10% of the world's population (Wittchen 2012). When compared to the normal population, people with mood disorder incline toward higher mortality rates and reduced life span (Angst et al. 1999; Kessing et al. 2015).

What's the connect? The human gut has trillions of bacterial which are known to play a critical role in communication between gut and brain via influencing neural, immune, and endocrine pathways (Dupont et al. 2020). Patients with a variety of psychiatric disordered such as BD, schizophrenia, depression, and autism spectrum disorder display a significant difference in the composition of the gut microbiome. Thus, there is renewed interest in treating the gut microbiome as a potential therapeutic target in psychiatry. Psychobiotics, a recently coined term, refers to favorable microbes that may provide value in diagnosed mental disorders cases. The term refers to '*a live microorganism that, when taken in adequate amounts, produces a beneficial health benefit in patients suffering from psychiatric illness*' (Dinan et al. 2013). The research on gut microbiome related to mood disorders is still in its nascent stage. Although growing evidence suggest that alteration in the gut microbiome plays a significant role in patients with mood disorders, the cause–effect relationship is yet to be determined. In major depressive disorder (MDD) and bioplar disorder, a consistent surge in the abundance of Actinobacteria and Enterobacteriaceae and a decline in Faecalibacterium were observed in different studies (Sowa-Kućma et al. 2018). Raised levels of Clostridium sp. have constantly been observed in patients with autism spectrum disorder. These discoveries imply that in patients with mood disorders, a decrease in bacterial genera which produces short-chain fatty acids and an increase in pro-inflammatory genera can be linked to chronic, low-grade systemic inflammation. On the other hand, till now, a clear relation between gut microbiota composition and anxiety disorders has been not been reported (Groen et al. 2018).

Exploring microbiome therapeutics: Changing the diet and incorporating probiotics/prebiotics in order to enhance the beneficial bacteria in the gut in both healthy and patient groups are likely to improve mood and decrease anxiety. Diet and food habit/pattern lead to the formation of gut microbiota which in turn regulate the host inflammation and thrombosis, causing brain disorders (Zhu et al. 2020a, b). For depression and other mental disorders, omega-3 fatty acids, ions such as zinc and magnesium, and plant phytochemicals are found to be more relevant nutrients (Kaplan et al. 2015; Trebatická and Ďuračková 2015; Haider et al. 2015; Bouayed et al. 2007; Solanki et al. 2015). Epidemiological studies have stated that traditional diet styles lead toward better mental health and hence reduce the risk of depression (Jacka et al. 2011; Jacka et al. 2010; Sánchez-Villegas et al. 2009; Skarupski et al. 2013; Rienks et al. 2013).

Food rich in polyphenol such as green tea, coffee, cocoa, curcumin, and other polyphenol-rich foods is connected with better mood, less fatigue, and hence reduced risk of depression in humans (Pham et al. 2014; Sathyapalan et al. 2010; Pase et al. 2013; Yu et al. 2015). These foods in preclinical settings prevent dysbiosis by inducing the growth of beneficial bacteria (Seo et al. 2015; Mills et al. 2015; Massot-Cladera et al. 2012; McFadden et al. 2015). It emerges that microbially transformed phytochemicals (e.g., quercetin after it has been subjected to fermentation) can alter the gut microbiota in healthy ways (i.e., growth of bifidobacteria and decrease in the ratio of Firmicutes to Bacteroidetes) (Parkar et al. 2013). Honey is an important diet constituent, high in phytochemicals such as phenolic acid and that of flavonoid families (Alvarez-Suarez et al. 2013), and has antidepressant and anxiolytic properties too (Azman et al. 2015; Mijanur Rahman et al. 2014). Honey flavonoids when stimulated by LPS hinder the release of pro-inflammatory cytokines, e.g., tumor necrosis factor alpha and interleukin-1 beta from microglia in a significant manner (Candiracci et al. 2012).

Free radicals or reactive oxygen species (ROS) lead to oxidative stress and trigger the pro-inflammatory avalanche, which involved IL-6 and CRP, both linked with depression. Several dietary antioxidants such as curcumin, ascorbic acid, carotenoids, and flavonoids are found to diminish depressive symptoms, either directly or linked to a lower occurrence of depression (Christy Harrison 2020).

Fermentation leads to transformation of bioavailable phytochemical structures and peptides which could lead to novel therapeutics (Selhub et al. 2014). It has been reported that consumption of fermented food results in reduced social anxiety, particularly in people with high neuroticism scores (Hilimire et al. 2015). Fermented food has heat-inactivated microorganisms, and their structural parts may also have significant effects on intestinal ecosystems (Yang et al. 2014).

10.7.12 Rheumatoid Arthritis

Rheumatoid arthritis (RA), an autoimmune disorder, is dependent on/controlled by/related to several factors, both genetic and environmental. Since it targets the self-antigens present in the synovium, cartilage, and bone, it leads to destruction of the joints and hence functional disability in most patient (Halpern et al. 2009; Tran et al. 2005). Majority of the drugs available for the treatment of RA today (including DMRDs or disease-modifying antirheumatic drugs) target cytokines, nonspecific immune suppression, or T-cell and B-cell activation (Isaacs 2008; Kumar and Banik 2013; Benjamin et al. 2018). Besides the genetic factors, diet, smoking, and stress are some of the environmental factors which influence the microbiota diversity/composition and hence the onset/outcome of RA.

What's the connect? The onset of autoimmunity may be associated with gastrointestinal tract. It is found that microbial composition in subjects with early rheumatoid arthritis differed from controls, with a reduction of bacteria from the family *Bifidobacterium* and *Bacteroides* (Vaahtovuo et al. 2008), and a noticeable increase of species from the genus *Prevotella* (Bernard 2014). Most of the studies indicated that treatment with MTX and HCQ leads to partial restoration of dysbiotic microbiomes to normal or alteration to increase abundance of beneficial microbial members (Zhang et al. 2015). Methotrexate reduced the abundance of Enterobacteriaceae and partially restored the gut microbiota in patients. Hydroxychloroquine increased the gut microbial species' richness and diversity. Hydroxychloroquine plus doxycycline treatment led to the reduction in abundance of phylum Bacteroidetes and Firmicutes (Bodkhe et al. 2019).

Recent studies in humans and mouse models of arthritis indicate that periodontal disease is linked with a heightened risk of RA (Scher et al. 2012; Arvikar et al. 2013; de Aquino et al. 2014). The presence of periodontitis is associated with anti-CCP antibody levels in RA patients (Dissick et al. 2010). Treatment of periodontitis enhances the disease activity of RA (Ortiz et al. 2009; Al-Katma et al. 2007). It is found that periodontal bacteria such as *P. Gingivalis* and *A. actinomycetemcomitans* induce the production of anti-CCP antibodies, which in turn leads to the arthritis thus suggesting the link of periodontal bacteria with RA pathogenesis. Another evidence comes from the finding that DNA of oral cavity bacteria such as *Fusobacterium nucleatum* and *Porphyromonas gingivalis* has been detected in the synovial fluid of RA patients (Reichert et al. 2013; Stephanie et al. 2012). In order to establish the mechanistic link of the oral cavity bacteria with that of development of RA in humans, more studies are needed. Oral microbiome and periodontal disease are described in another chapter in further detail.

Exploring microbiome therapeutics: Randomized controlled trials with probiotic bacteria like Lactobacillus rhamnosus, (Pineda et al. 2011; Hatakka et al. 2003), Lactobacillus casei (Vaghef-Mehrabany et al. 2014), Bacillus coagulans (Mandel et al. 2010), Lactobacillus reuteri, Lactobacillus acidophilus, and Bifidobacterium bifidum have been carried out to determine their ability to treat RA. Randomized controlled trials with probiotic bacteria have been studied for their ability to treat RA, like *Lactobacillus rhamnosus*, (Pineda et al. 2011; Hatakka et al. 2003)

Lactobacillus casei (Vaghef-Mehrabany et al. 2014), *Bacillus coagulans* (Mandel et al. 2010) *Lactobacillus reuteri, Lactobacillus acidophilus*, and *Bifidobacterium bifidum*. These probiotics have appeared to be effective for RA patients and are considered safe (Zamani et al. 2016). However, since there is limited data on the interaction between drug and microbiomes and probiotics for RA, further studies are needed for a probiotic to be used for immune homeostasis. This could be helpful in personalized medicine where a tailor-made probiotic (targeted organism/metabolite which is in low abundance) can be supplemented to create immune homeostasis in patients.

10.8 Conclusion

Acknowledging the impact that the microbiome has on human health and well-being by scientific and nonscientific communities has allowed us to get curious and explore this area in our own way. The existing approaches of microbiome diagnostics and therapeutics need to evolve keeping in pace with current technologies. New microbiome diagnostic marker panels developed based on Next-Gen sequencing can be used to diagnose (and potentially prognos) disease by clinicians. Predictive medicine using AI/ML algorithm-based analysis of patient's microbiome could predict the outcome of treatment options. To revolutionize microbiome therapeutics, the treatment options should move from general to personalized ones. Such personalized treatment strategy can be created based on the gut microbiome of an individual (patient-specific). The treatment options can range from 'precision probiotics' where a specific microbial cocktail can be administered to alter the species richness or 'precision prebiotics' where targeted microbial nutrition is given to promote growth of certain microbes or 'personalized dietary interventions' or 'targeted subtractive theory' (Fig. 10.5). Microbiome time series analysis to follow-up treatment success and establishment of normobiosis will allow early intervention and tweaking of treatment plans for better patient outcomes. Nevertheless, though the challenges are many, it is achievable. It will be thrilling to witness this microbiome-based therapeutic model being adopted in routine clinical practice and becoming a true translational discipline in the near future!

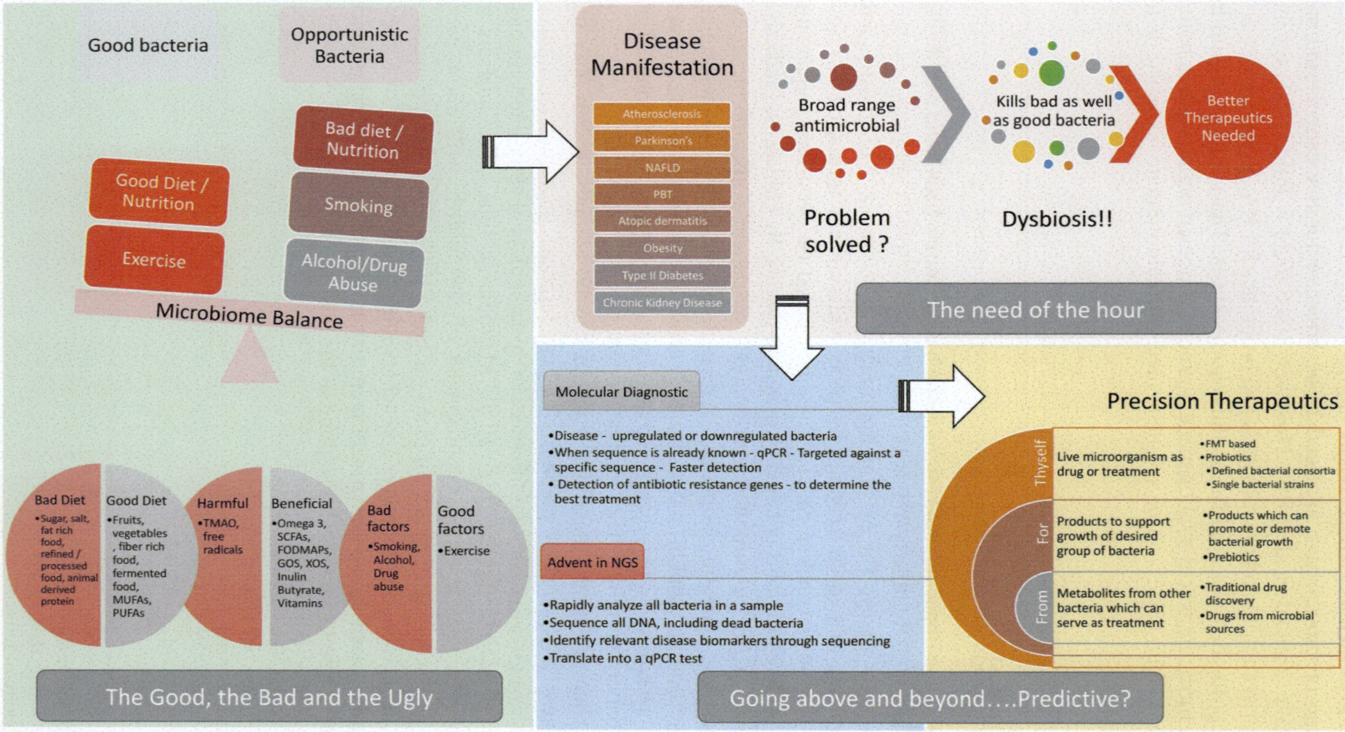

Fig. 10.5 A schematic illustration showing various aspects associated with microbiome. Microbiome balance is affected by many factors (left panel) and leads to disease condition (top-right panel). The use of broad range antimicrobial may lead to further disbalance the microbiome by killing the good bacteria along with the targeted bad ones. Microbiome diagnostic (bottom-left panel) can reveal the upregulated/downregulated microbes in disease condition. The advent in sequencing leads to better diagnostic, and the better the diagnostic, the better will be the therapeutics (bottom-left panel). Therapeutics involves the microbes either directly (live microbes) or indirectly (metabolites from bacteria or products to support the growth of good bacteria). (Allaband et al. 2019; Espinoza 2018; FitzGerald and Spek 2020; Raes 2016)

References

Aagaard K, Petrosino J, Keitel W, Watson M, Katancik J, Garcia N, Patel S, Cutting M, Madden T, Hamilton H, Harris E, Gevers D, Simone G, McInnes P, Versalovic J (2013) The human microbiome project strategy for comprehensive sampling of the human microbiome and why it matters. FASEB J 27:1012–1022

Alfano A, Perillo F, Fusco A, Savio V, Corsaro MM, Donnarumma G, Schiraldi C, Cimini D (2020) Lactobacillus brevis CD2: fermentation strategies and extracellular metabolites characterization. Probiotics Antimicrob Proteins 12(4):1542–1554

Al-Katma MK, Bissada NF, Bordeaux JM, Sue J, Askari AD (2007) Control of periodontal infection reduces the severity of active rheumatoid arthritis. JCR J Clin Rheumatol 13:134–137

Allaband C, McDonald D, Vázquez-Baeza Y, Minich JJ, Tripathi A, Brenner DA, Loomba R, Smarr L, Sandborn WJ, Schnabl B, Dorrestein P, Zarrinpar A, Knight R (2019) Microbiome 101: studying, analyzing, and interpreting gut microbiome data for clinicians. Clin Gastroenterol Hepatol 17(2):218–230

Alneberg J, Bjarnason BS, de Bruijn I, Schirmer M, Quick J, Ijaz UZ, Lahti L, Loman NJ, Andersson AF, Quince C (2014) Binning metagenomic contigs by coverage and composition. Nat Methods 11:1144–1146

Álvarez-Mercado AI, Navarro-Oliveros M, Robles-Sánchez C, Plaza-Díaz J, Sáez-Lara MJ, Muñoz-Quezada S, Fontana L, Abadía-Molina F (2019) Microbial population changes and their relationship with human health and disease. Microorganisms 7:1–27

Alvarez-Suarez J, Giampieri F, Battino M (2013) Honey as a source of dietary antioxidants: structures, bioavailability and evidence of protective effects against human chronic diseases. Curr Med Chem 20:621–638

Amar J, Burcelin R, Ruidavets JB, Cani PD, Fauvel J, Alessi MC, Chamontin B, Ferriéres J (2008) Energy intake is associated with endotoxemia in apparently healthy men. Am J Clin Nutr 87:1219–1223

Ang Z, Ding JL (2016) GPR41 and GPR43 in obesity and inflammation–protective or causative? Front Immunol 7

Angst J, Angst F, Stassen HH (1999) Suicide risk in patients with major depressive disorder. J Clin Psychiatry 60(2):57–62. discussion 75-76, 113–116

Aoki R, Kamikado K, Suda W, Takii H, Mikami Y, Suganuma N, Hattori M, Koga Y (2017) A proliferative probiotic Bifidobacterium strain in the gut ameliorates progression of metabolic disorders via microbiota modulation and acetate elevation. Sci Rep 7:43522

Aranda F, Bloy N, Pesquet J, Petit B, Chaba K, Sauvat A, Kepp O, Khadra N, Enot D, Pfirschke C, Pittet M, Zitvogel L, Kroemer G, Senovilla L (2015) Immune-dependent antineoplastic effects of cisplatin plus pyridoxine in non-small-cell lung cancer. Oncogene 34:3053–3062

Aronov PA, Luo FJ-G, Plummer NS, Quan Z, Holmes S, Hostetter TH, Meyer TW (2011) Colonic contribution to uremic solutes. J Am Soc Nephrol 22:1769–1776

Arvikar SL, Collier DS, Fisher MC, Unizony S, Cohen GL, McHugh G, Kawai T, Strle K, Steere AC (2013) Clinical correlations with Porphyromonas gingivalis antibody responses in patients with early rheumatoid arthritis. Arthritis Res Ther 15:R109

Asemi Z, Alizadeh S-A, Ahmad K, Goli M, Esmaillzadeh A (2016) Effects of beta-carotene fortified synbiotic food on metabolic control of patients with type 2 diabetes mellitus: a double-blind randomized cross-over controlled clinical trial. Clin Nutr 35:819–825

Asemi Z, Khorrami-Rad A, Alizadeh S-A, Shakeri H, Esmaillzadeh A (2014) Effects of synbiotic food consumption on metabolic status of diabetic patients: a double-blind randomized cross-over controlled clinical trial. Clin Nutr 33:198–203

Asemi Z, Zare Z, Shakeri H, Sabihi S, Esmaillzadeh A (2013) Effect of multispecies probiotic supplements on metabolic profiles, hs-CRP, and oxidative stress in patients with type 2 diabetes. Ann Nutr Metab 63:1–9

Aso Y, Akazan H (1992) Prophylactic effect of a Lactobacillus casei preparation on the recurrence of superficial bladder cancer. Urol Int 49:125–129

Azman K, Othman Z, Zakaria R, Abd Aziz C, Al-Rahbi B (2015) Tualang honey improves memory performance and decreases depressive-like behavior in rats exposed to loud noise stress. Noise Health 17:83

Bäckhed F, Fraser CM, Ringel Y et al (2012) Defining a healthy human gut microbiome: current concepts, future directions, and clinical applications. Cell Host Microbe 12(5):611–622

Baik I (2018) Forecasting obesity prevalence in Korean adults for the years 2020 and 2030 by the analysis of contributing factors. Nutr Res Pract 12:251

Balamurugan R, George G, Kabeerdoss J, Hepsiba J, Chandragunasekaran AMS, Ramakrishna BS (2010) Quantitative differences in intestinal Faecalibacterium prausnitzii in obese Indian children. Br J Nutr 103:335–338

Baldwin C, Millette M, Oth D, Ruiz MT, Luquet F-M, Lacroix M (2010) Probiotic Lactobacillus acidophilus and L. casei mix sensitize colorectal tumoral cells to 5-fluorouracil-induced apoptosis. Nutr Cancer 62:371–378

Baothman OA, Zamzami MA, Taher I, Abubaker J, Abu-Farha M (2016) The role of gut microbiota in the development of obesity and diabetes. Lipids Health Dis 15:108

Barathikannan K, Chelliah R, Rubab M, Daliri EB-M, Elahi F, Kim D-H, Agastian P, Oh S-Y, Oh DH (2019) Gut microbiome modulation based on probiotic application for anti-obesity: a review on efficacy and validation. Microorganisms 7:456

Barengolts E (2016) Gut microbiota, prebiotics, probiotics, and synbiotics in management of obesity and prediabetes: review of randomized controlled trials. Endocr Pract 22:1224–1234

Behzadi E, Mahmoodzadeh Hosseini H, Imani Fooladi AA (2017) The inhibitory impacts of lactobacillus rhamnosus GG-derived extracellular vesicles on the growth of hepatic cancer cells. Microb Pathog 110:1–6

Bell DSH (2015) Changes seen in gut bacteria content and distribution with obesity: causation or association? Postgrad Med 127:863–868

Bellini M, Gambaccini D, Stasi C, Urbano MT, Marchi S, Usai-Satta P (2014) Irritable bowel syndrome: a disease still searching for pathogenesis, diagnosis and therapy. World J Gastroenterol 20(27):8807–8820

Benjamin EJ, Virani SS, Callaway CW, Chamberlain AM, Chang AR, Cheng S, Chiuve SE, Cushman M, Delling FN, Deo R, de Ferranti SD, Ferguson JF, Fornage M, Gillespie C, Isasi CR, Jiménez MC, Jordan LC, Judd SE, Lackland D, Lichtman JH, Lisabeth L, Liu S, Longenecker CT, Lutsey PL, Mackey JS, Matchar DB, Matsushita K, Mussolino ME, Nasir K, O'Flaherty M, Palaniappan LP, Pandey A, Pandey DK, Reeves MJ, Ritchey MD, Rodriguez CJ, Roth GA, Rosamond WD, Sampson UKA, Satou GM, Shah SH, Spartano NL, Tirschwell DL, Tsao CW, Voeks JH, Willey JZ, Wilkins JT, Wu JH, Alger HM, Wong SS, Muntner P (2018) Heart disease and stroke statistics—2018 update: a report from the American Heart Association. Circulation 137

Bernard NJ (2014) Prevotellacopri associated with new-onset untreated RA. Nat Rev Rheumatol 10:2–2

Blaser MJ, Atherton JC (2004) Helicobacter pylori persistence: biology and disease. J Clin Invest 113:321–333

Bodkhe R, Balakrishnan B, Taneja V (2019) The role of microbiome in rheumatoid arthritis treatment. Ther Adv Musculoskelet Dis 11:1759720X1984463

Botschuijver et al (2017) Intestinal fungal dysbiosis is associated with visceral hypersensitivity in patients with irritable bowel syndrome and rats. Gasteroenterology 153(4):1026–1039

Bouayed J, Rammal H, Dicko A, Younos C, Soulimani R (2007) Chlorogenic acid, a polyphenol from Prunus domestica (Mirabelle), with coupled anxiolytic and antioxidant effects. J Neurol Sci 262:77–84

Boulanger C, Lüscher TF (1990) Release of endothelin from the porcine aorta inhibition by endothelium-derived nitric oxide. J Clin Invest 85:587–590

Boursier J, Mueller O, Barret M, Machado M, Fizanne L, Araujo-Perez F, Guy CD, Seed PC, Rawls JF, David LA, Hunault G, Oberti F, Calès P, Diehl AM (2016) The severity of nonalcoholic fatty liver disease is associated with gut dysbiosis and shift in the metabolic function of the gut microbiota. Hepatology 63:764–775

Brahe LK, Astrup A, Larsen LH (2016) Can we prevent obesity-related metabolic diseases by dietary modulation of the gut microbiota? Adv Nutr 7:90–101

Brandt LJ (2013) FMT: first step in a long journey. Am J Gastroenterol 108:1367–1368

Brantsaeter AL, Myhre R, Haugen M, Myking S, Sengpiel V, Magnus P, Jacobsson B, Meltzer HM (2011) Intake of probiotic food and risk of preeclampsia in primiparous women: the Norwegian mother and child cohort study. Am J Epidemiol 174:807–815

Brinkworth GD, Noakes M, Clifton PM, Bird AR (2009) Comparative effects of very low-carbohydrate, high-fat and high-carbohydrate, low-fat weight-loss diets on bowel habit and faecal short-chain fatty acids and bacterial populations. Br J Nutr 101:1493

Bullman S, Pedamallu CS, Sicinska E, Clancy TE, Zhang X, Cai D, Neuberg D, Huang K, Guevara F, Nelson T, Chipashvili O, Hagan T, Walker M, Ramachandran A, Diosdado B, Serna G, Mulet N, Landolfi S, Ramon Y, Cajal S, Fasani R, Aguirre AJ, Ng K, Elez E, Ogino S, Tabernero J, Fuchs CS, Hahn WC, Nuciforo P, Meyerson M (2017) Analysis of Fusobacterium persistence and antibiotic response in colorectal cancer. Science 358:1443–1448

Buti L, Spooner E, Van der Veen AG, Rappuoli R, Covacci A, Ploegh HL (2011) Helicobacter pylori cytotoxin-associated gene a (cag a) subverts the apoptosis-stimulating protein of p 53 (ASPP2) tumor suppressor pathway of the host. Proc Natl Acad Sci 108:9238–9243

Camerotto C, Cupisti A, D'Alessandro C, Muzio F, Gallieni M (2019) Dietary fiber and gut microbiota in renal diets. Nutrients 11:2149

Candiracci M, Piatti E, Dominguez-Barragán M, García-Antrás D, Morgado B, Ruano D, Gutiérrez JF, Parrado J, Castaño A (2012) Anti-inflammatory activity of a honey flavonoid extract on lipopolysaccharide-activated N13 microglial cells. J Agric Food Chem 60:12304–12311

Canfora EE, Jocken JW, Blaak EE (2015) Short-chain fatty acids in control of body weight and insulin sensitivity. Nat Rev Endocrinol 11:577–591

Canfora EE, Meex RCR, Venema K, Blaak EE (2019) Gut microbial metabolites in obesity, NAFLD and T2DM. Nat Rev Endocrinol 15:261–273

Cani PD, Amar J, Iglesias MA, Poggi M, Knauf C, Bastelica D, Neyrinck AM, Fava F, Tuohy KM, Chabo C, Waget A, Delmee E, Cousin B, Sulpice T, Chamontin B, Ferrieres J, Tanti J-F, Gibson GR, Casteilla L, Delzenne NM, Alessi MC, Burcelin R (2007) Metabolic endotoxemia initiates obesity and insulin resistance. Diabetes 56:1761–1772

Caporaso JG, Kuczynski J, Stombaugh J, Bittinger K, Bushman FD, Costello EK, Fierer N, Peña AG, Goodrich JK, Gordon JI, Huttley GA, Kelley ST, Knights D, Koenig JE, Ley RE, Lozupone CA, McDonald D, Muegge BD, Pirrung M, Reeder J, Sevinsky JR, Turnbaugh PJ, Walters WA, Widmann J, Yatsunenko T, Zaneveld J, Knight R (2010) QIIME allows analysis of high-throughput community sequencing data. Nat Methods 7:335–336

Carroll IM, Ringel-Kulka T, Keku TO, Chang Y-H, Packey CD, Sartor RB, Ringel Y (2011) Molecular analysis of the luminal- and mucosal-associated intestinal microbiota in diarrhea-predominant irritable bowel syndrome. Am J Physiol-Gastrointest Liver Physiol 301(5):G799–G807

Carroll IM, Ringel-Kulka T, Siddle JP, Ringel Y (2012) Alterations in composition and diversity of the intestinal microbiota in patients with diarrhea-predominant irritable bowel syndrome. Neurogastroenterol Motil 24(6):521–e248

Catry E, Bindels LB, Tailleux A, Lestavel S, Neyrinck AM, Goossens J-F, Lobysheva I, Plovier H, Essaghir A, Demoulin J-B, Bouzin C, Pachikian BD, Cani PD, Staels B, Dessy C, Delzenne NM (2018) Targeting the gut microbiota with inulin-type fructans: preclinical demonstration of a novel approach in the management of endothelial dysfunction. Gut 67:271–283

Cerdó T, García-Santos J, Bermúdez M, Campoy C (2019) The role of probiotics and prebiotics in the prevention and treatment of obesity. Nutrients 11:635

Chambers ES, Preston T, Frost G, Morrison DJ (2018) Role of gut microbiota-generated short-chain fatty acids in metabolic and cardiovascular health. Curr Nutr Rep 7:198–206

Chawla A, Nguyen KD, Goh YPS (2011) Macrophage-mediated inflammation in metabolic disease. Nat Rev Immunol 11:738–749

Chelakkot C, Choi Y, Kim D-K, Park HT, Ghim J, Kwon Y, Jeon J, Kim M-S, Jee Y-K, Gho YS, Park H-S, Kim Y-K, Ryu SH (2018) *Akkermansia muciniphila*-derived extracellular vesicles influence gut permeability through the regulation of tight junctions. Exp Mol Med 50:e450–e450

Chen C, Xun P, Tsinovoi C, He K (2017) Accumulated evidence on helicobacter pylori infection and the risk of asthma. Ann Allergy Asthma Immunol 119:137–145

Chen M, Yi L, Zhang Y, Zhou X, Ran L, Yang J, Zhu J, Zhang Q, Mi M (2016) Resveratrol attenuates trimethylamine- N-oxide (TMAO)-induced atherosclerosis by regulating TMAO synthesis and bile acid metabolism via remodeling of the gut microbiota. MBio 2016:7

Chen Q, Wang T, Li J, Wang S, Qiu F, Yu H, Zhang Y, Wang T (2017b) Effects of natural products on fructose-induced nonalcoholic fatty liver disease (NAFLD). Nutrients 9:96

Chong PP, Chin VK, Looi CY, Wong WF, Madhavan P, Yong VC (2019) The microbiome and irritable bowel syndrome - a review on the pathophysiology, current research and future therapy. Front Microbiol 10:1–23

Cluny NL, Eller LK, Keenan CM, Reimer RA, Sharkey KA (2015) Interactive effects of oligofructose and obesity predisposition on gut hormones and microbiota in diet-induced obese rats. Obesity 23:769–778

Coelho OGL, Cândido FG, de CG AR (2019) Dietary fat and gut microbiota: mechanisms involved in obesity control. Crit Rev Food Sci Nutr 59:3045–3053

Consoli MLD, da Silva RS, Nicoli JR, Bruña-Romero O, da Silva RG, de Vasconcelos GS, Correia MITD (2016) Randomized clinical trial. J Parenter Enteral Nutr 40:1114–1121

Cotillard A, Kennedy SP, Kong LC, Prifti E, Pons N, Le Chatelier E, Almeida M, Quinquis B, Levenez F, Galleron N, Gougis S, Rizkalla S, Batto J-M, Renault P, Doré J, Zucker J-D, Clément K, Ehrlich SD (2013) Dietary intervention impact on gut microbial gene richness. Nature 500:585–588

Cryan JF, Dinan TG (2012) Mind-altering microorganisms: the impact of the gut microbiota on brain and behaviour. Nat Rev Neurosci 13:701–712

Da Silva HE, Teterina A, Comelli EM, Taibi A, Arendt BM, Fischer SE, Lou W, Allard JP (2018) Nonalcoholic fatty liver disease is associated with dysbiosis independent of body mass index and insulin resistance. Sci Rep 8:1466

David LA, Maurice CF, Carmody RN, Gootenberg DB, Button JE, Wolfe BE, Ling AV, Devlin AS, Varma Y, Fischbach MA, Biddinger SB, Dutton RJ, Turnbaugh PJ (2014) Diet rapidly and reproducibly alters the human gut microbiome. Nature 505:559–563

de Aquino SG, Abdollahi-Roodsaz S, Koenders MI, van de Loo FAJ, Pruijn GJM, Marijnissen RJ, Walgreen B, Helsen MM, van den Bersselaar LA, de Molon RS, Campos MJA, Cunha FQ, Cirelli JA, van den Berg WB (2014) Periodontal pathogens directly promote autoimmune experimental arthritis by inducing a TLR2- and IL-1–driven Th17 response. J Immunol 192:4103–4111

De Leon LM, Watson JB, Kelly CR (2013) Transient flare of ulcerative colitis after fecal microbiota transplantation for recurrent Clostridium difficile infection. Clin Gastroenterol Hepatol 11:1036–1038

Del Chierico F, Nobili V, Vernocchi P, Russo A, De Stefanis C, Gnani D, Furlanello C, Zandonà A, Paci P, Capuani G, Dallapiccola B, Miccheli A, Alisi A, Putignani L (2017) Gut microbiota profiling of pediatric nonalcoholic fatty liver disease and obese patients unveiled by an integrated meta-omics-based approach. Hepatology 65:451–464

Delzenne NM, Neyrinck AM, Bäckhed F, Cani PD (2011) Targeting gut microbiota in obesity: effects of prebiotics and probiotics. Nat Rev Endocrinol 7:639–646

Demers M, Dagnault A, Desjardins J (2014) A randomized double-blind controlled trial: impact of probiotics on diarrhea in patients treated with pelvic radiation. Clin Nutr 33:761–767

Depommier C, Everard A, Druart C, Plovier H, Van Hul M, Vieira-Silva S, Falony G, Raes J, Maiter D, Delzenne NM, de Barsy M, Loumaye A, Hermans MP, Thissen J-P, de Vos WM,

Cani PD (2019) Supplementation with Akkermansia muciniphila in overweight and obese human volunteers: a proof-of-concept exploratory study. Nat Med 25:1096–1103

DeSantis TZ, Hugenholtz P, Larsen N, Rojas M, Brodie EL, Keller K, Huber T, Dalevi D, Hu P, Andersen GL (2006) Greengenes, a chimera-checked 16S rRNA gene database and workbench compatible with ARB. Appl Environ Microbiol 72:5069–5072

Devendran S, Méndez-García C, Ridlon JM (2017) Identification and characterization of a 20β-HSDH from the anaerobic gut bacterium Butyricicoccus desmolans ATCC 43058. J Lipid Res 58:916–925

Devendran S, Mythen SM, Ridlon JM (2018) The desA and desB genes from Clostridium scindens ATCC 35704 encode steroid-17, 20-desmolase. J Lipid Res 59:1005–1014

Dhakan DB, Maji A, Sharma AK et al (2019) The unique composition of Indian gut microbiome, gene catalogue, and associated fecal metabolome deciphered using multi-omics approaches. Gigascience 8(3):giz004

Dinan TG, Stanton C, Cryan JF (2013) Psychobiotics: a novel class of psychotropic. Biol Psychiatry 74:720–726

Dissick A, Redman RS, Jones M, Rangan BV, Reimold A, Griffiths GR, Mikuls TR, Amdur RL, Richards JS, Kerr GS (2010) Association of Periodontitis with rheumatoid arthritis: a pilot study. J Periodontol 81:223–230

Dridi B, Henry M, El Khéchine A, Raoult D, Drancourt M (2009) High prevalence of Methanobrevibacter smithii and Methanosphaera stadtmanae detected in the human gut using an improved DNA detection protocol. PLoS One 4(9):e7063

Duda-Chodak A, Tarko T, Satora P, Sroka P (2015) Interaction of dietary compounds, especially polyphenols, with the intestinal microbiota: a review. Eur J Nutr 54:325–341

Duncan SH, Barcenilla A, Stewart CS, Pryde SE, Flint HJ (2002) Acetate utilization and Butyryl coenzyme A (CoA): acetate-CoA transferase in butyrate-producing bacteria from the human large intestine. Appl Environ Microbiol 68:5186–5190

Dupont HL, Jiang Z-D, Dupont AW, Utay NS (2020) The intestinal microbiome in human health and disease. Trans Am Clin Climatol Assoc 131:178–197

Ejtahed HS, Mohtadi-Nia J, Homayouni-Rad A, Niafar M, Asghari-Jafarabadi M, Mofid V (2012) Probiotic yogurt improves antioxidant status in type 2 diabetic patients. Nutrition 28:539–543

El Kaoutari A, Armougom F, Gordon JI, Raoult D, Henrissat B (2013) The abundance and variety of carbohydrate-active enzymes in the human gut microbiota. Nat Rev Microbiol 11(7):497–504

Erickson M, Malin S, Wang Z, Brown J, Hazen S, Kirwan J (2019) Effects of lifestyle intervention on plasma trimethylamine N-oxide in obese adults. Nutrients 11:179

Espinoza JL (2018) Machine learning for tackling microbiota data and infection complications in immunocompromised patients with cancer. J Intern Med 284:189–192

Everard A, Belzer C, Geurts L, Ouwerkerk JP, Druart C, Bindels LB, Guiot Y, Derrien M, Muccioli GG, Delzenne NM, de Vos WM, Cani PD (2013) Cross-talk between Akkermansia muciniphila and intestinal epithelium controls diet-induced obesity. Proc Natl Acad Sci 110:9066–9071

Faith JJ, Guruge JL, Charbonneau M, Subramanian S, Seedorf H, Goodman AL, Clemente JC, Knight R, Heath AC, Leibel RL, Rosenbaum M, Gordon JI (2013) The Long-term stability of the human gut microbiota. Science 341:1237439

Faust K, Sathirapongsasuti JF, Izard J, Segata N, Gevers D, Raes J, Huttenhower C (2012) Microbial co-occurrence relationships in the human microbiome. PLoS Comput Biol 8: e1002606

Feng Z, Long W, Hao B, Ding D, Ma X, Zhao L, Pang X (2017) A human stool-derived Bilophila wadsworthia strain caused systemic inflammation in specific-pathogen-free mice. Gut Pathog 9:59

Fernández M, Reina-Pérez I, Astorga J, Rodríguez-Carrillo A, Plaza-Díaz J, Fontana L (2018) Breast cancer and its relationship with the microbiota. Int J Environ Res Public Health 15:1747

Ferreira AI, Garrido M, Castro-Poças F (2020) Irritable bowel syndrome: news from an old disorder. GE Port J Gastroenterol 27(4):255–268

FitzGerald MJ, Spek EJ (2020) Microbiome therapeutics and patent protection. Nat Biotechnol 38:806–810
Forslund K, Hildebrand F, Nielsen T, Falony G, Le Chatelier E, Sunagawa S, Prifti E, Vieira-Silva S, Gudmundsdottir V, Krogh Pedersen H, Arumugam M, Kristiansen K, Yvonne Voigt A, Vestergaard H, Hercog R, Igor Costea P, Roat Kultima J, Li J, Jørgensen T, Levenez F, Dore J, Bjørn Nielsen H, Brunak S, Raes J, Hansen T, Wang J, Dusko Ehrlich S, Bork P, Pedersen O (2015) Disentangling type 2 diabetes and metformin treatment signatures in the human gut microbiota. Nature 528:262–266
Franzosa EA, Huang K, Meadow JF, Gevers D, Lemon KP, Bohannan BJM et al (2015) Identifying personal microbiomes using metagenomic codes. Proc Natl Acad Sci 112:E2930–E2938
Frisan T (2016) Bacterial genotoxins: the long journey to the nucleus of mammalian cells. Biochim Biophys Acta Biomembr 1858:567–575
Frostegård J (2013) Immunity, atherosclerosis and cardiovascular disease. BMC Med 11:117
Furusawa Y, Obata Y, Fukuda S, Endo TA, Nakato G, Takahashi D, Nakanishi Y, Uetake C, Kato K, Kato T, Takahashi M, Fukuda NN, Murakami S, Miyauchi E, Hino S, Atarashi K, Onawa S, Fujimura Y, Lockett T, Clarke JM, Topping DL, Tomita M, Hori S, Ohara O, Morita T, Koseki H, Kikuchi J, Honda K, Hase K, Ohno H (2013) Commensal microbe-derived butyrate induces the differentiation of colonic regulatory T cells. Nature 504:446–450
Gamallat Y, Meyiah A, Kuugbee ED, Hago AM, Chiwala G, Awadasseid A, Bamba D, Zhang X, Shang X, Luo F, Xin Y (2016) Lactobacillus rhamnosus induced epithelial cell apoptosis, ameliorates inflammation and prevents colon cancer development in an animal model. Biomed Pharmacother 83:536–541
Gao R, Zhu C, Li H, Yin M, Pan C, Huang L, Kong C, Wang X, Zhang Y, Qu S, Qin H (2018) Dysbiosis signatures of gut microbiota along the sequence from healthy, young patients to those with overweight and obesity. Obesity 26:351–361
Garcia-Mantrana I, Selma-Royo M, Alcantara C, Collado MC (2018) Shifts on gut microbiota associated to mediterranean diet adherence and specific dietary intakes on general adult population. Front Microbiol 2018:9
Gauffin Cano P, Santacruz A, Moya Á, Sanz Y (2012) Bacteroides uniformis CECT 7771 ameliorates metabolic and immunological dysfunction in mice with high-fat-diet induced obesity. PLoS One 7:e41079
Gevers D, Knight R, Petrosino JF, Huang K, McGuire AL, Birren BW, Nelson KE, White O, Methé BA, Huttenhower C (2012a) The human microbiome project: a community resource for the healthy human microbiome. PLoS Biol 10:e1001377
Gevers D, Pop M, Schloss PD, Huttenhower C (2012b) Bioinformatics for the human microbiome project. PLoS Comput Biol 8:e1002779
Gibson GR, Hutkins R, Sanders ME, Prescott SL, Reimer RA, Salminen SJ, Scott K, Stanton C, Swanson KS, Cani PD et al (2017) Expert consensus document: the international scientific association for probiotics and prebiotics (isapp) consensus statement on the definition and scope of prebiotics. Nat Rev Gastroenterol Amp Hepatol 14:491
Gill SR, Pop M, Deboy RT et al (2006) Metagenomic analysis of the human distal gut microbiome. Science 312(5778):1355–1359
Glick-Bauer M, Yeh M-C (2014) The health advantage of a vegan diet: exploring the gut microbiota connection. Nutrients 6:4822–4838
Go AS, Mozaffarian D, Roger VL, Benjamin EJ, Berry JD, Blaha MJ, Dai S, Ford ES, Fox CS, Franco S, Fullerton HJ, Gillespie C, Hailpern SM, Heit JA, Howard VJ, Huffman MD, Judd SE, Kissela BM, Kittner SJ, Lackland DT, Lichtman JH, Lisabeth LD, Mackey RH, Magid DJ, Marcus GM, Marelli A, Matchar DB, McGuire DK, Mohler ER, Moy CS, Mussolino ME, Neumar RW, Nichol G, Pandey DK, Paynter NP, Reeves MJ, Sorlie PD, Stein J, Towfighi A, Turan TN, Virani SS, Wong ND, Woo D, Turner MB (2014) Heart disease and stroke statistics—2014 update. Circulation 129:5

Gomez-Arango LF, Barrett HL, McIntyre HD, Callaway LK, Morrison M, Dekker Nitert M (2016) Increased systolic and diastolic blood pressure is associated with altered gut microbiota composition and butyrate production in early pregnancy. Hypertension 68:974–981

Gómez-Guzmán M, Toral M, Romero M, Jiménez R, Galindo P, Sánchez M, Zarzuelo MJ, Olivares M, Gálvez J, Duarte J (2015) Antihypertensive effects of probiotics lactobacillus strains in spontaneously hypertensive rats. Mol Nutr Food Res 59:2326–2336

Goncalves MD, Lu C, Tutnauer J, Hartman TE, Hwang S-K, Murphy CJ, Pauli C, Morris R, Taylor S, Bosch K, Yang S, Wang Y, Van Riper J, Lekaye HC, Roper J, Kim Y, Chen Q, Gross SS, Rhee KY, Cantley LC, Yun J (2019) High-fructose corn syrup enhances intestinal tumor growth in mice. Science 363:1345–1349

Grabherr F, Grander C, Effenberger M, Adolph TE, Tilg H (2019) Gut dysfunction and non-alcoholic fatty liver disease. Front Endocrinol (Lausanne) 10:5

Graessler J, Qin Y, Zhong H, Zhang J, Licinio J, Wong M-L, Xu A, Chavakis T, Bornstein AB, Ehrhart-Bornstein M, Lamounier-Zepter V, Lohmann T, Wolf T, Bornstein SR (2013) Metagenomic sequencing of the human gut microbiome before and after bariatric surgery in obese patients with type 2 diabetes: correlation with inflammatory and metabolic parameters. Pharmacogenomics 13:514–522

Gravina AG, Zagari RM, De Musis C, Romano L, Loguercio C, Romano M (2018) Helicobacter pylori and extragastric diseases: a review. World J Gastroenterol 24:3204–3221

Green M, Arora K, Prakash S (2020) Microbial medicine: prebiotic and probiotic functional foods to target obesity and metabolic syndrome. Int J Mol Sci 21:5

Greenhill C (2015) Firmicutes and Bacteroidetes involved in insulin resistance by mediating levels of glucagon-like peptide 1. Nat Rev Endocrinol 11:254–254

Groen RN, de Clercq NC, Nieuwdorp M, Hoenders HJR, Groen AK (2018) Gut microbiota, metabolism and psychopathology: a critical review and novel perspectives. Crit Rev Clin Lab Sci 55:283–293

Gur C, Ibrahim Y, Isaacson B, Yamin R, Abed J, Gamliel M, Enk J, Bar-On Y, Stanietsky-Kaynan N, Coppenhagen-Glazer S, Shussman N, Almogy G, Cuapio A, Hofer E, Mevorach D, Tabib A, Ortenberg R, Markel G, Miklić K, Jonjic S, Brennan CA, Garrett WS, Bachrach G, Mandelboim O (2015) Binding of the Fap 2 protein of fusobacterium nucleatum to human inhibitory receptor TIGIT protects tumors from immune cell attack. Immunity 42:344–355

Gurung M, Li Z, You H, Rodrigues R, Jump DB, Morgun A, Shulzhenko N (2020) Role of gut microbiota in type 2 diabetes pathophysiology. EBioMedicine 51:102590

Haider S, Naqvi F, Batool Z, Tabassum S, Sadir S, Liaquat L, Naqvi F, Zuberi NA, Shakeel H, Perveen T (2015) Pretreatment with curcumin attenuates anxiety while strengthens memory performance after one short stress experience in male rats. Brain Res Bull 115:1–8

Halazonetis TD (2004) Constitutively active DNA damage checkpoint pathways as the driving force for the high frequency of p53 mutations in human cancer. DNA Repair (Amst) 3:1057–1062

Hall AB, Yassour M, Sauk J, Garner A, Jiang X, Arthur T, Lagoudas GK, Vatanen T, Fornelos N, Wilson R, Bertha M, Cohen M, Garber J, Khalili H, Gevers D, Ananthakrishnan AN, Kugathasan S, Lander ES, Blainey P, Vlamakis H, Xavier RJ, Huttenhower C (2017) A novel Ruminococcus gnavus clade enriched in inflammatory bowel disease patients. Genome Med 9:103

Hall J, Juncos L, Wang Z, Hall M, do Carmo J, da Silva A (2014) Obesity, hypertension, and chronic kidney disease. Int J Nephrol Renovasc Dis 2014:75

Halpern MT, Cifaldi MA, Kvien TK (2009) Impact of adalimumab on work participation in rheumatoid arthritis: comparison of an open-label extension study and a registry-based control group. Ann Rheum Dis 68:930–937

Hamady ZZR, Scott N, Farrar MD, Lodge JPA, Holland KT, Whitehead T, Carding SR (2010) Xylan-regulated delivery of human keratinocyte growth factor-2 to the inflamed colon by the human anaerobic commensal bacterium Bacteroides ovatus. Gut 59:461–469

Harrison C (2020) Of microbes and mental health: eating for mental wellness. https://asm.org/Articles/2020/February/Of-Microbes-and-Mental-Health-Eating-for-Mental-We#:~:text=Evidence%20suggests%20that%20probiotics%20mayto%20many%20mental%20health%20disorders

Hatakeyama M (2017) Structure and function of helicobacter pylori CagA, the first-identified bacterial protein involved in human cancer. Proc Japan Acad Ser 93:196–219

Hatakka K, Martio J, Korpela M, Herranen M, Poussa T, Laasanen T, Saxelin M, Vapaatalo H, Moilanen E, Korpela R (2003) Effects of probiotic therapy on the activity and activation of mild rheumatoid arthritis – a pilot study. Scand J Rheumatol 32:211–215

He Y, Wang J, Li F, Shi Y (2020) Main clinical features of COVID-19 and potential prognostic and therapeutic value of the microbiota in SARS-CoV-2 infections. Front Microbiol 11:1302

He Z, Hao W, Kwek E, Lei L, Liu J, Zhu H, Ma KY, Zhao Y, Ho HM, He W-S, Chen Z-Y (2019) Fish oil is more potent than flaxseed oil in modulating gut microbiota and reducing trimethylamine- N -oxide-exacerbated atherogenesis. J Agric Food Chem 67:13635–13647

Heshiki Y, Vazquez-Uribe R, Li J, Ni Y, Quainoo S, Imamovic L, Li J, Sørensen M, Chow BKC, Weiss GJ, Xu A, Sommer MOA, Panagiotou G (2020) Predictable modulation of cancer treatment outcomes by the gut microbiota. Microbiome 8:28

Hibberd AA, Lyra A, Ouwehand AC et al (2017) Intestinal microbiota is altered in patients with colon cancer and modified by probiotic intervention. BMJ Open Gastroenterol 4(1):e000145

Hilimire MR, DeVylder JE, Forestell CA (2015) Fermented foods, neuroticism, and social anxiety: an interaction model. Psychiatry Res 228:203–208

Hobby GP, Karaduta O, Dusio GF, Singh M, Zybailov BL, Arthur JM (2019) Chronic kidney disease and the gut microbiome. Am J Physiol Physiol 316:F1211–F1217

Hollister EB, Oezguen N, Chumpitazi BP, Luna RA, Weidler EM, Rubio-Gonzales M, Dahdouli M, Cope JL, Mistretta T-A, Raza S, Metcalf GA, Muzny DM, Gibbs RA, Petrosino JF, Heitkemper M, Savidge TC, Shulman RJ, Versalovic J (2019) Leveraging human microbiome features to diagnose and stratify children with irritable bowel syndrome. J Mol Diagn 21:449–461

Holscher HD (2017) Dietary fiber and prebiotics and the gastrointestinal microbiota. Gut Microbes 8:172–184

Hosseininasab Nodoushan S, Nabavi A (2019) The interaction of Helicobacter pylori infection and type 2 diabetes mellitus. Adv Biomed Res 8:15

Hoving LR, Katiraei S, Heijink M, Pronk A, van der Wee-Pals L, Streefland T, Giera M, Willems van Dijk K, van Harmelen V (2018) Dietary mannan oligosaccharides modulate gut microbiota, increase fecal bile acid excretion, and decrease plasma cholesterol and atherosclerosis development. Mol Nutr Food Res 62:1700942

Hu R, Zeng F, Wu L, Wan X, Chen Y, Zhang J, Liu B (2019) Fermented carrot juice attenuates type 2 diabetes by mediating gut microbiota in rats. Food Funct 10:2935–2946

Hulston CJ, Churnside AA, Venables MC (2015) Probiotic supplementation prevents high-fat, overfeeding-induced insulin resistance in human subjects. Br J Nutr 113:596–602

Human Microbiome Project Consortium (2012) Structure, function and diversity of the healthy human microbiome. Nature 486(7402):207–214. https://doi.org/10.1038/nature11234. PMID: 22699609; PMCID: PMC3564958

Huttenhower C, Gevers D, Knight R et al (2012) Structure, function and diversity of the healthy human microbiome. Nature 486:207–214

Inoue R, Ohue-Kitano R, Tsukahara T, Tanaka M, Masuda S, Inoue T, Yamakage H, Kusakabe T, Hasegawa K, Shimatsu A, Satoh-Asahara N (2017) Prediction of functional profiles of gut microbiota from 16S rRNA metagenomic data provides a more robust evaluation of gut dysbiosis occurring in Japanese type 2 diabetic patients. J Clin Biochem Nutr 61:217–221

Isaacs JD (2008) Therapeutic T-cell manipulation in rheumatoid arthritis: past, present and future. Rheumatology 47:1461–1468

Jacka FN, Mykletun A, Berk M, Bjelland I, Tell GS (2011) The association between habitual diet quality and the common mental disorders in community-dwelling adults. Psychosom Med 73:483–490

Jacka FN, Pasco JA, Mykletun A, Williams LJ, Hodge AM, O'Reilly SL, Nicholson GC, Kotowicz MA, Berk M (2010) Association of western and traditional diets with depression and anxiety in women. Am J Psychiatry 167:305–311

Jadhav P, Jiang Y, Jarr K, Layton C, Ashouri JF, Sinha SR (2020) Efficacy of dietary supplements in inflammatory bowel disease and related autoimmune diseases. Nutrients 12(7):2156

Jie Z, Xia H, Zhong S-L, Feng Q, Li S, Liang S, Zhong H, Liu Z, Gao Y, Zhao H, Zhang D, Su Z, Fang Z, Lan Z, Li J, Xiao L, Li J, Li R, Li X, Li F, Ren H, Huang Y, Peng Y, Li G, Wen B, Dong B, Chen J-Y, Geng Q-S, Zhang Z-W, Yang H, Wang J, Wang J, Zhang X, Madsen L, Brix S, Ning G, Xu X, Liu X, Hou Y, Jia H, He K, Kristiansen K (2017) The gut microbiome in atherosclerotic cardiovascular disease. Nat Commun 8:845

Jung CM, Heinze TM, Schnackenberg LK, Mullis LB, Elkins SA, Elkins CA, Steele RS, Sutherland JB (2009) Interaction of dietary resveratrol with animal-associated bacteria. FEMS Microbiol Lett 297:266–273

Kadooka Y, Sato M, Imaizumi K, Ogawa A, Ikuyama K, Akai Y, Okano M, Kagoshima M, Tsuchida T (2010) Regulation of abdominal adiposity by probiotics (Lactobacillus gasseri SBT2055) in adults with obese tendencies in a randomized controlled trial. Eur J Clin Nutr 64:636–643

Kang DD, Froula J, Egan R, Wang Z (2015) MetaBAT, an efficient tool for accurately reconstructing single genomes from complex microbial communities. Peer J 3:e1165

Kaplan BJ, Rucklidge JJ, Romijn A, McLeod K (2015) The emerging field of nutritional mental health. Clin Psychol Sci 3:964–980

Karlsson FH, Tremaroli V, Nookaew I, Bergström G, Behre CJ, Fagerberg B, Nielsen J, Bäckhed F (2013) Gut metagenome in European women with normal, impaired and diabetic glucose control. Nature 498:99–103

Kessing LV, Vradi E, McIntyre RS, Andersen PK (2015) Causes of decreased life expectancy over the life span in bipolar disorder. J Affect Disord 180:142–147

Khalesi S, Sun J, Buys N, Jayasinghe R (2014) Effect of probiotics on blood pressure. Hypertension 64:897–903

Kho ZY, Lal SK (2018) The human gut microbiome—a potential controller of wellness and disease. Front Microbiol 9:1–23

Kieffer DA, Piccolo BD, Vaziri ND, Liu S, Lau WL, Khazaeli M, Nazertehrani S, Moore ME, Marco ML, Martin RJ, Adams SH (2016) Resistant starch alters gut microbiome and metabolomic profiles concurrent with amelioration of chronic kidney disease in rats. Am J Physiol Physiol 310:F857–F871

Kikuchi K, Ben Othman M, Sakamoto K (2018) Sterilized bifidobacteria suppressed fat accumulation and blood glucose level. Biochem Biophys Res Commun 501:1041–1047

Kim B, Choi H-N, Yim J-E (2019) Effect of diet on the gut microbiota associated with obesity. J Obes Metab Syndr 28:216–224

Kim J, Yun JM, Kim MK, Kwon O, Cho B (2018) Lactobacillus gasseri BNR17 supplementation reduces the visceral fat accumulation and waist circumference in obese adults: a randomized, double-blind, placebo-controlled trial. J Med Food 21:454–461

Kim MH, Kang SG, Park JH, Yanagisawa M, Kim CH (2013a) Short-chain fatty acids activate GPR41 and GPR43 on intestinal epithelial cells to promote inflammatory responses in mice. Gastroenterology 145:396–406

Kim S-W, Park K-Y, Kim B, Kim E, Hyun C-K (2013b) Lactobacillus rhamnosus GG improves insulin sensitivity and reduces adiposity in high-fat diet-fed mice through enhancement of adiponectin production. Biochem Biophys Res Commun 431:258–263

King DE, Mainous AG, Lambourne CA (2012) Trends in dietary fiber intake in the United States, 1999-2008. J Acad Nutr Diet 112:642–648

Kinoshita M, Suzuki Y, Saito Y (2002) Butyrate reduces colonic paracellular permeability by enhancing PPARγ activation. Biochem Biophys Res Commun 293:827–831

Kira J, Isobe N (2019) Helicobacter pylori infection and demyelinating disease of the central nervous system. J Neuroimmunol 329:14–19

Kobyliak N, Abenavoli L, Mykhalchyshyn G, Kononenko L, Boccuto L, Kyriienko D, Dynnyk O (2018) A multi-strain probiotic reduces the fatty liver index, cytokines and aminotransferase levels in NAFLD patients: evidence from a randomized clinical trial. J Gastrointestin Liver Dis 27:41–49

Koeth RA, Wang Z, Levison BS, Buffa JA, Org E, Sheehy BT, Britt EB, Fu X, Wu Y, Li L, Smith JD, DiDonato JA, Chen J, Li H, Wu GD, Lewis JD, Warrier M, Brown JM, Krauss RM, Tang WHW, Bushman FD, Lusis AJ, Hazen SL (2013) Intestinal microbiota metabolism of l-carnitine, a nutrient in red meat, promotes atherosclerosis. Nat Med 19:576–585

Koh A, De Vadder F, Kovatcheva-Datchary P, Bäckhed F (2016) From dietary fiber to host physiology: short-Chain fatty acids as key bacterial metabolites. Cell 165:1332–1345

Koliada A, Syzenko G, Moseiko V, Budovska L, Puchkov K, Perederiy V, Gavalko Y, Dorofeyev A, Romanenko M, Tkach S, Sineok L, Lushchak O, Vaiserman A (2017) Association between body mass index and Firmicutes/Bacteroidetes ratio in an adult Ukrainian population. BMC Microbiol 17:120

Konishi H, Fujiya M, Tanaka H, Ueno N, Moriichi K, Sasajima J, Ikuta K, Akutsu H, Tanabe H, Kohgo Y (2016) Probiotic-derived ferrichrome inhibits colon cancer progression via JNK-mediated apoptosis. Nat Commun 7:12365

Kosti I, Lyalina S, Pollard KS, Butte AJ, Sirota M (2020) Meta-analysis of vaginal microbiome data provides new insights into preterm birth. Front Microbiol 11:1–13

Kumar P, Banik S (2013) Pharmacotherapy options in rheumatoid arthritis. Clin Med Insights Arthritis Musculoskelet Disord 6:5558

Kurdi P, Kawanishi K, Mizutani K, Yokota A (2006) Mechanism of growth inhibition by free bile acids in Lactobacilli and Bifidobacteria. J Bacteriol 188:1979–1986

Kwa M, Plottel CS, Blaser MJ, Adams S (2016) The intestinal microbiome and estrogen receptor-positive female breast cancer. J Natl Cancer Inst 108(8):djw029

Lagier J-C, Armougom F, Million M, Hugon P, Pagnier I, Robert C, Bittar F, Fournous G, Gimenez G, Maraninchi M, Trape J-F, Koonin EV, La Scola B, Raoult D (2012) Microbial culturomics: paradigm shift in the human gut microbiome study. Clin Microbiol Infect 18:1185–1193

Lagier J-C, Dubourg G, Million M, Cadoret F, Bilen M, Fenollar F, Levasseur A, Rolain J-M, Fournier P-E, Raoult D (2018) Culturing the human microbiota and culturomics. Nat Rev Microbiol 16:540–550

Lara-Tejero M (2000) A bacterial toxin that controls cell cycle progression as a deoxyribonuclease I-like protein. Science 290:354–357

Larsen N, Vogensen FK, van den Berg FWJ, Nielsen DS, Andreasen AS, Pedersen BK, Al-Soud WA, Sørensen SJ, Hansen LH, Jakobsen M (2010) Gut microbiota in human adults with type 2 diabetes differs from non-diabetic adults. PLoS One 5(e):9085

Lazar V, Ditu L-M, Pircalabioru GG, Picu A, Petcu L, Cucu N, Chifiriuc MC (2019) Gut microbiota, host organism, and diet trialogue in diabetes and obesity. Front Nutr 6:21

Le TKC, Hosaka T, Nguyen TT, Kassu A, Dang TO, Tran HB, Pham TP, Tran QB, Le THH, Da Pham X (2015) Bifidobacterium species lower serum glucose, increase expressions of insulin signaling proteins, and improve adipokine profile in diabetic mice. Biomed Res 36:63–70

Leber A, Hontecillas R, Abedi V, Tubau-Juni N, Zoccoli-Rodriguez V, Stewart C, Bassaganya-Riera J (2017) Modeling new immunoregulatory therapeutics as antimicrobial alternatives for treating Clostridium difficile infection. Artif Intell Med 78:1–13

Lee J-W, Shin J-G, Kim EH, Kang HE, Yim IB, Kim JY, Joo H-G, Woo HJ (2004) Immunomodulatory and antitumor effects in vivo by the cytoplasmic fraction of lactobacillus casei and Bifidobacterium longum. J Vet Sci 5:41–48

Lenoir M, del Carmen S, Cortes-Perez NG, Lozano-Ojalvo D, Muñoz-Provencio D, Chain F, Langella P, de Moreno de LeBlanc A, LeBlanc JG, Bermúdez-Humarán LG (2016) Lactobacillus casei BL23 regulates Treg and Th17 T-cell populations and reduces DMH-associated colorectal cancer. J Gastroenterol 51:862–873

Levy SE, Myers RM (2016) Advancements in next-generation sequencing. Annu Rev Genomics Hum Genet 17:95–115

Li D, Liu C-M, Luo R, Sadakane K, Lam T-W (2015) MEGAHIT: an ultra-fast single-node solution for large and complex metagenomics assembly via succinct de Bruijn graph. Bioinformatics 31:1674–1676

Li D, Luo R, Liu C-M, Leung C-M, Ting H-F, Sadakane K, Yamashita H, Lam T-W (2016) MEGAHIT v1.0: a fast and scalable metagenome assembler driven by advanced methodologies and community practices. Methods 102:3–11

Li K, Zhang L, Xue J, Yang X, Dong X, Sha L, Lei H, Zhang X, Zhu L, Wang Z, Li X, Wang H, Liu P, Dong Y, He L (2019a) Dietary inulin alleviates diverse stages of type 2 diabetes mellitus via anti-inflammation and modulating gut microbiota in db/db mice. Food Funct 10:1915–1927

Li X, Gan, Sun, Meng, Shang, Mao & Li (2019b) Targeting gut microbiota for the prevention and management of diabetes mellitus by dietary natural products. Foods 8, 440

Liang Q, Chiu J, Chen Y, Huang Y, Higashimori A, Fang J, Brim H, Ashktorab H, Ng SC, Ng SSM, Zheng S, Chan FKL, Sung JJY, Yu J (2017) Fecal bacteria act as novel biomarkers for noninvasive diagnosis of colorectal cancer. Clin Cancer Res 23:2061–2070

Lim MY, Rho M, Song Y-M, Lee K, Sung J, Ko G (2015) Stability of gut enterotypes in Korean monozygotic twins and their association with biomarkers and diet. Sci Rep 4:7348

Limaye SA, Haddad RI, Cilli F, Sonis ST, Colevas AD, Brennan MT, Hu KS, Murphy BA (2013) Phase 1b, multicenter, single blinded, placebo-controlled, sequential dose escalation study to assess the safety and tolerability of topically applied AG013 in subjects with locally advanced head and neck cancer receiving induction chemotherapy. Cancer 119:4268–4276

Liu G, Liang L, Yu G, Li Q (2018a) Pumpkin polysaccharide modifies the gut microbiota during alleviation of type 2 diabetes in rats. Int J Biol Macromol 115:711–717

Liu Q, Tang G-Y, Zhao C-N, Feng X-L, Xu X-Y, Cao S-Y, Meng X, Li S, Gan R-Y, Li H-B (2018b) Comparison of antioxidant activities of different grape varieties. Molecules 23:2432

Liu RT, Walsh RFL, Sheehan AE (2019) Prebiotics and probiotics for depression and anxiety: a systematic review and meta-analysis of controlled clinical trials. Neurosci Biobehav Rev 102:13–23

Lloyd-Price J, Mahurkar A, Rahnavard G, Crabtree J, Orvis J, Hall AB, Brady A, Creasy HH, McCracken C, Giglio MG, McDonald D, Franzosa EA, Knight R, White O, Huttenhower C (2017) Strains, functions and dynamics in the expanded human microbiome project. Nature 550:61–66

Loguercio C, Federico A, Tuccillo C, Terracciano F, Auria MV, De Simone C, Blanco CDV (2005) Beneficial effects of a probiotic VSL#3 on parameters of liver dysfunction in chronic liver diseases. J Clin Gastroenterol 39:540–543

Loomba R, Seguritan V, Li W, Long T, Klitgord N, Bhatt A, Dulai PS, Caussy C, Bettencourt R, Highlander SK, Jones MB, Sirlin CB, Schnabl B, Brinkac L, Schork N, Chen C-H, Brenner DA, Biggs W, Yooseph S, Venter JC, Nelson KE (2017) Gut microbiome-based metagenomic signature for non-invasive detection of advanced fibrosis in human nonalcoholic fatty liver disease. Cell Metab 25:1054–1062

Louis P, Flint HJ (2017) Formation of propionate and butyrate by the human colonic microbiota. Environ Microbiol 19:29–41

Lu R, Wu S, Zhang Y, Xia Y, Liu X, Zheng Y, Chen H, Schaefer KL, Zhou Z, Bissonnette M, Li L, Sun J (2014) Enteric bacterial protein Avr a promotes colonic tumorigenesis and activates colonic beta-catenin signaling pathway. Oncogene 3:e105–e105

Ma C, Han M, Heinrich B, Fu Q, Zhang Q, Sandhu M, Agdashian D, Terabe M, Berzofsky JA, Fako V, Ritz T, Longerich T, Theriot CM, McCulloch JA, Roy S, Yuan W, Thovarai V, Sen SK,

Ruchirawat M, Korangy F, Wang XW, Trinchieri G, Greten TF (2018) Gut microbiome–mediated bile acid metabolism regulates liver cancer via NKT cells. Science 360:5931

Ma FX, Zhou B, Chen Z, Ren Q, Lu SH, Sawamura T, Han ZC (2006) Oxidized low density lipoprotein impairs endothelial progenitor cells by regulation of endothelial nitric oxide synthase. J Lipid Res 47:1227–1237

Malaguarnera M, Vacante M, Antic T, Giordano M, Chisari G, Acquaviva R, Mastrojeni S, Malaguarnera G, Mistretta A, Li Volti G, Galvano F (2012) Bifidobacterium longum with Fructo-oligosaccharides in patients with non-alcoholic steatohepatitis. Dig Dis Sci 57:545–553

Mandel DR, Eichas K, Holmes J (2010) Bacillus coagulans: a viable adjunct therapy for relieving symptoms of rheumatoid arthritis according to a randomized, controlled trial. BMC Complement Altern Med 10:1

Marchesi JR, Adams DH, Fava F, Hermes GDA, Hirschfield GM, Hold G, Quraishi MN, Kinross J, Smidt H, Tuohy KM, Thomas LV, Zoetendal EG, Hart A (2016) The gut microbiota and host health: a new clinical frontier. Gut 65:330–339

Mariat D, Firmesse O, Levenez F, Guimarães V, Sokol H, Doré J, Corthier G, Furet J-P (2009) The Firmicutes/Bacteroidetes ratio of the human microbiota changes with age. BMC Microbiol 9:123

Markowiak P, Ślizewska K (2017) Effects of probiotics, prebiotics, and synbiotics on human health. Nutrients 9:1021

Markowitz VM, Chen I-MA, Chu K, Szeto E, Palaniappan K, Jacob B, Ratner A, Liolios K, Pagani I, Huntemann M, Mavromatis K, Ivanova NN, Kyrpides NC (2012) IMG/M-HMP: a metagenome comparative analysis system for the human microbiome project. PLoS One 7: e40151

Marques FZ, Nelson E, Chu P-Y, Horlock D, Fiedler A, Ziemann M, Tan JK, Kuruppu S, Rajapakse NW, El-Osta A, Mackay CR, Kaye DM (2017) High-fiber diet and acetate supplementation change the gut microbiota and prevent the development of hypertension and heart failure in hypertensive mice. Circulation 135:964–977

Massot-Cladera M, Pérez-Berezo T, Franch A, Castell M, Pérez-Cano FJ (2012) Cocoa modulatory effect on rat faecal microbiota and colonic crosstalk. Arch Biochem Biophys 527:105–112

McBurney MI, Davis C, Fraser CM, Schneeman BO, Huttenhower C, Verbeke K, Walter J, Latulippe ME (2019) Establishing what constitutes a healthy human gut microbiome: state of the science, regulatory considerations, and future directions. J Nutr 149:1882–1895

McFadden R-MT, Larmonier CB, Shehab KW, Midura-Kiela M, Ramalingam R, Harrison CA, Besselsen DG, Chase JH, Caporaso JG, Jobin C, Ghishan FK, Kiela PR (2015) The role of curcumin in modulating colonic microbiota during colitis and colon cancer prevention. Inflamm Bowel Dis 21:2483–2494

Mego M, Chovanec J, Vochyanova-Andrezalova I, Konkolovsky P, Mikulova M, Reckova M, Miskovska V, Bystricky B, Beniak J, Medvecova L, Lagin A, Svetlovska D, Spanik S, Zajac V, Mardiak J, Drgona L (2015) Prevention of irinotecan induced diarrhea by probiotics: a randomized double blind, placebo controlled pilot study. Complement Ther Med 23:356–362

Mei X, Zhang X, Wang Z, Gao Z, Liu G, Hu H, Zou L, Li X (2016) Insulin sensitivity-enhancing activity of Phlorizin is associated with lipopolysaccharide decrease and gut microbiota changes in obese and type 2 diabetes (db/db) mice. J Agric Food Chem 64:7502–7511

Menni C, Lin C, Cecelja M, Mangino M, Matey-Hernandez ML, Keehn L, Mohney RP, Steves CJ, Spector TD, Kuo C-F, Chowienczyk P, Valdes AM (2018) Gut microbial diversity is associated with lower arterial stiffness in women. Eur Heart J 39:2390–2397

Menni C, Mangino M, Cecelja M, Psatha M, Brosnan MJ, Trimmer J, Mohney RP, Chowienczyk P, Padmanabhan S, Spector TD, Valdes AM (2015) Metabolomic study of carotid–femoral pulse-wave velocity in women. J Hypertens 33:791–796

Meroni M, Longo M, Dongiovanni P (2019) The role of probiotics in nonalcoholic fatty liver disease: a new insight into therapeutic strategies. Nutrients 11:2642

Mijanur Rahman M, Gan SH, Khalil MI (2014) Neurological effects of honey: current and future prospects. Evid-Based Complement Altern Med 2014:1–13

Mills CE, Tzounis X, Oruna-Concha M-J, Mottram DS, Gibson GR, Spencer JPE (2015) In vitro colonic metabolism of coffee and chlorogenic acid results in selective changes in human faecal microbiota growth. Br J Nutr 113:1220–1227

Mimee M, Citorik RJ, Lu TK (2016) Microbiome therapeutics — advances and challenges. Adv Drug Deliv Rev 105:44–54

Miyamoto J, Kasubuchi M, Nakajima A, Irie J, Itoh H, Kimura I (2016) The role of short-chain fatty acid on blood pressure regulation. Curr Opin Nephrol Hypertens 25:379–383

Mohajeri MH, La Fata G, Steinert RE, Weber P (2018) Relationship between the gut microbiome and brain function. Nutr Rev 76:481–496

Mohamadshahi M, Veissi M, Haidari F, Shahbazian H, Kaydani G-A, Mohammadi F (2014) Effects of probiotic yogurt consumption on inflammatory biomarkers in patients with type 2 diabetes. Bioimpacts 4:83–88

Morgan XC, Tickle TL, Sokol H, Gevers D, Devaney KL, Ward DV et al (2012) Dysfunction of the intestinal microbiome in inflammatory bowel disease and treatment. Genome Biol 13:R79

Motta J-P, Bermudez-Humaran LG, Deraison C, Martin L, Rolland C, Rousset P, Boue J, Dietrich G, Chapman K, Kharrat P, Vinel J-P, Alric L, Mas E, Sallenave J-M, Langella P, Vergnolle N (2012) Food-grade bacteria expressing Elafin protect against inflammation and restore colon homeostasis. Sci Transl Med 4:158ra144

Moya-Pérez A, Neef A, Sanz Y (2015) Bifidobacterium pseudocatenulatum CECT 7765 reduces obesity-associated inflammation by restoring the lymphocyte-macrophage balance and gut microbiota structure in high-fat diet-fed mice. PLoS One 10:e0126976

Munukka E, Pekkala S, Wiklund P, Rasool O, Borra R, Kong L, Ojanen X, Cheng SM, Roos C, Tuomela S, Alen M, Lahesmaa R, Cheng S (2014) Gut-adipose tissue axis in hepatic fat accumulation in humans. J Hepatol 61:132–138

Munukka E, Rintala A, Toivonen R, Nylund M, Yang B, Takanen A, Hänninen A, Vuopio J, Huovinen P, Jalkanen S, Pekkala S (2017) Faecalibacterium prausnitzii treatment improves hepatic health and reduces adipose tissue inflammation in high-fat fed mice. ISME 11:1667–1679

Murata-Kamiya N, Kurashima Y, Teishikata Y, Yamahashi Y, Saito Y, Higashi H, Aburatani H, Akiyama T, Peek RM, Azuma T, Hatakeyama M (2007) Helicobacter pylori CagA interacts with E-cadherin and deregulates the β-catenin signal that promotes intestinal transdifferentiation in gastric epithelial cells. Oncogene 26:4617–4626

Murphy R, Tsai P, Jüllig M, Liu A, Plank L, Booth M (2017) Differential changes in gut microbiota after gastric bypass and sleeve gastrectomy bariatric surgery vary according to diabetes remission. Obes Surg 27:917–925

Natividad JM, Lamas B, Pham HP, Michel M-L, Rainteau D, Bridonneau C, da Costa G, van Hylckama VJ, Sovran B, Chamignon C, Planchais J, Richard ML, Langella P, Veiga P, Sokol H (2018) Bilophila wadsworthia aggravates high fat diet induced metabolic dysfunctions in mice. Nat Commun 9:2802

Navab-Moghadam F, Sedighi M, Khamseh ME, Alaei-Shahmiri F, Talebi M, Razavi S, Amirmozafari N (2017) The association of type II diabetes with gut microbiota composition. Microb Pathog 110:630–636

Newburg DS (2000) Oligosaccharides in human milk and bacterial colonization. J Pediatr Gastroenterol Nutr 30:8–17

Nicolucci AC, Reimer RA (2017) Prebiotics as a modulator of gut microbiota in paediatric obesity. Pediatr Obes 12:265–273

NIH Human Microbiome Portfolio Analysis Team (2019) A review of 10 years of human microbiome research activities at the US National Institutes of Health, Fiscal Years 2007-2016. Microbiome 7(1):31

Nihei N, Okamoto H, Furune T, Ikuta N, Sasaki K, Rimbach G, Yoshikawa Y, Terao K (2018) Dietary α-cyclodextrin modifies gut microbiota and reduces fat accumulation in high-fat-diet-fed obese mice. Bio Factors 44:336–347

Ning L, Liu R, Lou X, Du H, Chen W, Zhang F, Li S, Chen X, Xu G (2019) Association between helicobacter pylori infection and nonalcoholic fatty liver disease: a systemic review and meta-analysis. Eur J Gastroenterol Hepatol 31:735–742

Nouri Z, Karami F, Neyazi N, Modarressi MH, Karimi R, Khorramizadeh MR, Taheri B, Motevaseli E (2016) Dual anti-metastatic and anti-proliferative activity assessment of two probiotics on HeLa and HT-29 cell lines. Cell J 18:127–134

Nurk S, Meleshko D, Korobeynikov A, Pevzner PA (2017) Meta SPAdes: a new versatile metagenomic assembler. Genome Res 27:824–834

Ohman, Simrén (2013) Intestinal microbiota and its role in irritable bowel syndrome (IBS). Curr Gastroenterol Rep 15(5):323

Orlando A, Linsalata M, Russo F (2016) Antiproliferative effects on colon adenocarcinoma cells induced by co-administration of vitamin K1 and Lactobacillus rhamnosus GG. Int J Oncol 48:2629–2638

Ortiz P, Bissada NF, Palomo L, Han YW, Al-Zahrani MS, Panneerselvam A, Askari A (2009) Periodontal therapy reduces the severity of active rheumatoid arthritis in patients treated with or without tumor necrosis factor inhibitors. J Periodontol 80:535–540

Ott SJ (2004) Reduction in diversity of the colonic mucosa associated bacterial microflora in patients with active inflammatory bowel disease. Gut 53(5):685–693

Paavonen J, Naud P, Salmerón J, Wheeler C, Chow S-N, Apter D, Kitchener H, Castellsague X, Teixeira J, Skinner S, Hedrick J, Jaisamrarn U, Limson G, Garland S, Szarewski A, Romanowski B, Aoki F, Schwarz T, Poppe W, Bosch F, Jenkins D, Hardt K, Zahaf T, Descamps D, Struyf F, Lehtinen M, Dubin G (2009) Efficacy of human papillomavirus (HPV)-16/18 AS04-adjuvanted vaccine against cervical infection and precancer caused by oncogenic HPV types (PATRICIA): final analysis of a double-blind, randomised study in young women. Lancet 374:301–314

Packer CS, Rice AE, Johnson TC, Pelaez NJ, Temm CJ, Potter GV, White WA, Roth AH, Dominguez JH, Peterson RG (2014) Oxidized low density lipoprotein (OX-LDL) induced arterial muscle contraction signaling mechanisms. Open Hypertens 6:20–26

Pandey KR, Naik SR, Vakil BV (2015) Probiotics, prebiotics and synbiotics- a review. J Food Sci Technol 52:7577–7587

Parkar SG, Trower TM, Stevenson DE (2013) Fecal microbial metabolism of polyphenols and its effects on human gut microbiota. Anaerobe 23:12–19

Pase MP, Scholey AB, Pipingas A, Kras M, Nolidin K, Gibbs A, Wesnes K, Stough C (2013) Cocoa polyphenols enhance positive mood states but not cognitive performance: a randomized, placebo-controlled trial. J Psychopharmacol 27:451–458

Paulos CM, Wrzesinski C, Kaiser A, Hinrichs CS, Chieppa M, Cassard L, Palmer DC, Boni A, Muranski P, Yu Z, Gattinoni L, Antony PA, Rosenberg SA, Restifo NP (2007) Microbial translocation augments the function of adoptively transferred self/tumor-specific CD8+ T cells via TLR4 signaling. J Clin Invest 117:2197–2204

Pedret A, Valls RM, Calderón-Pérez L, Llauradó E, Companys J, Pla-Pagà L, Moragas A, Martín-Luján F, Ortega Y, Giralt M, Caimari A, Chenoll E, Genovés S, Martorell P, Codoñer FM, Ramón D, Arola L, Solà R (2019) Effects of daily consumption of the probiotic Bifidobacterium animalis subsp. lactis CECT 8145 on anthropometric adiposity biomarkers in abdominally obese subjects: a randomized controlled trial. Int J Obes (Lond) 43:1863–1868

Pédron T, Nigro G, Sansonetti PJ (2016) From homeostasis to pathology: decrypting microbe-host symbiotic signals in the intestinal crypt. Philos Trans R Soc Lond B Biol Sci 371 (1707):20150500

Peluso I, Morabito G, Urban L, Ioannone F, Serafi M (2012) Oxidative stress in atherosclerosis development: the central role of LDL and oxidative burst. Endocrine, Metab Immune Disord Targets 12:351–360

Petrof EO, Claud EC, Gloor GB, Allen-Vercoe E (2013) Microbial ecosystems therapeutics: a new paradigm in medicine? Benefic Microbes 4(1):53–65

Pham NM, Nanri A, Kurotani K, Kuwahara K, Kume A, Sato M, Hayabuchi H, Mizoue T (2014) Green tea and coffee consumption is inversely associated with depressive symptoms in a Japanese working population. Public Health Nutr 17:625–633

Pieczynska MD, Yang Y, Petrykowski S, Horbanczuk OK, Atanasov AG, Horbanczuk JO (2020) Gut microbiota and its metabolites in atherosclerosis development. Molecules 25:594

Pineda ML, Thompson SF, Summers K et al (2011) A randomized, double-blinded, placebo-controlled pilot study of probiotics in active rheumatoid arthritis. Med Sci Monit 17:CR347–CR3354

Piovani D, Danese S, Peyrin-Biroulet L, Nikolopoulos GK, Lytras T, Bonovas S (2019) Environmental risk factors for inflammatory bowel diseases: an umbrella review of meta-analyses. Gastroenterology 157:647–659

Plovier H, Everard A, Druart C, Depommier C, Van Hul M, Geurts L, Chilloux J, Ottman N, Duparc T, Lichtenstein L, Myridakis A, Delzenne NM, Klievink J, Bhattacharjee A, van der Ark KCH, Aalvink S, Martinez LO, Dumas M-E, Maiter D, Loumaye A, Hermans MP, Thissen J-P, Belzer C, de Vos WM, Cani PD (2017) A purified membrane protein from Akkermansia muciniphila or the pasteurized bacterium improves metabolism in obese and diabetic mice. Nat Med 23:107–113

Pluznick JL, Protzko RJ, Gevorgyan H, Peterlin Z, Sipos A, Han J, Brunet I, Wan L-X, Rey F, Wang T, Firestein SJ, Yanagisawa M, Gordon JI, Eichmann A, Peti-Peterdi J, Caplan MJ (2013) Olfactory receptor responding to gut microbiota-derived signals plays a role in renin secretion and blood pressure regulation. Proc Natl Acad Sci 110:4410–4415

Pruesse E, Quast C, Knittel K, Fuchs BM, Ludwig W, Peplies J, Glockner FO (2007) SILVA: a comprehensive online resource for quality checked and aligned ribosomal RNA sequence data compatible with ARB. Nucleic Acids Res 35:7188–7196

Qiao Y, Sun J, Xia S, Tang X, Shi Y, Le G (2014) Effects of resveratrol on gut microbiota and fat storage in a mouse model with high-fat-induced obesity. Food Funct 5:1241

Qin J, Li R, Raes J, Arumugam M, Burgdorf KS, Manichanh C, Nielsen T, Pons N, Levenez F, Yamada T, Mende DR, Li J, Xu J, Li S, Li D, Cao J, Wang B, Liang H, Zheng H, Xie Y, Tap J, Lepage P, Bertalan M, Batto J-M, Hansen T, Le Paslier D, Linneberg A, Nielsen HB, Pelletier E, Renault P, Sicheritz-Ponten T, Turner K, Zhu H, Yu C, Li S, Jian M, Zhou Y, Li Y, Zhang X, Li S, Qin N, Yang H, Wang J, Brunak S, Doré J, Guarner F, Kristiansen K, Pedersen O, Parkhill J, Weissenbach J, Bork P, Ehrlich SD, Wang J (2010) A human gut microbial gene catalogue established by metagenomic sequencing. Nature 464:59–65

Qiu L, Tao X, Xiong H, Yu J, Wei H (2018) Lactobacillus plantarum ZDY04 exhibits a strain-specific property of lowering TMAO via the modulation of gut microbiota in mice. Food Funct 9:4299–4309

Qiu L, Yang D, Tao X, Yu J, Xiong H, Wei H (2017) Enterobacter aerogenes ZDY01 attenuates choline-induced trimethylamine N-oxide levels by remodeling gut microbiota in mice. J Microbiol Biotechnol 27:1491–1499

Raes J (2016) Microbiome-based companion diagnostics: no longer science fiction? Gut 65:896–897

Ramezani A, Raj DS (2014) The gut microbiome, kidney disease, and targeted interventions. J Am Soc Nephrol 25:657–670

Redman MG, Ward EJ, Phillips RS (2014) The efficacy and safety of probiotics in people with cancer: a systematic review. Ann Oncol 25:1919–1929

Reichert S, Haffner M, Keyßer G, Schäfer C, Stein JM, Schaller H-G, Wienke A, Strauss H, Heide S, Schulz S (2013) Detection of oral bacterial DNA in synovial fluid. J Clin Periodontol 40:591–598

Reimer RA, Grover GJ, Koetzner L, Gahler RJ, Lyon MR, Wood S (2014) Combining sitagliptin/metformin with a functional fiber delays diabetes progression in Zucker rats. J Endocrinol 220:361–373

Rey FE, Faith JJ, Bain J, Muehlbauer MJ, Stevens RD, Newgard CB, Gordon JI (2010) Dissecting the in vivo metabolic potential of two human gut acetogens. J Biol Chem 285:22082–22090

Rezac S, Kok CR, Heermann M, Hutkins R (2018) Fermented foods as a dietary source of live organisms. Front Microbiol 9:1785

Rezaei Nejad H, Oliveira BCM, Sadeqi A, Dehkharghani A, Kondova I, Langermans JAM, Guasto JS, Tzipori S, Widmer G, Sonkusale SR (2019) Ingestible osmotic pill for in vivo sampling of gut microbiomes. Adv Intell Syst 1(5):1900053

Ridlon JM, Kang D-J, Hylemon PB (2006) Bile salt biotransformations by human intestinal bacteria. J Lipid Res 47:241–259

Rienks J, Dobson AJ, Mishra GD (2013) Mediterranean dietary pattern and prevalence and incidence of depressive symptoms in mid-aged women: results from a large community-based prospective study. Eur J Clin Nutr 67:75–82

Riley LW, Raphael E, Faerstein E (2013) Obesity in the United States–dysbiosis from exposure to low-dose antibiotics? Front Public Health 1:69

Rousseaux C, Thuru X, Gelot A, Barnich N, Neut C, Dubuquoy L, Dubuquoy C, Merour E, Geboes K, Chamaillard M, Ouwehand A, Leyer G, Carcano D, Colombel J-F, Ardid D, Desreumaux P (2007) Lactobacillus acidophilus modulates intestinal pain and induces opioid and cannabinoid receptors. Nat Med 13:35–37

Rubinstein MR, Wang X, Liu W, Hao Y, Cai G, Han YW (2013) Fusobacterium nucleatum promotes colorectal carcinogenesis by modulating E-cadherin/β-catenin signaling via its Fad A adhesin. Cell Host Microbe 14:195–206

Salamon D, Sroka-Oleksiak A, Kapusta P, Szopa M, Mrozińska S, Ludwig-Słomczyńska AH, Wołkow PP, Bulanda M, Klupa T, Małecki MT, Gosiewski T (2018) Characteristics of the gut microbiota in adult patients with type 1 and 2 diabetes based on the analysis of a fragment of 16S rRNA gene using next-generation sequencing. Polish Arch Intern Med 128:336

Sanchez-Rodriguez E, Egea-Zorrilla A, Plaza-Díaz J, Aragón-Vela J, Muñoz-Quezada S, Tercedor-Sánchez L, Abadia-Molina F (2020) The gut microbiota and its implication in the development of atherosclerosis and related cardiovascular diseases. Nutrients 12:605

Sánchez-Villegas A, Delgado-Rodríguez M, Alonso A, Schlatter J, Lahortiga F, Majem LS, Martínez-González MA (2009) Association of the Mediterranean Dietary Pattern with the incidence of depression. Arch Gen Psychiatry 66:1090

Sathyapalan T, Beckett S, Rigby AS, Mellor DD, Atkin SL (2010) High cocoa polyphenol rich chocolate may reduce the burden of the symptoms in chronic fatigue syndrome. Nutr J 9:55

Scher JU, Ubeda C, Equinda M, Khanin R, Buischi Y, Viale A, Lipuma L, Attur M, Pillinger MH, Weissmann G, Littman DR, Pamer EG, Bretz WA, Abramson SB (2012) Periodontal disease and the oral microbiota in new-onset rheumatoid arthritis. Arthritis Rheum 64:3083–3094

Schiffrin EL (2014) Immune mechanisms in hypertension and vascular injury. Clin Sci (Lond) 126:267–274

Schlaberg R (2020) Microbiome diagnostics. Clin Chem 66(1):68–76

Schloss PD, Westcott SL, Ryabin T, Hall JR, Hartmann M, Hollister EB, Lesniewski RA, Oakley BB, Parks DH, Robinson CJ, Sahl JW, Stres B, Thallinger GG, Van Horn DJ, Weber CF (2009) Introducing mothur: open-source, platform-independent, community-supported software for describing and comparing microbial communities. Appl Environ Microbiol 75:7537–7541

Seldin MM, Meng Y, Qi H, Zhu W, Wang Z, Hazen SL, Lusis AJ, Shih DM (2016) Trimethylamine N-oxide promotes vascular inflammation through signaling of mitogen-activated protein kinase and nuclear factor-κB. J Am Heart Assoc 5(2):e002767

Selhub EM, Logan AC, Bested AC (2014) Fermented foods, microbiota, and mental health: ancient practice meets nutritional psychiatry. J Physiol Anthropol 33:2

Seo D-B, Jeong HW, Cho D, Lee BJ, Lee JH, Choi JY, Bae I-H, Lee S-J (2015) Fermented green tea extract alleviates obesity and related complications and alters gut microbiota composition in diet-induced obese mice. J Med Food 18:549–556

Seppo L, Jauhiainen T, Poussa T, Korpela R (2003) A fermented milk high in bioactive peptides has a blood pressure–lowering effect in hypertensive subjects. Am J Clin Nutr 77:326–330

Shen T-CD, Albenberg L, Bittinger K, Chehoud C, Chen Y-Y, Judge CA, Chau L, Ni J, Sheng M, Lin A, Wilkins BJ, Buza EL, Lewis JD, Daikhin Y, Nissim I, Yudkoff M, Bushman FD, Wu GD (2015) Engineering the gut microbiota to treat hyperammonemia. J Clin Invest 125:2841–2850

Simon M-C, Strassburger K, Nowotny B, Kolb H, Nowotny P, Burkart V, Zivehe F, Hwang J-H, Stehle P, Pacini G, Hartmann B, Holst JJ, Mac Kenzie C, Bindels LB, Martinez I, Walter J, Henrich B, Schloot NC, Roden M (2015) Intake of Lactobacillus reuteri improves incretin and insulin secretion in glucose-tolerant humans: a proof of concept. Diabetes Care 38:1827–1834

Sirich TL, Plummer NS, Gardner CD, Hostetter TH, Meyer TW (2014) Effect of increasing dietary fiber on plasma levels of colon-derived solutes in hemodialysis patients. Clin J Am Soc Nephrol 9:1603–1610

Skarupski KA, Tangney CC, Li H, Evans DA, Morris MC (2013) Mediterranean diet and depressive symptoms among older adults over time. J Nutr Health Aging 17:441–445

Solanki N, Alkadhi I, Atrooz F, Patki G, Salim S (2015) Grape powder prevents cognitive, behavioral, and biochemical impairments in a rat model of posttraumatic stress disorder. Nutr Res 35:65–75

Song J-X, Ren H, Gao Y-F, Lee C-Y, Li S-F, Zhang F, Li L, Chen H (2017) Dietary capsaicin improves glucose homeostasis and alters the gut microbiota in obese diabetic Ob/Ob mice. Front Physiol 8:602

Sowa-Kućma M, Styczeń K, Siwek M, Misztak P, Nowak RJ, Dudek D, Rybakowski JK, Nowak G, Maes M (2018) Are there differences in lipid peroxidation and immune biomarkers between major depression and bipolar disorder: effects of melancholia, atypical depression, severity of illness, episode number, suicidal ideation and prior suicide attempts. Prog Neuro-Psychopharmacol Biol Psychiatry 81:372–383

Statovci D, Aguilera M, Mac Sharry J, Melgar S (2017) The impact of Western diet and nutrients on the microbiota and immune response at mucosal interfaces. Front Immunol 8:838

Stephanie T, Alia C, Ali A, Ahmed EH, Steven F (2012) Identification of oral bacterial DNA in synovial fluid of arthritis patients with native and failed prosthetic joints. J Clin Rheumatol 18 (3):117–121

Subah Packer C (2007) Estrogen protection, oxidized LDL, endothelial dysfunction and vasorelaxation in cardiovascular disease: new insights into a complex issue. Cardiovasc Res 73:6–7. https://doi.org/10.1016/j.cardiores.2006.11.013

Suceveanu AI, Stoian AP, Parepa I, Voinea C, Hainarosie R, Manuc D, Nitipir C, Mazilu L, Suceveanu AP (2018) Gut microbiota patterns in obese and type 2 diabetes (T2D) patients from Romanian Black Sea coast region. Rev Chim 69:2260–2267

Suzuki C, Aoki-Yoshida A, Aoki R, Sasaki K, Takayama Y, Mizumachi K (2017) The distinct effects of orally administered lactobacillus rhamnosus GG and Lactococcus lactis subsp. lactis C59 on gene expression in the murine small intestine. PLoS One 12:e0188985

Taherian-Esfahani Z, Abedin-Do A, Nouri Z, Mirfakhraie R, Ghafouri-Fard S, Motevaseli E (2016) Lactobacilli differentially modulate mTOR and Wnt/β-catenin pathways in different cancer cell lines. Iran J Cancer Prev 9(3):e5369

Tajabadi-Ebrahimi M, Sharifi N, Farrokhian A, Raygan F, Karamali F, Razzaghi R, Taheri S, Asemi Z (2016) A randomized controlled clinical trial investigating the effect of Synbiotic administration on markers of insulin metabolism and lipid profiles in overweight type 2 diabetic patients with coronary heart disease. Exp Clin Endocrinol Diabetes 125:21–27

Takagi A, Ikemura H, Matsuzaki T, Sato M, Nomoto K, Morotomi M, Yokokura T (2008) Relationship between the in vitro response of dendritic cells to lactobacillus and prevention of tumorigenesis in the mouse. J Gastroenterol 43:661–669

Takayama F, Taki K, Niwa T (2003) Bifidobacterium in gastro-resistant seamless capsule reduces serum levels of indoxyl sulfate in patients on hemodialysis. Am J Kidney Dis 41:S142–S145

Tan JK, McKenzie C, Mariño E, Macia L, Mackay CR (2017) Metabolite-sensing G protein–coupled receptors—facilitators of diet-related immune regulation. Annu Rev Immunol 35:371–402

Tap et al (2017) Identification of an intestinal microbiota signature associated with severity of irritable bowel syndrome. Gastroenterology 152(1):111–123

Taranu I, Marin D, Braicu C, Pistol G, Sorescu I, Pruteanu L, Neagoe IB, Vodnar D (2018) In vitro transcriptome response to a mixture of lactobacilli strains in intestinal porcine epithelial cell line. Int J Mol Sci 19(7):1923

Terpou A, Papadaki A, Lappa IK, Kachrimanidou V, Bosnea LA, Kopsahelis N (2019) Probiotics in food systems: significance and emerging strategies towards improved viability and delivery of enhanced beneficial value. Nutrients 11:01591

The Integrative HMP (iHMP) Research Network Consortium (2019) The integrative human microbiome project. Nature 569:641–648

Tomova A, Bukovsky I, Rembert E, Yonas W, Alwarith J, Barnard ND, Kahleova H (2019) The effects of vegetarian and vegan diets on gut microbiota. Front Nutr 6:47

Tong X, Xu J, Lian F, Yu X, Zhao Y, Xu L, Zhang M, Zhao X, Shen J, Wu S, Pang X, Tian J, Zhang C, Zhou Q, Wang L, Pang B, Chen F, Peng Z, Wang J, Zhen Z, Fang C, Li M, Chen L, Zhao L (2018) Structural alteration of gut microbiota during the amelioration of human type 2 diabetes with hyperlipidemia by metformin and a traditional chinese herbal formula: a multicenter, randomized, open label clinical trial. MBio 9:e02392

Topçuoğlu BD, Lesniak NA, Ruffin MT, Wiens J, Schloss PD (2020) A framework for effective application of machine learning to microbiome-based classification problems. MBio 1:1–13

Tran CN, Lundy SK, Fox DA (2005) Synovial biology and T cells in rheumatoid arthritis. Pathophysiology 12:183–189

Trebatická J, Ďuračková Z (2015) Psychiatric disorders and polyphenols: can they be helpful in therapy? Oxid Med Cell Longev 2015:1–16

Truong DT, Franzosa EA, Tickle TL, Scholz M, Weingart G, Pasolli E, Tett A, Huttenhower C, Segata N (2015) MetaPhlAn2 for enhanced metagenomic taxonomic profiling. Nat Methods 12 (10):902–903

Turnbaugh PJ, Ley RE, Hamady M, Fraser-Liggett CM, Knight R, Gordon JI (2007) The human microbiome project. Nature 449(7164):804–810

Uribe G, Villéger R, Bressollier P, Dillard RN, Worthley DL, Wang TC, Powell DW, Urdaci MC, Pinchuk IV (2018) Cell Microbiol 20(11):e12871

Vaahtovuo J, Munukka E, Korkeamäki M, Luukkainen R, Toivanen P (2008) Fecal microbiota in early rheumatoid arthritis. J Rheumatol 35:1500–1505

Vaghef-Mehrabany E, Alipour B, Homayouni-Rad A, Sharif S-K, Asghari-Jafarabadi M, Zavvari S (2014) Probiotic supplementation improves inflammatory status in patients with rheumatoid arthritis. Nutrition 30:430–435

Van Herck M, Vonghia L, Francque S (2017) Animal models of nonalcoholic fatty liver disease—a Starter's guide. Nutrients 9:1072

Van Hul M, Geurts L, Plovier H, Druart C, Everard A, Ståhlman M, Rhimi M, Chira K, Teissedre P-L, Delzenne NM, Maguin E, Guilbot A, Brochot A, Gérard P, Bäckhed F, Cani PD (2018) Reduced obesity, diabetes, and steatosis upon cinnamon and grape pomace are associated with changes in gut microbiota and markers of gut barrier. Am J Physiol Metab 314:E334–E352

van Nood E, Speelman P, Nieuwdorp M, Keller J (2014) Fecal microbiota transplantation: facts and controversies. Curr Opin Gastroenterol 30(1):34–39

Vanderhoof JA, Young R (2008) Probiotics in the United States. Clin Infect Dis 46:S67–S72

Vaziri ND, Liu S-M, Lau WL, Khazaeli M, Nazertehrani S, Farzaneh SH, Kieffer DA, Adams SH, Martin RJ (2014) High amylose resistant starch diet ameliorates oxidative stress, inflammation, and progression of chronic kidney disease. PLoS One 9:e114881

Vaziri ND, Wong J, Pahl M, Piceno YM, Yuan J, DeSantis TZ, Ni Z, Nguyen T-H, Andersen GL (2013) Chronic kidney disease alters intestinal microbial flora. Kidney Int 83:308–315

Versalovic J, Dore J, Guarner F, Luna RA, Ringel Y (2017) Microbiome-based diagnostics: ready for applications in laboratory medicine? Clin Chem 63:1674–1679

Vieira-Silva S, Falony G, Belda E, Nielsen T, Aron-Wisnewsky J, Chakaroun R, Forslund SK, Assmann K, Valles-Colomer M, Nguyen TTD, Proost S, Prifti E, Tremaroli V, Pons N, Le Chatelier E, Andreelli F, Bastard J-P, Coelho LP, Galleron N, Hansen TH, Hulot J-S, Lewinter C, Pedersen HK, Quinquis B, Rouault C, Roume H, Salem J-E, Søndertoft NB, Touch S, Dumas M-E, Ehrlich SD, Galan P, Gøtze JP, Hansen T, Holst JJ, Køber L, Letunic I, Nielsen J, Oppert J-M, Stumvoll M, Vestergaard H, Zucker J-D, Bork P, Pedersen O, Bäckhed F, Clément K, Raes J (2020) Statin therapy is associated with lower prevalence of gut microbiota dysbiosis. Nature 581:310–315

Vitetta L, Llewellyn H, Oldfield D (2019) Gut dysbiosis and the intestinal microbiome: *Streptococcus thermophilus* a key probiotic for reducing uremia. Microorganisms 7(8):228. https://doi.org/10.3390/microorganisms7080228

Vivarelli S, Salemi R, Candido S, Falzone L, Santagati M, Stefani S, Torino F, Banna GL, Tonini G, Libra M (2019) Gut microbiota and cancer: from pathogenesis to therapy. Cancers (Basel) 11:38

Wang I-K, Lai H-C, Yu C-J, Liang C-C, Chang C-T, Kuo H-L, Yang Y-F, Lin C-C, Lin H-H, Liu Y-L, Chang Y-C, Wu Y-Y, Chen C-H, Li C-Y, Chuang F-R, Huang C-C, Lin C-H, Lin H-C (2012) Real-time PCR analysis of the intestinal microbiotas in peritoneal dialysis patients. Appl Environ Microbiol 78:1107–1112

Wang J, Tang H, Zhang C, Zhao Y, Derrien M, Rocher E, van Hylckama Vlieg JE, Strissel K, Zhao L, Obin M, Shen J (2015a) Modulation of gut microbiota during probiotic-mediated attenuation of metabolic syndrome in high fat diet-fed mice. ISME 9:1–15

Wang Z, Roberts AB, Buffa JA, Levison BS, Zhu W, Org E, Gu X, Huang Y, Zamanian-Daryoush-M, Culley MK et al (2015b) Non-lethal inhibition of gut microbial trimethylamine production for the treatment of atherosclerosis. Cell 163:1585–1595

Watzl B, Girrbach S, Roller M (2005) Inulin, oligofructose and immunomodulation. Br J Nutr 93: S49–S55

Wegh C, Geerlings SY, Knol J, Roeselers G, Belzer C (2019) Postbiotics and their potential applications in early life nutrition and beyond. Int J Mol Sci 20(19):4673

Wehmeyer MH, Zyriax B-C, Jagemann B, Roth E, Windler E, Schulze zur Wiesch J, Lohse AW, Kluwe J (2016) Nonalcoholic fatty liver disease is associated with excessive calorie intake rather than a distinctive dietary pattern. Medicine (Baltimore) 95:e3887

WHO (1946) Preamble to the constitution of WHO as adopted by the international health conference, New York, 19 June–22 July 1946; signed on 22 July 1946 by the representatives of 61 States (Official Records of WHO, no. 2, p 100) and entered into force on 7 April 1948

Wijarnpreecha K, Thongprayoon C, Panjawatanan P, Manatsathit W, Jaruvongvanich V, Ungprasert P (2018) Helicobacter pylori and risk of nonalcoholic fatty liver disease. J Clin Gastroenterol 52:386–391

Wittchen H-U (2012) The burden of mood disorders. Science 338:15

Woldeamlak B, Yirdaw K, Biadgo B (2019) Role of gut microbiota in type 2 diabetes mellitus and its complications: novel insights and potential intervention strategies. Korean J Gastroenterol 74:314

Wong J, Piceno YM, DeSantis TZ, Pahl M, Andersen GL, Vaziri ND (2014) Expansion of urease- and uricase-containing, indole- and p-cresol-forming and contraction of short-Chain fatty acid-producing intestinal microbiota in ESRD. Am J Nephrol 39:230–237

Wood DE, Salzberg SL (2014) Kraken: ultrafast metagenomic sequence classification using exact alignments. Genome Biol 15:R46

Wrzosek L, Miquel S, Noordine M-L, Bouet S, Chevalier-Curt M, Robert V, Philippe C, Bridonneau C, Cherbuy C, Robbe-Masselot C, Langella P, Thomas M (2013) Bacteroides thetaiotaomicron and Faecalibacterium prausnitzii influence the production of mucus glycans and the development of goblet cells in the colonic epithelium of a gnotobiotic model rodent. BMC Biol 11:61

Wu GD, Chen J, Hoffmann C, Bittinger K, Chen Y-Y, Keilbaugh SA, Bewtra M, Knights D, Walters WA, Knight R, Sinha R, Gilroy E, Gupta K, Baldassano R, Nessel L, Li H, Bushman

FD, Lewis JD (2011) Linking Long-term dietary patterns with gut microbial enterotypes. Science 334:105–108

Wu H, Esteve E, Tremaroli V, Khan MT, Caesar R, Mannerås-Holm L, Ståhlman M, Olsson LM, Serino M, Planas-Fèlix M, Xifra G, Mercader JM, Torrents D, Burcelin R, Ricart W, Perkins R, Fernàndez-Real JM, Bäckhed F (2017) Metformin alters the gut microbiome of individuals with treatment-naive type 2 diabetes, contributing to the therapeutic effects of the drug. Nat Med 23:850–858

Wu S, Rhee K-J, Zhang M, Franco A, Sears CL (2007) Bacteroides fragilis toxin stimulates intestinal epithelial cell shedding and -secretase-dependent E-cadherin cleavage. J Cell Sci 120:1944–1952

Xavier JB, Young VB, Skufca J, Ginty F, Testerman T, Pearson AT, Macklin P, Mitchell A, Shmulevich I, Xie L, Caporaso JG, Crandall KA, Simone NL, Godoy-Vitorino F, Griffin TJ, Whiteson KL, Gustafson HH, Slade DJ, Schmidt TM, Walther-Antonio MRS, Korem T, Webb-Robertson B-JM, Styczynski MP, Johnson WE, Jobin C, Ridlon JM, Koh AY, Yu M, Kelly L, Wargo JA (2020) The cancer microbiome: distinguishing direct and indirect effects requires a systemic view. Trend Cancer 6:192–204

Yang H, Xia H, Ye Y, Zou W, Sun Y (2014) Probiotic Bacillus pumilus SE5 shapes the intestinal microbiota and mucosal immunity in grouper Epinephelus coioides. Dis Aquat Organ 111:119–127

Yang J-Y, Lee Y-S, Kim Y, Lee S-H, Ryu S, Fukuda S, Hase K, Yang C-S, Lim HS, Kim M-S, Kim H-M, Ahn S-H, Kwon B-E, Ko H-J, Kweon M-N (2017a) Gut commensal Bacteroides acidifaciens prevents obesity and improves insulin sensitivity in mice. Mucosal Immunol 10:104–116

Yang Y, Weng W, Peng J, Hong L, Yang L, Toiyama Y, Gao R, Liu M, Yin M, Pan C, Li H, Guo B, Zhu Q, Wei Q, Moyer M-P, Wang P, Cai S, Goel A, Qin H, Ma Y (2017b) Fusobacterium nucleatum increases proliferation of colorectal cancer cells and tumor development in mice by activating toll-like receptor 4 signaling to nuclear factor−κB, and up-regulating expression of micro RNA-21. Gastroenterology 152:851–866

Yasuda G, Shibata K, Takizawa T, Ikeda Y, Tokita Y, Umemura S, Tochikubo O (2002) Prevalence of constipation in continuous ambulatory peritoneal dialysis patients and comparison with hemodialysis patients. Am J Kidney Dis 39:1292–1299

Yoshida N, Emoto T, Yamashita T, Watanabe H, Hayashi T, Tabata T, Hoshi N, Hatano N, Ozawa G, Sasaki N, Mizoguchi T, Amin HZ, Hirota Y, Ogawa W, Yamada T, Hirata K (2018) Bacteroides vulgatus and Bacteroides dorei reduce gut microbial lipopolysaccharide production and inhibit atherosclerosis. Circulation 138:2486–2498

Yu J-J, Pei L-B, Zhang Y, Wen Z-Y, Yang J-L (2015) Chronic supplementation of curcumin enhances the efficacy of antidepressants in major depressive disorder. J Clin Psychopharmacol 35(4):406–410

Yuan J, Chen C, Cui J, Lu J, Yan C, Wei X, Zhao X, Li N, Li S, Xue G, Cheng W, Li B, Li H, Lin W, Tian C, Zhao J, Han J, An D, Zhang Q, Wei H, Zheng M, Ma X, Li W, Chen X, Zhang Z, Zeng H, Ying S, Wu J, Yang R, Liu D (2019) Fatty liver disease caused by high-alcohol-producing Klebsiella pneumoniae. Cell Metab 30:675–688

Zamani B, Golkar HR, Farshbaf S, Emadi-Baygi M, Tajabadi-Ebrahimi M, Jafari P, Akhavan R, Taghizadeh M, Memarzadeh MR, Asemi Z (2016) Clinical and metabolic response to probiotic supplementation in patients with rheumatoid arthritis: a randomized, double-blind, placebo-controlled trial. Int J Rheum Dis 19:869–879

Zhang W, Li J, Lu S et al (2019) Gut microbiota community characteristics and disease-related microorganism pattern in a population of healthy Chinese people. Sci Rep 9:1594

Zhang X, Shen D, Fang Z, Jie Z, Qiu X, Zhang C, Chen Y, Ji L (2013) Human gut microbiota changes reveal the progression of glucose intolerance. PLoS One 8:e71108

Zhang X, Zhang D, Jia H, Feng Q, Wang D, Liang D, Wu X, Li J, Tang L, Li Y, Lan Z, Chen B, Li Y, Zhong H, Xie H, Jie Z, Chen W, Tang S, Xu X, Wang X, Cai X, Liu S, Xia Y, Li J, Qiao X, Al-Aama JY, Chen H, Wang L, Wu Q, Zhang F, Zheng W, Li Y, Zhang M, Luo G,

Xue W, Xiao L, Li J, Chen W, Xu X, Yin Y, Yang H, Wang J, Kristiansen K, Liu L, Li T, Huang Q, Li Y, Wang J (2015) The oral and gut microbiomes are perturbed in rheumatoid arthritis and partly normalized after treatment. Nat Med 21:895–905

Zhao S, Liu W, Wang J, Shi J, Sun Y, Wang W, Ning G, Liu R, Hong J (2017) Akkermansia muciniphila improves metabolic profiles by reducing inflammation in chow diet-fed mice. J Mol Endocrinol 58:1–14

Zheng J, Li H, Zhang X, Jiang M, Luo C, Lu Z, Xu Z, Shi J (2018) Prebiotic Mannan-oligosaccharides augment the hypoglycemic effects of metformin in correlation with modulating gut microbiota. J Agric Food Chem 66:5821–5831

Zhu L, Zhang D, Zhu H, Zhu J, Weng S, Dong L, Liu T, Hu Y, Shen X (2018) Berberine treatment increases Akkermansia in the gut and improves high-fat diet-induced atherosclerosis in Apoe −/− mice. Atherosclerosis 268:117–126

Zhu S, Jiang Y, Xu K, Cui M, Ye W, Zhao G, Jin L, Chen X (2020a) The progress of gut microbiome research related to brain disorders. J Neuroinflammation 17:25. https://doi.org/10.1186/s12974-020-1705-z

Zhu Y, Li Q, Jiang H (2020b) Gut microbiota in atherosclerosis: focus on trimethylamine N-oxide. APMIS 128:353–366

Zimmermann M, Zimmermann-Kogadeeva M, Wegmann R, Goodman AL (2019) Mapping human microbiome drug metabolism by gut bacteria and their genes. Nature 570:462–467

Zitvogel L, Ma Y, Raoult D, Kroemer G, Gajewski TF (2018) The microbiome in cancer immunotherapy: diagnostic tools and therapeutic strategies. Science 359:1366–1370

Microbiome Therapeutics: Emerging Concepts and Challenges

11

J. Sarada, S. Anju, Y. Aparna, and K. Anuradha

Abstract

The human microbiome has become a thrust area for researchers and industrial scientists in understanding microbial roles in health and disease. With the advent of NGS methods, good progress is made in analyzing the composition of the microbiome and their key metabolites produced, so many researchers are now trying to explore the potential of the microbiome as an important therapeutic tool. Microbes in the human gut are identified as key contributors in host metabolism and could be explored as novel therapeutics. A large number of research papers reviewed different aspects of the microbiome and its potential role in cardiometabolic disorders, inflammatory bowel diseases, neuropsychiatric diseases, and cancer. Research studies made in the last 5 years proved that the microbiome disruptions play a vital role in malnutrition, obesity and also modulate associations between diet and disease. Beyond the gut, the human microbiome likely affects all organs through the immune, circulatory, and nervous systems, affecting human behavior and cognitive function. Researchers have made efforts to clarify the role of the microbiome in autoimmune disease development and find new therapeutic approaches to treat immune-mediated diseases. For example, bacterium *A. muciniphila* is viewed as a next-generation beneficial microbe as supplementation of it protects against several cardiometabolic features and decreases several pathological conditions, such as obesity, type 2 diabetes, hypertension, hypercholesterolemia, and liver disease. Hence microbiome-based therapeutics are being identified as an integral part of the precision medicine approach because of the contribution to inter-individual variability in diseases condition and also as a modifiable factor leading to the

J. Sarada (✉) · S. Anju · Y. Aparna · K. Anuradha
Department of Microbiology, Bhavan's Vivekananda College of Science Humanities and Commerce, Sainikpuri, Secunderabad, Telangana, India

P. V. Bramhachari (ed.), *Microbiome in Human Health and Disease*,
https://doi.org/10.1007/978-981-16-3156-6_11

development of future therapeutics. They have the potential to influence the therapeutic strategies in the treatment of several diseases, generate new economic opportunities, and benefit humans around the globe.

Keywords

Microbiome · Therapeutics · Human gut · Health and disease

11.1 Introduction

The microbiome has become the central concept among academic researchers and biotechnology-based companies to explore the hidden microbial roles in health, disease and to understand microbiome-based interventions. Different microbial communities are found throughout the mucosal surfaces and cavities present in the human body and contribute to diverse physiological processes like immunity, metabolism, nutritional homeostasis, and neuronal activity in a healthy individual (Fig. 11.1). The human body in return provides a suitable environment for stable colonization of this commensal microbiome.

With the advent of Metagenomic and Gnotobiotic techniques, microbiome contribution to human health is being established. Variations in gut microbiome composition and function are connected to the etiology of several diseases. Loss or changes in the microbiome and its homeostatic function could be due to infection, inflammation, improper diet, and or antibiotics. The microbiota of each organ in the human body is unique and the microbial population in different organs gives important information about the occurrence of disease symptoms. Therefore, the microbiome is found responsible for the clinical symptoms and drug response of individuals. The population of beneficial microbes in healthy individuals is more

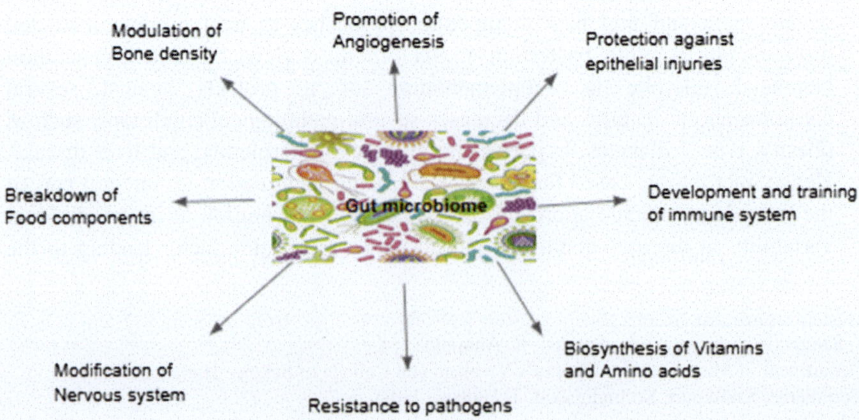

Fig. 11.1 Role of Gut Microbiome in Human Health

diverse and can withstand physiological changes than the disease-associated microbiota which causes inflammation.

Several studies were made to prove that interventions to modulate the microbiome help in improving the health of a person. Current microbiome-based therapeutics (Table 11.1) aim at altering the gut microbiome by using live microbes as probiotics. Instead of probiotics, metabolite-based approach or prebiotics are also used to affect microbiome composition and/or function in a beneficial way.

The Human Microbiome Project (HMP) was launched by the National Institute of Health in 2007, to provide expertise and knowledge to characterize the human microbiome and understand its role in the health of a person. HMP provides guidance and direction to unravel the role microbes play in human health, disease, nutrition, and immunity.

Metagenomic projects assisted in the detection of the pathobionts in Colorectal Cancer (CRC), Inflammatory Bowel Diseases (IBD), Irritable Bowel Syndrome (IBS), *Clostridium difficile infection* (CDI), Cardiovascular Diseases (CVD), autoimmune diseases, and several Non-Communicable Diseases (NCDs) like Obesity, Asthma, Type 1 Diabetes, Eczema, etc. (Fig. 11.2). These studies help to prevent and even cure by modulating the human microbiota. As a biomarker, the human microbiome can be used to detect diseased conditions and has the potential for the diagnostic and therapeutic role. This approach supports diagnosis and aids in novel therapeutic strategies.

Microbiome-based personalized medicine has become a new therapeutic approach especially for the treatment of cancer. Since microbiota is identified to modulate carcinogenesis through mechanisms like inflammation and immunity, the microbiome can be used to develop anticancer therapies.

11.1.1 Colon Rectal Cancer (CRC) Therapy

Any changes of the normal microbiome that could interrupt host–microbial interactions lead to dysbiosis which in turn result in diseases. Dysbiotic condition is commonly observed in several human diseases especially cancer. Colon harbors 70% of the human microbiome and thus more prone to cancer development than the small intestine in our body. Origin and development of Colorectal Cancer (CRC) involve alteration in the gut microbiota. Hausen reported that 20% of the cancers are attributed to intestinal microbiota changes from normal flora to the infectious state (Hausen 2009). Mutations in proto-oncogene, tumor suppressor genes, and genes involved in DNA repair lead to induction of CRC. Environmental factors like smoking, alcohol, processed foods, animal fat, and low intake of fiber and fruits are known to trigger the development of CRC.

Gut microbiota help to maintain mucosal homeostasis and epithelial barriers. Wu et al. and Yoshioka et al. reported that microbial metabolite butyrate initiates dysfunction in the gut epithelia by activating pro-inflammatory mediators like cytokines, interleukin-6, and tumor necrosis factor-α (Yoshioka 2009; Wu 2006). Goodwin et al. stated that *Bacteriodes fragilis* and *Enterococcus faecalis* stimulate

Table 11.1 Microbiome therapeutics in various diseases

Disease	Types of microbiota present in disease condition or dysbiotic microbiota	Depletion of microbiota	Biomarker strains for diagnosis	Therapeutic microflora
Colorectal cancer	*Streptococcus gallolyticus*, *F. nucleatum*, *E. coli*, E. *fragilis*, and *E. faecalis*	*Roseburi, Clostridium, Faecalibacterium,* and *Bifidobacterium*	*F. Nucleatum*	*Bifidobacteria spp.; Akkermansia muciniphila, Enterococcus hirae, Bacteroides spp.*
Irritable bowel syndrome (IBS)	No information	*Lactobacillus, Bifidobacterium, and Faecalibacterium prausnitzii*	No information	*Bifidobacterium, Lactobacillus,* and *acidophilus, L helveticus,* and *Bifidobacterium infantis Pediococcus acidilactici* and two *L plantarum* strains
Inflammatory bowel diseases (IBDs)	*Enterobacteriaceae*	*Firmicutes: Clostridium clade,* and *Bacteroidetes*	*Enterobacteriaceae*	*Bacteroides fragilis Segmented Filamentous Bacteria* (SFB) and *Mucispirillum*, engineered *Lactococcus lactis*
Clostridium difficile infection (CDI)	*C. difficile*	No information	*C. difficile*	*Lactobacillus casei, Lactobacillus bulgaricus,* and *Streptococcus thermophilus,* purified *Firmicutes* spores
Food allergy	*Bacteroides, Propionibacterium,* and *Klebsiella*	No information	No information	*Lactobacillus rhamnosus GG* (LGG)
Type 1 diabetes (T1D)	*Bacteroidetes, Bacteroides dorei, Bacteroides vulgatus*	No information	No information	*Lactobacillaceae-*

Multiple sclerosis	*Methanobrevibacter, ermansia, Pseudomonas, Mycoplasma, Haemophilus, Blautia* and *Dorei*	*Butyricimonas*	No information	*Lactobacillus spp., Pediococcus acidilactici, Bifidobacterium bifidum, Bifidobacterium animalis,* and *Bacteroides fragilis*
Systemic lupus erythematosus	*Bacteroidetes* members, and *Rhodococcus, Eggerthella, Klebsiella, Prevotella, Eubacterium, Flavonifractor, and Incertae sedis*	*Firmicutes*	No information	Supplementation with Treg-inducing bacteria

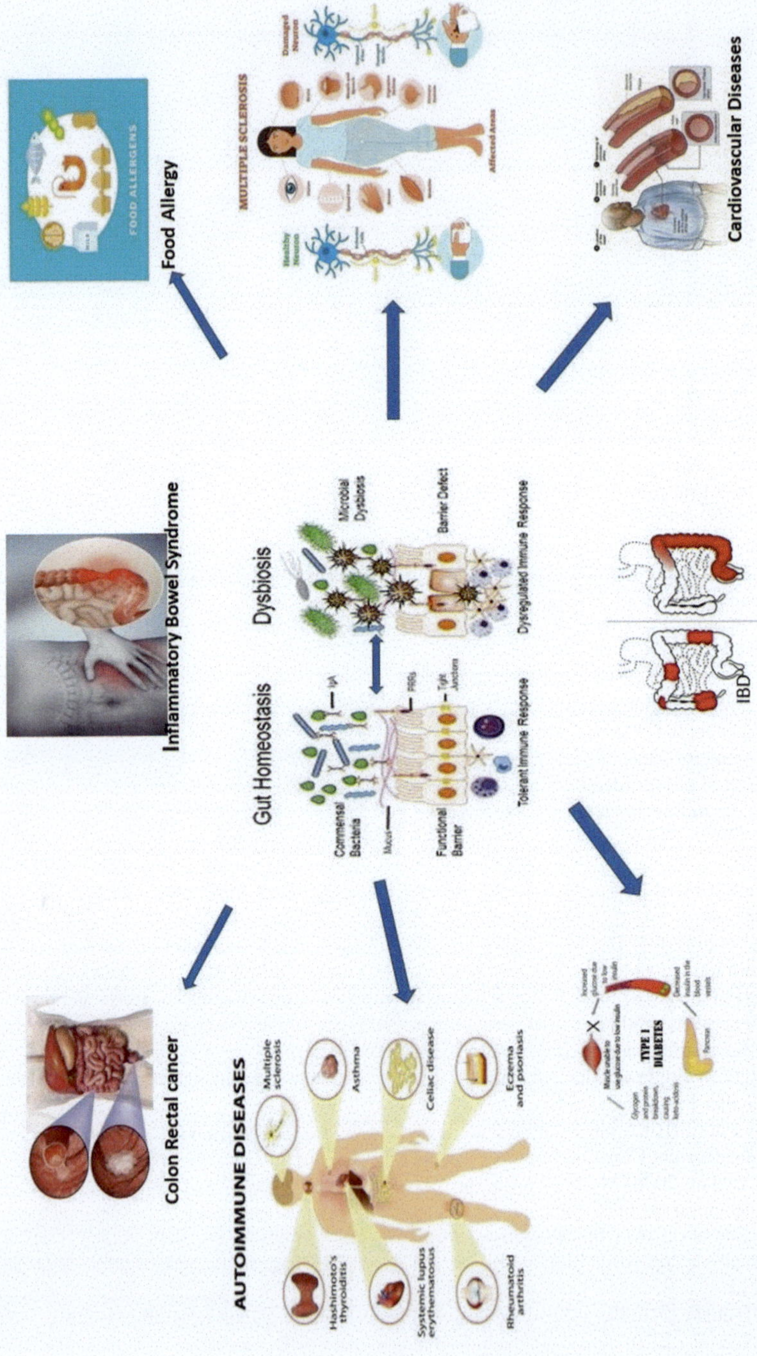

Fig. 11.2 Dysbiosis related diseases

initiating preliminary responses for inflammation and transformation of epithelia and make cancer indigenous bacteria participate in cancer induction (Goodwin et al. 2011). Dysbiosis gut microbiota causes functional imbalance by initiating pro-inflammatory responses and epithelial cell transformation and make driver bacteria (cancer indigenous intestinal bacteria) participate in cancer initiation. During the tumor developing stage, microenvironmental alterations favor the growth of bacterial passengers or opportunistic bacteria. Comparison of fecal microbiota from CRC patients with healthy individuals shows enriched and depleted microorganisms differentiating CRC patients from control populations. The occurrence of *Streptococcus gallolyticus, F. nucleatum, Escherichia coli, B. fragilis,* and *E. faecalis* was more in CRC patients than in healthy people. Interestingly bacteria like *Roseburia, Clostridium, Faecalibacterium,* and *Bifidoba*cterium are found to be depleted in CRC patients.

Diagnosis at the initial stages can prevent CRC either by surgeries or by therapy which has a higher success rate. Generally, FOBT (Fecal occult blood test) and FIT (fecal immunochemical test) are used along with colonoscopy for diagnosis. Microbiome signatures in stool samples were used as microbial markers in the early detection of CRC. Yu et al. reported that adenoma and CRC patient's stool samples had *F. nucleatum* (Yu 2017). In a meta-analysis research study, Zhang et al. also reported that *F. nucleatum* becomes a biomarker for a non-invasive screening in CRC and colorectal adenoma (Zhang 2019). Another strategy is to screen for fecal metabolome metabolite markers like SCFA, fructose, linoleic acid, and nicotinic acid in CRC patients.

Microbiome modulation stands as an alternative approach to prevent or treat CRC. Strategies so far in use are fecal microbiota-transplantation (FMT), pre−/ probiotics, and diet. Among all, FMT has become a hopeful strategy for CRC patients. Since the gut microbiome stimulates the immune system, they possess anticancer effects and become key contributors to tumor immunotherapy. *Bifidobacteria spp.; Akkermansia muciniphila, Enterococcus hirae, Bacteroides spp* exhibit anti-tumoral response by activating T-cells. The role for *A. muciniphila* is identified to be prominent in anticancer immunotherapy, especially with anti-PD-1 treatment.

Singh and his group observed that diet influences the human gut microbiome (Singh 2017). In a recent report, it was observed that the mouse model when induced by deoxycholic acid and fed with a western-type diet initiated intestinal carcinogenesis. However, with supplementation of a diet rich in fiber content and butyrate-producers colon tumor growth could be drastically reduced (Cao 2017; Pandey 2015). Probiotics were known to inhibit CRC by producing detoxifying agents, anti-inflammatory factors, anticancer compounds, and short-chain fatty acids (SCFAs) which improve the intestinal barrier function.

A combination of chemotherapy and/or immunotherapy with supplementation with diet, probiotics, and prebiotics targeting the gut microbiota is a promising strategy. A study made by Ding et al. has shown that oral administration of *Bifidobacteria* along with PD-L1 inhibitors had a synergistic effect inhibiting tumor growth, against the individual on chemotherapy alone (Ding 2018).

Pandey et al. reported the use of synbiotics like OAT fiber/*L. plantarum* and FOS/L. sporogens for therapy (Pandey 2015). Bozkurt et al. genetically modified *Bifidobacterium animalis* species for butyrate and mycosporin-like amino acids production for prebiotics effects (Bozkurt 2019). This genetically altered strain could modulate host immunity by regulating cytokine production, macrophages and lymphocyte proliferation. Ding et al. stated that pro−/synbiotics therapies were useful for CRC patients. *Bifidobacteria spp., Akkermansia muciniphila, Enterococcus hirae, Bacteroides spp* were reported to cause an anticancer immune response by T-cell activation.

11.1.2 Gut Microbiome and Chimeric Antigen Receptor (CAR) T-Cell Therapy.

The targeted immunotherapeutic approach is found to be more effective and less toxic in cancer treatment than conventional chemotherapy. Immunotherapy includes targeting programmed cell death 1 (PD1) receptor, programmed cell death ligand 1 (PD-L1) or cytotoxic T-lymphocyte antigen-4 (CTLA-4) inhibitors, and chimeric antigen receptor (CAR) T-cells which could control the natural solid and hematological malignancies (HM). Chimeric Antigen Receptor (CAR) T-cells are autologous T-cells targeted to a tumor-specific antigen and seem to be more promising for patients suffering from unmanageable hematological malignancies.

CD19-directed CAR T-cells therapy was approved by FDA for cancer patients. However, variations in clinical responses to immuno and CAR T-cells therapies call for deeper investigations on factors influencing the therapy. *Bifidobacteria longum* is known to boost response in Immune Checkpoint Inhibition (ICIs) while Bifidobacterium *bifidum* is immunosuppressive. Usage of antibiotics during immunotherapy and CAR T-cells therapy lead to dysbiosis.

Gut microbiota negotiates with both systemic immunity and immune checkpoint inhibition (ICI) responses and demarcates the native T-cells either into pro-inflammatory (Th17) or anti-inflammatory (Tregs) effector cells. Then these cells migrate to systemic circulation. Several strategies like interference in pre-CAR conditioning, usage of antibiotics with the narrow spectrum, suppression of CAR T-cells by Treg cells could be explored by using specific gut microbes.

In recent research investigations, it was observed that gut microbiome diversity and composition influence response to immunotherapy. Muhammad Bilal Abid et al. hypothesized that the modulation of the gut microbiome could enhance the CAR T-cell response (Muhammad Bilal Abid 2019).

Bacteria like *Faecalibacterium prausnitzii, Ruminococcaceae, Clostridia* when used in therapy gave superior response and survival to patients while *Enterobacteriaceae members* yielded low response and decreased survival. *Akkermansia muciniphila* is known to boost the immune response and increase survival. Many clinical studies had proved that manipulation of gut microbiota results in enhancement of immunotherapy responses.

11.1.3 Irritable Bowel Syndrome (IBS) Therapy

IBD is also known as a functional gastrointestinal disease (FGID), has a global prevalence of 10%, and is known to affect the quality of life among the patients suffering from it. Pathogenic symptoms like dysmotility, visceral hypersensitivity, abnormal responses to stress play a role in the development of IBS. Diagnosis is based on symptoms like constant abdominal pain during defecation and change in stool frequency and form. Microbiota mediated changes in bile acid deconjugation affect stool volume and consistency and changes in microbial fermentation affect gas volume in these patients. IBS diagnosis poses a challenge as it lacks a universally reliable biomarker, and heterogeneity in its progression, and nonspecific symptoms.

Enteric Dysbiosis and Small Intestine Bacterial Overgrowth (SIBO) are identified as the main factors associated with the IBS altering mucosal permeability. Enteric bacteria like *Lactobacillus, Bifidobacterium, and Faecalibacterium prausnitzii* were fewer in these patients than in healthy individuals. For therapy presently prebiotics, probiotics, antibiotics, diet, and FMT are in use, all targeting gut microbiota.

11.1.3.1 Prebiotics

Prebiotics like fructooligosaccharides (FOS) or galactooligosaccharides (GOS) are identified to support the growth of *Bifidobacteria and Lactobacilli* and give health benefits for the host. Supplementation of trans-GOS at 3.5 g/day improved stool consistency and flatulence, while supplementation at 7 g/day only improved subjective universal estimation. Interestingly at two doses *Bifidobacteria* and *Lactobacilli abundance* significantly increased. Vulevic, et al. reported that supplementation of β-GOS improved flatulence, stool consistency, bloating, and growth of *Bifidobacteria* at low (3.5 g/day) and high (7 g/day) dose groups (Vulevic et al. 2018).

11.1.3.2 Probiotics

Probiotics treatment includes supplementation of *Bifidobacterium, Lactobacillus*, or *Saccharomyces* alone could cure the IBS-based symptoms. Meta-analysis of probiotic treatment identified that a treatment period of fewer than 8 weeks could improve the overall symptoms and Quality of life (QoL). Multiple Probiotic mixtures—*VSL #3* has *Bifidobacterium, Lactobacillus,* and *Streptococcus* strains and when supplemented decreased abdominal bloating in patients with IBS-Diarrhea. Supplementation in IBS patients alone reduced colonic transit time and flatulence. Another probiotic mixture *SCM-III,* containing *L. acidophilus, L. helveticus* and *Bifidobacterium* species, was effective and reduced bloating, abdominal pain, and improved bowel habit. *Pediococcus acidilactici* and two *L. plantarum* strains as the probiotic mixture could significantly improve gut-specific anxiety. However, continuous use of probiotics in susceptible patients may cause infection, sepsis, and endocarditis.

11.1.3.3 Dietary Therapy

Dietary therapy recommends the low consumption of Fermentable Oligosaccharides, Disaccharides, Monosaccharides, and Polyols also known as FODMAPs. As the gut microbiome quickly ferments FODMAPs and starts water secretion and gas production, a diet low in FODMAP (LFD) helps to control IBS symptoms. LFD with probiotic supplementation also prevents the reduction in *Bifidobacteria* population. Diets rich in FODMAPs and excessive prebiotics usage lead to dysbiosis of enteric bacteria resulting in the formation of excessive gas. Several research groups are working to treat IBS by restoring homeostatic balance in enteric bacteria with a suitable strategy.

11.1.3.4 FMT

FMT tries to restore the microflora in the intestine by injecting fecal suspension from a healthy donor into the gastrointestine of the patient. A review of FMT studies revealed that it was beneficial up to 58% of patients treated and success depended on host–microbiome characteristics. Donors rich *with Bifidobacteria species* induce symbiosis in IBS patients more efficiently. In a study made by Halkjaer et al. with oral FMT and placebo capsules, they observed enteric biodiversity in both groups (Halkjaer et al. 2017). However, an investigation made by Johnsen and group revealed that prominent improvement was obtained with freshly prepared and frozen FMT than the placebo (Johnsen et al. 2018). These observations indicate that the efficacy of FMT depends upon the factors like bowel preparation, route of delivery.

Even though the above-mentioned strategies were found to be promising, several limitations were also observed in these treatments. Proof of concept for a direct relationship between bacterial overgrowth in the small intestine with IBS symptoms exists, the linkage between the microbiome and IBS symptoms is yet to be evaluated and needs more validated experimental designs. Prebiotic usage is often linked with unwanted abdominal symptoms and their efficacy yet needs to be proved. Probiotics have heterogeneous methodologies but to a greater extent have benefits for IBS symptoms. The efficiency of probiotics requires further research on the type of probiotic to be used, measured quantity, treatment duration, etc. FMT needs improvisation in course of therapy and method of administration.

11.1.4 Inflammatory Bowel Diseases (IBDs) Therapy

IBD is an intestinal disorder resulting in extended inflammation of the gastrointestinal region and caused because of unusual immune reactions to intestinal bacteria. IBD has two diseases, namely Crohn's disease (CD) and ulcerative colitis (UC). Inflammation of the large intestine is seen in UC wherein surface mucosal layers are affected. Patchy transmural inflammation in any part of the digestive tract is found in Crohn's disease. Many countries across the world are found suffering from IBD. Dysbiosis of the gut microbiome resulting in a decrease in anti-inflammatory microbes like *Firmicutes: Clostridium clade*, and *Bacteroidetes* supports patient susceptibility to colonization by colitogenic strains, like *Escherichia species*. Four effective therapeutic strategies for treatment are (1) manipulation gut flora, (2) gut immunity, (3) maintain gut homeostasis, and (4) decrease inflammation.

11.1.4.1 Gut Microbiota Mediated Immunomodulation Therapy

Hosts immunological tolerance to commensal microbiota comes from immunological responses triggered by gut microbes. Polysaccharide A capsule of *B. fragilis* induces Treg cells and aids in developing tissue homeostasis. *B. fragilis* stimulates dendritic cells by secreting cytokine and interleukin-10 to curb inflammation of mucosa and activate CD4+ T-cells to release transforming growth factor-β. Protective immunomodulatory signals are elicited to immune cells by *B. fragilis* through outer membrane vesicles. Commensal *Segmented Filamentous Bacteria* (SFB) and *Mucispirillum* elicit T-dependent IgA production. These investigations provide a deeper understanding of the role of gut microbes in triggering or inhibiting host immune response, which could be used later to modulate inflammatory responses in clinical IBD (Chu et al. 2016; Lécuyer 2014; Bunker et al. 2015).

11.1.4.2 Precision Edited Gut Microbiota Mediated Therapy

This method was investigated by research Zhu et al. with a colitis mouse model (Zhu et al. 2018). Gut inflammation caused by *Enterobacteriaceae* was specifically inhibited by targeting molybdenum-dependent pathways by Tungsten (W) treatment because tungsten can replace molybdenum, and prevents its consumption in pathways during inflammation. This treatment reverses the dysbiotic state of the gut to normal composition in colitic mice by diminishing the pro-inflammatory markers and pathological changes. Colonization of *Clostridium scindens* which makes 7α-dehydroxylation of bile acid was observed to give resistance to *C. difficile* infection during precision microbiome editing experiments carried out by Buffie et al. (2014). Multiprotein oligomers responsible for activating inflammatory responses are known as inflammasomes.

Inflammasomes work on the composition of gut microbiota with pre-designed microbiota induced T-reg cells which increase the anti-inflammatory activity in the intestine and thus control IBD. The potential of remodeled microbiota was also investigated in experimental colitis, by using mutant strains of *Candida glabrata* lacking chitin synthase-3 had a less inflammatory response when compared with control yeast strains. Thus in the future remodeled microbiota could be explored as an effective treatment for colitis.

11.1.4.3 Probiotics Mediated Therapy

Two major bacterial genera, *Lactobacillus* and *Bifidobacterium,* were identified to maintain the equilibrium between modulations in gut microbiota and repress the overgrowth of pathogens. In the mouse model of Ulcerative Colitis, probiotics induced T-reg cell production by regulating the growth of TH2 cells and the ratio of Treg/TH1/TH2 cell populations. Overall protection against mucosal inflammation in the intestine is achieved by changing TLRs and signaling pathways in a probiotic-dependent manner. Depending on the disease context probiotic strains differ in mechanism of action and have diverse pathways in regulation. Further studies are required in dose, treatment duration, and strains used, to recommend probiotics in clinical therapy of IBD.

11.1.4.4 Engineered Probiotics Mediated Therapy

Employing genetically engineered probiotic strain—*Lactococcus lactis* an IL10 (anti-inflammatory molecule) producer, 50% of colitis was attenuated in the mouse model. Protective role engineered strain of L. *lactis* was understood when given orally to trinitrobenzene sulfonic acid-treated mice (Steidler et al. 2000). In a DSS colitis model, administration of modified *L. lactis* to produce an IL-7 like cytokine could attenuate the inflammation by lowering the secretions of DCs. This approach of manipulating the probiotic strain to produce anti-inflammatory cytokines seems to be promising to treat inflammatory diseases effectively.

Enhanced levels of reactive oxygen species would develop inflammation and IBD leading to tissue damage. *Citrobacter rodentium* growth results in better oxygenation to mucosal surface. The effects of ROS are neutralized by superoxide dismutase (SODs) secreted by intestinal mucosa making diffusible hydrogen peroxide from superoxide anions. Liu et al. demonstrated that engineered *Bifidobacterium longum* containing recombinant human manganese SOD fusion protein successfully expressed in mice model and successfully stopped colitis by decreasing pro-inflammatory cytokines (Liu et al. 2018). In another study, *B. longum* expressing α melanocyte-stimulating hormone an anti-inflammatory peptide is known to differentially control the anti-inflammatory cytokines in experimental colitis.

Attempts were also made to utilize H47 producing EcN, i.e., bacteria's narrow spectrum antibiotics to kill pathogens responsible for inflammatory bowel diseases. Spisni and colleagues demonstrated that RNAi technology targeting COX2, a pro-inflammatory enzyme in a pathogenic *E.coli* strain could successfully decrease the activity of pro-inflammatory cytokines when administered in a mouse model (Spisni et al. 2015). Treatment of IBD was also carried out by using systemic antibodies against cytokines. Various other strategies like engineering *Lactobacilli* to recognize pro-inflammatory cytokines, cloning natural protease inhibitor, and lipoteichoic acid-mediated secretion of IL10 by mutant *L. plantarum* to decrease pro-inflammatory cytokines were proved to be effective in mouse models; however, such methods need evaluation in humans for application.

11.1.4.5 Microbial Metabolites Mediated Therapy

In IBD patients, butyrate-producing bacteria are found to be washed out supporting the growth of pathobionts and thus paving the way to inflammation. SCFAs produced by gut bacteria are important mediators of "protective" signals. The host system identifies and responds to SCFA which are known to stimulate various physiological functions, thus reducing inflammation. SCFAs control excessive intestinal inflammation by promoting T-cell responses through T-reg cells expressing SCFA-GPCR (G protein-coupled receptors) mechanisms or by histone deacetylase (HDACs) inhibition. Therefore, SCFAs produced by gut microbiota are proved to be beneficial as they promote mediators to resist inflammation and inhibit cytokines using different mechanisms and establish homeostasis in the gut.

Administration of butyrate to colitis-induced mice showed resistance towards migration of enteropathogens. Butyrate could induce segregation of homeostatic macrophages and increase antimicrobial activity in macrophages. This was possible by making shifts in macrophages metabolism, e.g., reduced glycolysis, mTOR kinase activity, and increased bacterial autophagy protein. SCFAs, for example, acetate was produced by fermentation of dietary fiber in the gut and triggered IL-18 production and restored gut homeostasis in the animal colitis model (Macia et al. 2015). Propionate—a less well-studied SCFA could reduce the expression of pro-inflammatory cytokines. Non-SCFAs microbial-derived products like vitamin B6 when supplemented along with *Bacteroides* spp. could clear pathogens. These studies highlight the mechanisms of microbial metabolites in modulating the host responses. These strategies look promising for clinical application in the future.

11.1.4.6 Intestinal Stem Cells-Gut Microbiota Mediated Therapy

Epithelial integrity and tissue homeostasis are maintained by pluripotent Lgr5+ stem cells. Intestinal stem cells (ISCs) interact with gut microflora and change epithelial regeneration during inflammation. When these cells are exposed to microbial muramyl dipeptide (MDP), cytosolic NOD2 expression is increased causing cytoprotection of stem cells against oxidative stress. In the intestine, microbes indirectly raise stem cell numbers and influence host barrier function. Even probiotics and diet also influence the gut microbiome and present alternate mechanisms for enhancing the stem cell population. There is a scope for further investigation on the interplay between ISC, probiotic, and dietary as therapeutics for IBD treatment. Mesenchymal stem cells which are heterogeneous non-hematopoietic, can differentiate into various cell types and participate significant roles in gut immunity and inflammation.

Since IBD is an immune-mediated disease, therapies-based MSCs were tried in pre-clinical models for tissue regeneration, restoring immune balance, and microbiome. The potential role of ISCs as therapeutics was discussed in detail in these studies (Shamoon 2019).

11.1.5 Microbiome-Based CDI Therapy

Clostridium difficile infection (CDI) has caught the attention of clinicians and researchers alike because of its prominence and prevalence. CDI is identified with symptoms like mild to moderate non-bloody diarrhea, abdominal cramping, and tenderness. In severe cases, watery diarrhea with abdominal pain, fever, nausea, anorexia, and malaise are noted. Besides, leukocytosis, elevated C-reactive protein, and low albumin levels are present. CDIs are frequently attributed to antibiotics altering the normal gut flora thus allowing *C. difficile* to nourish (Katz 2006). The majority of the therapeutic agents designed for CDI control are antimicrobial agents, toxin binding agents, immune modifying agents, probiotics, and FMT.

11.1.5.1 Probiotics

Probiotics are considered an effective method for restoring homeostasis in the gastrointestinal tract and to prevent or treat CDIs. Commonly used probiotics for CDI therapy are *Lactobacillus* and *Bifidobacterium* and the yeast *Saccharomyces* (Boyle et al. 2006; Katz 2006).

In a clinical study conducted by Hickson et al., the effectiveness of *Lactobacillus casei*, *Lactobacillus bulgaricus,* and *Streptococcus thermophilus* as prophylactic agents with 115 patients of age 50 years who were on antibiotics were investigated (Hickson et al. 2007). Within 2 days of antibiotic therapy, probiotic administration was initiated and continued for 7 days. They observed a 17% reduction in the prevention of CDI. However, probiotics usage has limitations because they do not undergo rigorous testing, unlike pharmaceutical agents (Boyle et al. 2006). Probiotics while eliminating the infectious bacteria may show certain side effects for e.g. bloating and flatulence caused by *Lactobacillus and Bifidobacterium.*

11.1.5.2 FMT

FMT comes under live biotherapeutics drugs for CDI treatment and is regulated by the FDA just like any other drug. It is an option for patients with recurrent CDI as 80% of clinical response rates were recorded. Since FMT includes the administration of live feces-derived mixtures of microorganisms, potential safety issues are involved. The impact of long-term manipulation of gut microbiota by FMT is unknown and needs consent from both the receivers and donors. Limitations in present delivery routes encourage researchers to design alternative drugs with reduced risk and ease of delivery without compromising on the efficacy of FMT.

Another approach is the use of freeze-thawed and encapsulated material instead of fresh feces for FMT. When compared with controls, freeze-thawed versus fresh feces delivered through enema was good without recurrence up to 13 weeks. However, in each treatment group about one-third of the patients required two FMTs to prevent recurrences. *RBX2660* is a standardized product of live microbiota suspension developed for the treatment of recurrent CDI and proved to be effective as FMT (Orenstein et al. 2016). *SER-109* or *Seres* is another biotherapeutics prepared by encapsulation of *Firmicutes* spores mixture obtained from human feces. Ethanol treatment was used to reduce the spread of infectious agents in the product. Seres have completed two clinical phase trials. In the first noncomparative clinical trial 26 out of 30 patients (86.7%) could be recovered and 96.7% of patients had a good clinical resolution. SER-109 could turn around the loss of gut microbiota after the treatment. In the second clinical study, SER-109 could not reduce CDI recurrence but was found effective in aged patients. Keeping because of these limitations, alternative approaches are under clinical trials (Khanna et al. 2016; Wilcoxb 2016).

11.1.6 Microbiome Therapeutics for Food Allergy (FA)

Food allergies are estimated to affect around 4%–6% of children and 4% of adults in a survey made by the Center for Disease Control and Prevention. It occurs because of the hyper-reactivity of the immune response to any component in food. FA is most frequent in young children but can appear at any age. Food allergy correlates dysbiosis during childhood and immune conditions later part of life. During epidemiological surveys, it was identified that environmental factors like caesarian births, no breastfeeding, drugs, and diet with low-fiber content and high-fat are important contributors to dysbiosis.

In a research investigation, the fecal microbiota of infants suffering from allergy was compared with healthy infants revealed alterations in gut microflora with the development of allergy (Nakayama and Kobayashi 2011). Allergic group infants had more Bacteroides, *Propionibacterium,* and *Klebsiella* in first & second months, respectively, while *Acinetobacter* and *Clostridium* were found in the non-allergic group of similar age. Non-allergic infants have SCFA producing *Clostridium* species which prevent the growth and proliferation of *Bacteroides* which are most abundant in FA. SCFAs are vital metabolites of gut flora that protect from a food allergy and regulate the immune system in epigenetic mode. Depletion of butyrate-producers could lead to dysbiosis and initiate FA. Therefore SCFA supplementation could be explored further to cure food allergy.

Microbiota-directed therapy has the potential in decreasing the occurrence of FA. One research study had demonstrated that dietary patterns have an impact on the progress of FA and influence the gut microbiota population. Breast milk provides a primary source of commensal microbiota in the infant's gut and later helps in the maturation of the microbiota. An abundance of oligosaccharides and TGF-β2 in breast milk promotes the growth of SCFA producing *Bifidobacterium* and *Streptococcaceae* and *Ruminococcaceae* members, respectively.

Modulation of gut flora is more effective through diet as well. An infant diet enriched with fruits, vegetables, and homemade foods could lessen the FA. A high-fiber rich diet promotes the growth of *Bifidobacterium* and *Lactobacillus* which ferment the fiber foods and thereby increase serum SCFA levels. Studies on neonatal prebiotic supplementation could control eczema.

Supplementation of *Lactobacillus rhamnosus GG* (LGG) to breastfeeding mothers and infants could control the risk of eczema. *L. rhamnosus* supplementation along with oral immunotherapy led to tolerance to peanuts in allergic children. Similarly, administration of LGG containing extensively hydrolyzed casein formula provided tolerance to an allergy of cow milk. Comparison of the fecal microbiota of infants receiving EHFC + LGG with the infants receiving EHCF alone revealed enrichment of butyrate-producing flora and high fecal butyrate levels. Butyrate is reported as an epigenetic regulator and the use of EHCF+LGG induces methylation of the promoter of cytokines genes. Such epigenetic regulation was observed in children supplemented with EHCF+LGG rather than children treated with other methods. Long-term protection was possible by targeting gut flora in children suffering from FA. Many more trials are necessary to assess the roles of probiotic strains in manipulating gut flora. Integration of clinical trials with epigenetics and metabolomics data may lead to novel therapeutics (Rosita Aitoro 2017).

11.1.7 Microbiome Therapy for Autoimmune Diseases

Autoimmune diseases are caused due to intestinal dysbiosis leading to impaired gut barrier function, increased inflammation, and immunity. Dysbiotic flora induces neo-epitope formation by imperfect post-translational modification of luminal proteins. This in turn aggravates systemic autoimmunity and provokes autoimmune diseases (Lerner et al. 2016).

11.1.7.1 Type 1 Diabetes (T1D)

Type 1 diabetes (T1D) is caused due to the destruction of insulin-secreting cells, therefore cells start depending on exogenous insulin. Several research studies tried to establish the relationship between dysbiosis and T1D. The role of the gut flora in humans was first investigated in four T1D children in Finland by examining their stool samples and reported a low ratio of the *Firmicutes: Bacteroidetes* than controls (Giongo et al. 2011; De Goffau et al. 2013). Later, De Goffau also reported that children with autoantibodies against b-cells had more *Bacteroidetes* and less lactate and butyrate producers in their feces samples (Li et al. 2015). In a study with seroconverted T1D patients, Davis-Richardson et al. found plenty of *Bacteroides dorei* and *Bacteroides vulgatus*. From these research studies, it could be possible to consider early dysbiosis as an indication to predict T1D in genetically predisposed individuals (Davis-Richardson 2014).

Administration of *Lactobacillaceae*-enriched probiotic orally protected NOD mice from T1D by suppressing autoimmune response in the gut (Dolpady et al. 2016). In humans, a TEDDY group evaluated probiotic supplementation for 7473 children from 4 to 10 years in the age group with genetic risk for T1D during their first year of life. Early probiotic administration was correlated with a decreased risk of islet autoimmunity when compared with the group that received probiotics after 27 days of life or no supplementation (Uusitalo et al. 2016).

11.1.7.2 Multiple Sclerosis

Multiple sclerosis (MS) is a seriously disabling disease of the Central Nervous System in which the immune system attacks the protective myelin of nerve fibers and causes communication problems between the brain and the rest of the body. This disease can cause permanent deterioration of the nerves and thus disability in young adults. Environmental factors like a viral infection, a hypocaloric diet, vitamin D deficiency, and dysbiosis trigger MS along with susceptible HLA alleles (Dendrou et al. 2015). Research studies indicate that intestinal dysbiosis is one of the factors of extraintestinal disease development.

A recent research study made on 60 Relapsing-Remitting MS (RRMS) patients, observed an increased number of *Methanobrevibacter* which is involved in recruiting macrophages and activating dendritic cells, *Akkermansia* playing a role in degrading the mucus layer and promoting inflammation and decrease in *Butyricimonas*—butyrate-producing bacteria which has immunomodulatory properties in the gut of untreated patients (Cantarel et al. 2015). However, in patients treated with IFN-b and glatiramer acetate, there was an increase in the number of *Prevotella* which is associated with high-fiber ingestion and has regulatory roles via butyrate generation. Dysbiosis in RRMS patients was also reported by Chen et al., in their study while comparing stool samples from patients with healthy controls. They observed a plethora of bacterial strains like *Pseudomonas, Mycoplasma, Haemophilus, Blautia,* and *Dorei* in patients when compared with healthy counterparts who had only *Prevotella* and *Parabacteroides* in abundance (Chen et al. 2016a, b). Current research findings point to the gut–brain axis connection during dysbiosis in animal models and MS patients. The relationship between immunity in the gastrointestinal mucosa and commensal bacteria seems to promote important physiological homeostasis for the host (Colpitts et al. 2017).

Several studies established the immunoregulatory functions of probiotic administration in MS. Treatment with *Lactobacillus spp., Pediococcus acidolactici, Bifidobacterium bifidum, Bifidobacterium animalis,* and *Bacteroides fragilis* improved CNS inflammation by inducing Treg cells in the gut mucosa thus promoting the secretion of IL-10, transforming growth factor-b and inducing decreased Th1/Th17 inflammatory subsets in mice model (Lavasani et al. 2010). Kouchaki et al. reported a decrease in inflammatory markers in MS patients treated with probiotic supplementation containing *Lactobacillus acidophilus, Lactobacillus casei, Lactobacillus fermentum,* and *Bifidobacterium bifidum* (Kouchaki et al. 2016). This randomized double-blind placebo-controlled clinical trial analyzed probiotic intake for 12 weeks in 60 MS patients. Although these studies give hope for the treatment of MS, future studies are still required to determine the real role of the gut microbiota in CNS demyelinating diseases.

11.1.7.3 Rheumatoid Arthritis

Another systemic autoimmune disorder is Rheumatoid Arthritis (RA) which is characterized by persistent inflammation of joints. Autoantibodies like anti-cyclic citrullinated peptide and/or rheumatoid factor are detected before RA occurrence (Van de Wiele et al. 2017) and initiated with factors like smoking, infections, and dysbiosis (Klareskog et al. 2006). Chen et al. identified decreased species richness (alpha-diversity) in gut microbiota in RA patients and a positive connection among rheumatoid factor levels and disease progression. In RA patients along with the pro-inflammatory cytokine *Eggerthella, Actinomyces, Turicibacter, Streptococcus,* and *Collinsella* species were more in gut microbiota (Chen et al. 2016a, b). Beta-diversity found in the gut microbiota of RA patients was observed to have a link with the levels of rheumatoid factor, C-reactive protein, suggesting that these parameters play role in gut microbiota modulation (Yatsunenko et al. 2012). The establishment of intestinal dysbiosis signature will thus help to develop therapeutic tools (Ciccia et al. 2016; Di Paola et al. 2016).

11.1.7.4 Systemic Lupus Erythematosus (SLE)

SLE is a heterogeneous disease wherein autoantibodies mainly bind with nuclear and cytoplasm antigens damaging the vital parts of the body (Apostolidis et al. 2011). Genetic and external factors like viral infections, and exposure to UV-B rays are involved in pathogenesis. Several research studies provide the role of intestinal dysbiosis in SLE development (Van de Wiele et al. 2017). Hevia et al. found decreased Firmicutes*:* *Bacteroidetes* ratios in stool samples of SLE when analyzed with the healthy counterparts (Hevia et al. 2016) Another study made with stool sample analysis of SLE patients with control subjects reported low numbers of *Firmicutes* and an enhanced population of *Bacteroidetes* members. The gut microbiota profile of SLE patients has a predominance of *Rhodococcus, Eggerthella, Klebsiella, Prevotella, Eubacterium, Flavonifractor,* and *Incertae sedis* genera. Based on these reports, investigation on in vitro differentiation of T-reg, Th1, and Th17 cells in SLE patients was carried out (Lopez et al. 2016). Treg-inducing bacterial supplementation decreased the balance of Th17/Th1 in SLE patients, indicating that these strains could be used as therapeutic probiotics.

11.1.8 Cardiovascular Disease (CVD)

Modulations in gut microbiome composition are known to be associated with CVD including atherosclerosis, dyslipidemia, hypertension, and heart failure. Gut microbiota persuades the host signaling and immunomodulatory effects through microbial metabolites. For example, SCFA, bile acids, and trimethylamine-N-oxide (TMAO) are contributing factors of CVD risk.

Patients with coronary artery disease contain DNA from bacterial species in the atherosclerotic plaques (Kubinak et al. 2016). Modulation of SCFA production and inflammatory signaling played a prominent role in inducing atherosclerosis when experimented within mice with gut microbial transplants. Similar experiments also demonstrated that TMAO is proatherogenic, prothrombotic, and an important underlying contributor to CAD risk (Hamada et al. 2002). In an investigation with animal models, TMAO is observed to decrease the mechanism of removal of cholesterol transport by macrophages. Intestinal microbiota could also regulate host lipid metabolism.

Gut bacteria with their microbial enzymes make different bile acid metabolites which activate host receptors and interfere with lipid metabolism. Short-chain fatty acids and sodium-dependent inflammation are microbiome-mediated contributors to hypertension (Fagarasan et al. 2010). Though the evidence suggests a fundamental role for the microbiota in CVD, many questions remain in the mechanism they cause. Drug pharmacokinetics and pharmacodynamics are observed to be influenced by gut microbiota (Fukata et al. 2013). Microbiome-based precision medicine could be designed or developed by gaining knowledge in molecular mechanisms contributing CVD risk and drug response.

11.1.9 Microbiome-Based Therapy for Non-Communicable Diseases (NCD)

NCDs occur in different organs of the body and become systemic diseases. Unlike infectious diseases which are transient, latent, or lethal, NCDs are seen over a longer duration. For example—allergy, asthma, and some autoimmune diseases come under early-onset NCD, while CVD, metabolic disease, and neurodegenerative disorders fall under later-onset NCDs. Generally, NCDs were thought of as diseases of more developed countries alone, but now they are global problems affecting entire human health. NCDs combined with microbial resistance to antibiotics had posed a major threat to the integrity of healthcare systems and to the global economy.

Interestingly all NCDs have common environmental risk factors with altered gut colonization patterns. A decrease in the microbiota, increase in immunologic and metabolic deregulation are very common attributes seen in all NCDs. Fluctuations in the gut microbiome occur because of complex changes in nutritional patterns and environment. Early environmental changes are involved in the increase in early-onset inflammatory NCDs like an allergic disease. The gut microbial deprivation hypothesis demonstrates a link between reduced gut microbiota diversity and early-onset NCDs, including atopy, eczema, and asthma. Microbiome modulation with strategies like microbiota seeding, feeding, and rebiosis appear as promising methods for sustainable healthcare. These methods try to treat emerging NCDs, reduce later-life health risks, and reduce recurrent infections with personalized medicine approaches (Fukata et al. 2013).

Recently, researchers and clinicians are also making efforts to adopt systems biology-approach to manage NCDs. New therapeutics could be designed by integrating metagenomic and metabolomic information which would extend effective healthcare and preventative strategies to the patients. This approach majorly focused on the early phases of conception to birth where epigenetic programming influenced over the life course and helped researchers to integrate human gut microbiota information with epigenetic. These combined approaches are ultimately direct to better strategies to control the diseases.

11.2 Conclusion

Although dysbiosis and decrease in microbiota diversity are common in almost all diseases, yet whether a microbial deviation is the reason or result needs to be investigated. To address the challenges faced in microbiome modulation, researchers need to focus on microbiota signatures as biomarker development for diagnosis. Instead of looking at microbiome composition, it is most important to study the role of the microbiome in a disease condition. Deeper insight into the role of a healthy gut microbiome is possible through Human Microbiome Project and MetaHit projects to design or plan dietary modulation. Gut colonization patterns need to be developed with prebiotics, nonconventional indigenous gut bacteria, and FMT because they might have continuous influences on health. Therefore, research on microbiome modulation should enhance the assessment of risk in diseases, discovery, and efficacy of drugs.

Microbiome-based therapeutics are being identified as precision medicine because of the contribution to inter-individual variability in disease conditions and also as a variable factor for the improvement of potential therapeutics. They have the potential to influence the therapeutic strategies in the treatment of several diseases, generate new economic opportunities, and benefit humans around the globe.

Acknowledgement Authors thank the Principal and Management of Bhavan's Vivekananda College, for the encouragement and support.

Conflict of Interest The authors declare that there is no conflict of interest.

References

Apostolidis SA, Crispín JC, Tsokos GC (2011) IL-17-producing T cells in lupus nephritis. Lupus 20 (2):120–124

Boyle RJ, Robins-Browne RM, Tang MLK (2006) Probiotic use in clinical practice: what are the risks? Am J Clin Nutr 83:1256–1264

Bozkurt HQ (2019) Bifidobacterium animalis subspecies lactis engineered to produce mycosporine-like amino acids in colorectal cancer prevention. SAGE Open Med 7:2050312119825784

Buffie CG et al (2014) Precision microbiome reconstitution restore s bile acid mediated resistance to Clostridium difficile. Nature 517:205–208

Bunker JJ et al (2015) Innate and adaptive humoral responses coat distinct commensal bacteria with immunoglobulin A. Immunity 43:541–553

Cantarel BL et al (2015) Gut microbiota in multiple sclerosis: possible influence of immunomodulators. Invest Med 63:729–734

Cao HX (2017) Secondary bile acid-induced dysbiosis promotes intestinal carcinogenesis. Int J Cancer 140:2545–2556

Chen J et al (2016a) Multiple sclerosis patients have a distinct gut microbiota compared to healthy controls. Sci Rep 6:28484

Chen J et al (2016b) An expansion of rare lineage intestinal microbes characterizes rheumatoid arthritis. Genoma Med 8:43

Chu H et al (2016) Gene-microbiota interactions contribute to the pathogenesis of inflammatory bowel disease. Science 352:1116–1120

Ciccia F et al (2016) The role of the gastrointestinal tract in the pathogenesis of rheumatic diseases. Best Pract Res Clin Rheumatol 30:889–900

Colpitts SL et al (2017) A bidirectional association between the gut microbiota and CNS disease in a biphasic murine model of multiple sclerosis. Gut Microbes 8(6):561–573

Davis-Richardson AG (2014) Bacteroides dorei dominates gut microbiome prior to autoimmunity in Finnish children at high risk for type 1 diabetes. Front Microbiol 5:678

De Goffau MC et al (2013) Fecal microbiota composition differs between children with beta-cell autoimmunity and those without. Diabetes 62:1238–1244

Dendrou CA et al (2015) Immunopathology of multiple sclerosis. Nat Rev Immunol 15(9):545–558

Di Paola M et al (2016) Alteration of fecal microbiota profiles in juvenile idiopathic arthritis. Associations with HLA-B27 allele and disease status. Front Microbiol 7:979

Ding CT (2018) Intestinal microbiota: a novel perspective in colorectal cancer biotherapeutics. OncoTargets Ther 11:4797–4810

Dolpady J et al (2016) Oral probiotic VSL#3 prevents autoimmune diabetes by modulating microbiota and promoting indoleamine 2,3-dioxygenase-enriched tolerogenic intestinal environment. J Diabetes Res 2016:7569431

Fagarasan S et al (2010) Adaptive immune regulation in the gut: T cell-dependent and T cell-independent IgA synthesis. Annu Rev Immunol 28:243–273

Fukata M et al (2013) The role of pattern recognition receptors in intestinal inflammation. Mucosal Immunol 6:451–463

Giongo A et al (2011) Toward defining the autoimmune microbiome for type 1 diabetes. ISME J 5:82–91

Goodwin AC et al (2011) Polyamine catabolism contributes to enterotoxigenic Bacteroides fragilis-induced colon tumorigenesis. Proc Natl Acad Sci 108:15354–15359

Halkjaer SI et al (2017) Can fecal microbiota transplantation cure irritable bowel syndrome? World J Gastroenterol 23:4112–4120

Hamada H et al (2002) Identification of multiple isolated lymphoid follicles on the antimesenteric wall of the mouse small intestine. J Immunol 168:57–64

Hausen Z (2009) The search for infectious causes of human cancers: where and why. Virology 392:1–10

Hevia A et al (2016) Intestinal dysbiosis associated with systemic lupus erythematosus. MBio 5: e01548

Hickson M et al (2007) Use of probiotic lactobacillus preparation to prevent diarrhoea associated with antibiotics: randomized double blind placebo controlled trial. BMJ 335:80–83

Johnsen PH et al (2018) Faecal microbiota transplantation versus placebo for moderate-to-severe irritable bowel syndrome: a double-blind, randomised, placebo-controlled, parallel-group, single-centre trial. Lancet Gastroenterol Hepatol 3:17–24

Katz JA (2006) Probiotics for the prevention of antibiotic-associated diarrhea and Clostridium difficile diarrhea. J Clin Gastroenterol 40:249–255

Khanna S et al (2016) A novel microbiome therapeutic increases && gut microbial diversity and prevents recurrent Clostridium difficile infection. J Infect Dis 214:173–181

Klareskog L et al (2006) Mechanisms of disease: genetic susceptibility and environmental triggers in the development of rheumatoid arthritis. Nat Clin Pract Rheumatol 2:425–433

Kouchaki E et al (2016) Clinical and metabolic response to probiotic supplementation in patients with multiple sclerosis: a randomized, double-blind, placebo-controlled trial. Clin Nutr 16:30214

Kubinak JL et al (2016) Do antibodies select a healthy microbiota. Nat Rev Immunol 26:767–774

Lavasani S et al (2010) A novel probiotic mixture exerts a therapeutic effect on experimental autoimmune encephalomyelitis mediated by IL-10 producing regulatory T cells. PLoS One 5: e9009

Lécuyer E (2014) Segmented filamentous bacterium uses secondary and tertiary lymphoid tissues to induce gut IgA and specific T helper 17 cell responses. Immunity 40:608–620

Lerner A et al (2016) Dysbiosis may trigger autoimmune diseases via inappropriate post-translational modification of host proteins. Front Microbiol 7:84

Li X et al (2015) The role for gut permeability in the pathogenesis of type 1 diabetes – a solid or leaky concept? Pediatr Diabetes 16:485–492

Liu M et al (2018) Oral engineered Bifidobacterium longum expressing rhMnSOD to suppress experimental colitis. Int Immunopharmacol 57:25–32

Lopez P et al (2016) Th17 responses and natural IgM antibodies are related to gut microbiota composition in systemic lupus erythematosus patients. Sci Rep 6:24072

Macia L et al (2015) Metabolite-sensing receptors GPR43 and GPR109A facilitate dietary fibre-induced gut homeostasis through regulation of the inflammation. Nat Commun 6:6734

Muhammad Bilal Abid NN (2019) Gut microbiome and CAR-T therapy. Exp Hematol Oncol 8:31

Nakayama J, Kobayashi T (2011) Aberrant structures of fecal bacterial community in allergic infants profiled by 16S rRNA gene pyrosequencing. FEMS Immunol Med Microbiol 63:397–406

Orenstein R et al (2016) Safety && and durability of RBX2660 (microbiota suspension) for recurrent Clostridium difficile infection: results of the PUNCH CD study. Clin Infect Dis 62:596–602

Pandey KN (2015) Probiotics, prebiotics and synbiotics- a review. J Food Sci Technol 52:7577–7587

Rosita Aitoro LP (2017) Gut microbiota as a target for preventive and therapeutic intervention against food allergy. Nutrients 9:672

Shamoon M (2019) Recent advances in gut microbiota mediated therapeutic targets in inflammatory bowel diseases: emerging modalities for future pharmacological implications. Pharmacol Res 148:104344

Singh RC-W (2017) Influence of diet on the gut microbiome and implications for human health. J Transl Med 15:73

Spisni E et al (2015) Cyclooxygenase-2 silencing for the treatment of colitis: a combined in vivo strategy based on RNA interference and engineered Escherichia coli. Mol Ther 23:278–289

Steidler L et al (2000) Treatment of murine colitis by Lactococcus lactis secreting Interleukin-10. Science 289:1352–1355

Uusitalo U et al (2016) Association of early exposure of probiotics and islet autoimmunity in the TEDDY study. AMA Pediatr 170:20–28

Van de Wiele TV et al (2017) How the microbiota shapes rheumatic diseases. Nat Rev Rheumatol 12:398–411

Vulevic J et al (2018) Effect of a prebiotic galacto oligosaccharide mixture on gastrointestinal symptoms in adults selected from a general population who suffer with bloating, abdominal pain, or flatulence. Neurogastroenterol Motil 30:e13440

Wilcoxb JM (2016) New and emerging therapies for Clostridium difficile infection. Curr Opin Infect Dis 29:546–554

Wu NW (2006) Dysbiosis signature of fecal matter microbiota in colorectal cancer patients. Microb Ecol 66:462–470

Yatsunenko T et al (2012) Human gut microbiomes are viewed across age and geography. Nature 486:222–227

Yoshioka NT-I (2009) Intestinal macrophages involved in the homeostasis of the intestine have the potential for responding to LPS. Anticancer Res 29:4861–4865

Yu JF (2017) Metagenomic analysis of faecal microbiome as a tool towards targeted non-invasive biomarkers for colorectal cancer. Gut 66:70–78

Zhang XZ-Y (2019) Fecal Fusobacterium nucleatum for the diagnosis of colorectal tumor: a systematic review and meta-analysis. Cancer Med 8:480–491

Zhu W et al (2018) Precision editing of the gut microbiota ameliorates colitis. Nature 553:208–211

Recent Advancements in Microbiome–Immune Homeostasis and their Involvement in Cancer Immunotherapy

12

Anusha Konatala, Fain Paracket, and Pola Sudhakar

Abstract

The microbial genome of bacteria, archaea, protists, fungi, and viruses which colonize in humans is known as the microbiome. The population of microbes in the human body is known as microbiota, its composition may differ with concerning host factors like sex, age, obesity, physical inactivity, alcohol consumption, smoking, diet, and polymorphisms in dominant human oncogenes. The current chapter is ascertaining the connection between microbiota and cancer, the role of the microbiota in cancer immunotherapy, which leads to significant advances and scope in the etiology of cancer. Different processes are studied and have been promising to conclude the role of microbiota in tumorigenesis and progression, processes like genotoxicity, induction of chronic inflammation, bacteria-mediated cell proliferation, and activation of procarcinogens show the interference of microbiota with the tumors. More research studies must focus on microbiota interaction with the host to define its contribution to the growth and development of cancers and identify microbiome as a potential cancer marker and develop personalized medicine to treat malignancies. This chapter outlines various researches, explaining how the microbiota itself enclose a novel paradigm in the prevention of cancer and its management. Paramount to develop microbiota-based immunotherapy for treating cancer, few challenges in microbiome research are to identify individual microbial species such as viruses, protozoans, archaea, protists, and fungi that causally affect cancer phenotypes and unravel the underlying mechanisms. Here, we discuss a few relevant technologies and few

A. Konatala · F. Paracket
University of Windsor, Windsor, ON, Canada

P. Sudhakar (✉)
Department of Biotechnology, AUCST, Andhra University, Visakhapatnam, Andhra Pradesh, India
e-mail: sudhakar@andhrauniversity.edu.in

P. V. Bramhachari (ed.), *Microbiome in Human Health and Disease*,
https://doi.org/10.1007/978-981-16-3156-6_12

challenges in studying the microbiome and their involvement in cancer immunotherapy.

Keywords

Microbiome · Immunotherapy · Carcinogenesis · Epidemiology · CTLA-4 Cytotoxic T-lymphocyte Antigen-4

12.1 Introduction

The human microbiota, collective of primary bacteria and other microbes, like archaea, fungi, protozoa, protozoan viruses, helminthic worms that reside in the human body, the genetic material of microbiota is known as the microbiome. It is an inhabitant in the human body, within the surface of the epithelial barrier, 99% occupied in the gut and 1% in skin, vagina, nasal, and mouth. The fluctuation of the equilibrium, which is detrimental due to the loss of beneficial microorganisms, is known as dysbiosis. Physiology factors, lifestyle changes in the diet all affect the health of microbiomes in the host. This dysbiosis causes inflammation in epithelial cells and is known to cause tumor development. There are corrective treatments for dysbiosis mentioned in this chapter, which showed promising results in the microbiome caused by diseases. Moreover, multiple techniques and models used in the characterization of microbes are discussed briefly in this chapter.

The gut microbiome has shown significant importance in immune cell development and maintains equilibrium with commensal microbes. These microbes have regulatory roles in the development of mucosal immune systems. Intestinal microbes produce short-chain fatty acids, which play a crucial role in tumor prevention and activation for apoptosis. Besides, they have an impact on efficacy immunotherapy in cancer patients by blocking cytotoxic T-lymphocyte-associated antigen 4 (CTLA-4) and programmed cell death ligand 1 (PD-L1), which is explained below in detail (Temraz et al. 2019).

12.2 Microbiome

The Human Microbiome Research observed 11,174 primary biological specimens in a study conducted in 242 healthy adults (Methé et al. 2012). The microbiota has a considerable impact on the host's immunity, physiological functions, particularly metabolism, cognitive and neurological function, inflammation, hematopoiesis (Roy and Trinchieri 2017). The microbiome is first acquired by vertical transmission from mothers during delivery and lactation period. Newborns birth by cesarean has abundant skin microbiota of mothers compared to birth by vaginal delivery, later possess more maternal vaginal microbiota (Dominguez-Bello et al. 2016). Total microbial counts about ten times more than the human cells, with more genes, concerning the human genome (Shahanavaj et al. 2015). Primarily dominant bacteria found in the healthy human gut are *Bacteroidetes* and *Firmicutes;* their percentage range varies from 10% to 90% (Allaband et al. 2019). Some research studies detailed, secretions by the bacteria make it dominant in the interbacterial competition with a high population of *B. fragilis* strain (Garud and Pollard 2020). Microbiota of the gut is known as the "second genome" as it shows a significant impact on the internal activities of the host; thus, it is also called "forgotten organ." The gastrointestinal microbiome has been observed to have a crucial effect on overall health and serves as the best model to understand microbiota and host interactions (Schwabe and Jobin 2013).

Gut bacteria is categorized into three types based on their function in the host:

- *Symbionts*: (90% in the gut) Microbes in mutualistic benefit to the host.
- *Conditioned Pathobionts*: Usually harmless, causes disease in unfavorable condition.
- *Pathobionts*: Disease-causing microorganisms.

The pathobionts can be harmful when there is a disturbance in equilibrium between the gut microbiota and the host due to altered dietary habits, exposure of pathogens, the action of antibiotics, and other environmental factors like change of weather or disturbance in the circadian clock. Alterations in the homeostasis in the microbial communities are known as "dysbiosis" (Helmink et al. 2019a). It is affected by the physiological and pathological changes that take place in the host (Shui et al. 2020). Some recent finding suggests colorectal cancer (CRC), inflammatory bowel disease (IBD), celiac disease, obesity may have been caused by pathobionts. Dysbiosis leads to a leaky gut by exposure to pathogens, increasing the intestinal permeability, promoting translocation of gut bacteria, and dysplasia of the immune system affecting the homeostasis of the gut (Zhou et al. 2020). To treat the above condition caused by dysbiosis through fecal microbiota transplantation (FMT) and other novel therapies can be introduced. Gut microbiome study is in a preliminary stage of the investigation to know the functional properties of commensal bacteria, and its mechanism involved to interact with the host is not completely understood. Moreover, multiple promising types of research have suggested gut microbiota showcase great potential towards medical treatments of cancer and other diseases (Kho and Lal 2018).

12.2.1 Esophageal Microbiota

The esophageal microbiome has been studied in healthy and diseased conditions with the help of recent gene sequencing tools. It is observed that a healthy human esophagus contains abundant *Streptococcus* (gram +ve), compared to the infected esophagus, with a high amount of gram-negative bacteria (Rajagopala et al. 2017). Microorganisms inhabiting the esophagus are *Bacteroidetes, fusobacteria, proteobacteria,* and *spirochetes*. These microbes produce lipopolysaccharides (LPS), which acts as an immune-activating agent in stimulating innate immune responses that can directly treat malignancies. LPS interact with the innate immune system by binding to toll-like receptor-4 (TLR4), resulting in the activation of nuclear factor kappa B (NF-κB). High levels of NF-κB are observed in esophageal adenocarcinoma patients cascading levels of inflammatory cytokines like IL-1b, Il-6, IL-8, and TNF-a. In some research studies in mice, LPS prolongs the time of gastric emptying, which helps in increased gastric reflux to the esophagus. Therefore, using NF-κB host cell pathway inhibitors, probiotics, antibiotics, and microbiome in the esophagus can prevent cancer development (Shahanavaj et al. 2015). Some detailed studies are required to find the target in different diseases in the esophagus, diagnosing, therapeutics, and prevention (Lv et al. 2019).

12.3 Healthy and the Unhealthy Microbiome

12.3.1 Healthy Microbiome

Characterization of microbes as healthy and unhealthy is necessary to understand their functions and their roles in healthy and diseased conditions are critical. The gut microbiota is diverse compared to other host sites; a healthy microbiome considered in one host may not be healthy for others. Although there have been some patterns found in a study conducted in patients from different zones. Healthy hosts have rich microbiota, which harbors 1000 species of bacteria belonging to *Firmicutes* and *Bacteroidetes*. Different proposals from researchers say that a host with favorable gut microbiota has elicited an immune response against cancers due to antigen presentation and enhanced T cell function. Research studies in mice explain T cell response is defined for *B. fragilis or B. thetaiotaomicron* in microbiota promoting CTL-4 blockade seen in patients. Gopalakrishnan found responses to anti-PD-1 in skin cancer affected by gut microbiota by changing CD4+ IL-17+ cells and CD4+ FoxP3 + T cells. Restoring the efficacy of anti-PD-1 through T cells recruitment by *Akkermansia* in the gut microbiota sensitizing the cancer cells was studied (Chen et al. 2020). *Bacillus polyfermenticus*, a probiotic bacterium observed to affect the development of colon cancer cells by obstructing receptors like ErbB2 and ErbB3 by immune suppression, chronic inflammation, immune evasion (Shahanavaj et al. 2015). Moreover, probiotics destroying hepatocellular carcinoma is through SCFA production (Zhou et al. 2020).

Symbionts inhabiting the gut of the host play role of cancer transforming agents in distal and local carcinogenesis and involve indirectly causing induction of inflammation and immune suppression. Due to disturbed equilibrium, some microbes tend to act as a part of an unhealthy microbiome being involved in altering host physiology and metabolism (Li et al. 2019).

12.3.2 Unhealthy Microbiome

The gut microbiota of the host affects its immune system indirectly by suppressing and inducing inflammation leading to cancer development (Li et al. 2019). Few observed that some bacteria promote chronic inflammation to activate macrophages, increase reactive oxygen species (ROS) generation leading to DNA breakage and mutations (Zhou et al. 2020). Obesity leads to dysbiosis with a high volume of Clostridia, which produces secondary bile acid, deoxycholic acid (DCA) and promotes hepatic cellular carcinoma (HCC) (Schwabe and Jobin 2013).

Human tissues tightly regulate growth and death promoting signals to maintain homeostatic cell densities, tissue function, and architecture of the tissue or organ. Disruption in these signals results in uncontrolled cellular proliferation. E-cadherin and intercellular adhesion molecule have been a target for the intestinal bacteria to promote epithelial cell proliferation by activation of Wnt/u-catenin pathway (Fulbright et al. 2017).

Cyclophosphamide was found effective in translocation of the *Enterococcus hirae* small intestine bacteria to spleen and colonization of *Barnesiella intestinihominis* in the colon of the host; these microbes together contribute to the antitumor immunity (Li et al. 2019). Some examples of an unhealthy and healthy microbiota, affecting the physiological and metabolic activities of the host are listed in Fig. 12.1.

12.4 Techniques and Tools for Microbiome Analysis

Gut microbiota is well understood in recent years with the help of advanced gene sequencing tools and humanized gnotobiotic models (Kho and Lal 2018). These advanced sequencing tools have helped researchers to generate millions of sequences to study different microbial communities. Conventional techniques used to unravel the gut microbiome are 16S ribosomal RNA, metabolic characterization of the microbiome, gene amplicon sequencing, shotgun, single-cell RNA sequencing by CRISPR–Cas technology, metagenomic sequencing (Elinav et al. 2019), and next-generation sequencing tools. Of all, the composition of host–microbiota can be defined by 16S RNA ribosome amplicon sequencing and whole-genome shotgun (WGS) sequencing. Through WGS appropriate detection of the species, strains with diversity within the samples can be determined, which are concluded in 16S rRNA amplicon sequencing. The primary disadvantage of 16SrRNA sequencing is it lacks taxonomic resolution. In either case, microorganism DNA sequence samples are studied by next-generation sequencing technologies in comparison with known database sequences to analyze the presence and abundance of taxa (Saus et al. 2019). Microbial community analysis can be achieved with genomic databases and tools such as the quantitative insights into microbial ecology (QIIME), ribosomal database project (RDP) pyrosequencing, procrustes analysis, taxonomy, and ecology of ribosomal sequences (W.A.T.E.R.S) (Ursell et al. 2012).

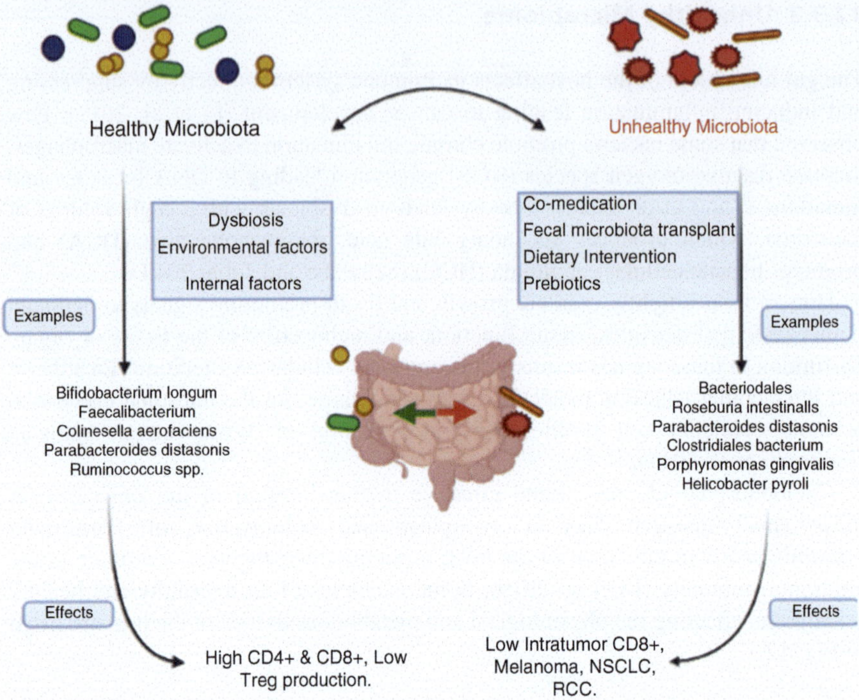

Fig. 12.1 Healthy and unhealthy microbiota inhabiting host: Different factors like dysbiosis, internal and external factors of host result in loss of healthy microbiota leading to unhealthy or harmful microbiota promoting uncontrolled cell proliferation and other clinical disorders

Human-compatible preclinical models, humanizing mice, organ-chips, and human-derived organoids are used in the study of gut microbiota and their reactions to other tissues. Fecal transplant trials have been conducted in patients, to unwrap the composition of essential microorganisms to help in immunotherapy for cancers (Elinav et al. 2019). Advanced tools and genomic database consortium are most popularly used to identify bacteria and study their effects on the host. Standardization in this research study helps to compare the various preclinical and clinical studies and understand how microbiota with different genomes have been involved in the development of malignancies.

12.5 Microbiome Therapies

The microbiota damaged due to antibiotics, drugs, changes in diet, and dysbiosis can be restored with the addition of new microbes in the gut that would mimic healthy gut composition. New therapeutic strategies are involved in altering gut microbe to mimic gut microbiota found in healthy humans to restore the resistance capacity of gut microbe towards the disease (Cerdó et al. 2019). Different microbiome-targeted therapies such as prebiotic-resistant starches (fiber), probiotics, and fecal microbiota transplantation are in play to replenish the microbiota composition in aid to treat diseases (Zhou et al. 2020). The role of the prebiotic and its influence depends on the existing bacteria in the host. The combined approach of prebiotics and growing specific bacteria is known as synbiotics, may be promising in treating diseases (Li et al. 2019).

- *Prebiotics* (*Food Components*): Edible substance that helps to promote the growth of defined microbes to enhance the host health by restoring the stability of microbiota and decrease proinflammatory pathways.
- *Probiotics* (*live Microorganism*): Beneficial and active microbes are composed mainly of yeast, Lactobacillus, Actinomycetes, Bifidobacterium, Clostridium which assist in inhibiting harmful bacteria growth by colonizing in human reproductive systems and intestines through maintaining microecology of the host.
- *Fecal Microbiota Transplantation*: It is the transplantation of healthy human feces to the gastrointestinal tract (GIT) of the patient to develop healthy functional microbiota to treat extraintestinal and intestinal diseases.

In late 300 AD, in Eastern Jinn Dynasty, China Ge Hong's "Elbow Reserve Emergency" has a record in the treatment of patients with human waste for treating conditions like diarrhea, food poisoning, fever, and death (Zhou et al. 2020). Although prebiotic and probiotics have shown promising effects in several treatments, their molecular mechanism is still unknown (Vieira et al. 2013). In an FMT study conducted inpatient with *C. difficile* infection, followed by transplantation, there is increased *Bacteroidetes* in the gut. The microbiota composition after FMT was like the donor and differences were observed in the metagenomic profile in recipients. Further studies are needed to identify a specific colony that can modulate the immunity of the host and prevent tumor development and also restore the balance between gut microbiota and host (Seekatz et al. 2014).

12.6 Microbiome in Mice and Humans

In gut microbiome study, researchers mostly use mice, as they share similar digestive tract. However, animal models like Zebrafish, drosophila, fruit fly, and the Hawaiian bobtail squid have also widely been used in the study of host–microbiota interactions. Mice genes share 99% of similarities with human genes and also have a close resemblance with microbiome phylum as in humans (Kostic et al. 2013a).

In mice, the gut is different from the human due to low pH and oxygen tension in the intestine affecting the fidelity of human microbiota. The changes in the glycan profile of the mucus and around 4% of microbial gene sequences were found to be shared between humans and mice (Fessler et al. 2019a), leaving less scope of research using mice to mimic the human environment. Sequencing technologies revealed that microbiota exhibit varied genetic sequences between hosts and within a host over time. This variation of microbiota can be like one nucleotide variant, short insertions and deletions, and more substantial structural variations like deletions, insertions, duplications, inversions, and gene copy-number variants (Garud and Pollard 2020). Microbiome composition in humans evolves in the first 3 years after birth and then stabilizes and tends to remain constant until being affected by external and internal factors. Among microbiota, the gut has most of it, with approximately 3×10^{13} bacterial cells count ten times to the number of human cells (Roy and Trinchieri 2017).

12.7 Role of the Microbiome in Healthy Individuals

Microbiota is potential enough in transforming a variety of metabolites like some proteins, impacting the immune response in protecting the host from cancer generation and progression (Prosperi 2020). Gut microbiota and its products have been observed to be influencing the anticancer effect by modulating the immune system of the host through the immunological cell death pathway (Chen et al. 2020). Eubiosis, a rich and diverse microbiota regulating micro-ecological balance within the host, helps to maintain immunity by activating TLR signaling pathway acting as an adjuvant enhancing the immune response (Li et al. 2019). The gut microbiome regulates the homeostasis of the host intestine by processing the dietary fiber ingredients consumed by the host into digestible byproducts and plays an important role in eliciting an immune response against invading microbes and resisting them (Shui et al. 2020). It is demonstrated that the microbiome has been involved in the maturation of immune cells like CD4+ T cells, CD8+ T cells, and dendritic cells (Zhou et al. 2020). SCFA, such as butyrate, is generated by the dietary fibers through microbial fermentation, the primary energy source for the colonocytes (Fulbright et al. 2017). Butyrate is sensed by the dendritic cells, T cells expressing G protein-coupled receptors, GPR41 or 43 (Zitvogel et al. 2018). Glucose obtained from glycolysis becomes the primary carbon source for the cancer cells; this is known as the "Warburg Effect." In a diet with high fiber, content butyrate is produced and due to its impaired metabolism, there is a high percentage of the butyrate making the cancer cells starve.

Also, butyrate promotes apoptosis, inhibits histone deacetylase, regulates immunogenic cell death (ICD), cellular proliferation through epigenetic modifications (Fulbright et al. 2017). *Bifidobacterium infantis* involve in the differentiation of immune cells Tregs and dendritic cells and promote Foxp3+ regulatory T cells. *Clostridium butyricum* maintains intestinal immune homeostasis by regulating pancreatic T cells (Chen et al. 2020). Table 12.1 shows the systemic effects of gut microbiota on the host: The following functions in the human body are affected due to the gut microbiome impacting homeostasis leading to clinical disorders (Roy and Trinchieri 2017).

A good understanding of the bacteria gut microbiome is established. More research and details about the action of the virus and fungi composition and their

Table 12.1 Systemic effects of gut microbiota on the host physiology

Physiological functions	Non-neoplastic pathology
• Cardiovascular and musculoskeletal.Functions • Metabolism. • Neurological and cognitive functions. • Hematopoiesis and myeloid cell functions. • Inflammation and immunity. • Aging.	• Insulin resistance. • Obesity. • Autoimmune. • Non-alcoholic steatohepatitis.

interaction and their effect on the host would give complete handling of the human microbiota and utilize them efficiently as anticancer agents (Saus et al. 2019).

12.8 Microbiota Effects on Immune System Development

Initial colonization of intestinal bacteria depends upon gut-associated with lymphoid tissues (GALT); similarly, *Bacillus subtilis* and *Bacteroides fragilis* gut microbes have shown to promote GALT development (Rhee et al. 2004). GALT is a component of mucosa-associated lymphoid and they are divided into three sections (Cebra 1999):

(a) Payer patches (containing B cell and T cell),
(b) Lamina propria (consisting of immunoglobulins, dendritic cells, mast cells),
(c) Intraepithelial leukocyte spaces (NK cells, T cells).

Gut microbiota is involved in the maintenance of the mucosal immune system, during myelopoiesis, and the function of dendritic cells, macrophages, and neutrophils (Gorjifard and Goldszmid 2016; Fessler et al. 2019b). The mucosal immune system undergoes significant changes once bacterial colonization establishes in the intestinal tract. The gut microbiome has been shown to play a significant role in promoting NK cell differentiation, dendritic cells (Wu and Wu 2012). *B. fragilis* and *Clostridia* shape the polarity of macrophages and are observed to be coordinated mutualistic relationships between macrophages and microbes (Mezouar et al. 2018). Moreover, microbe-derived luminal ATP molecule activates $CD70^{high}$ $CD11c^{low}$ cells, which promotes TH17 cell differentiation (Atarashi et al. 2008).

Multiple diseases and chronic disorders had common intestinal dysbiosis that may have contributed to the pathogenicity of these diseases. Symbiotic bacteria are essential for lymphoid tissue development. Germ-free mice have shown gut-associated lymphoid tissues (GALT), developmental issues, and impaired lymphoid follicles compared to a pathogen-free mouse (Kim et al. 2017).

12.9 Microbiome Role in Epithelial Barrier

The epithelial barriers contain goblet cells, Paneth cells, on damage to these cells microbes infiltrate into the blood, few acting as the procarcinogenic agents to spread carcinogenesis and inflammation (Rajagopala et al. 2017). The gut epithelium and its tight junctions act as a barrier for a wide variety of bacteria and internal gut milieu, "at density up to 10^{12} organisms/ml intestinal content" were observed from a unicellular layer of epithelium (Sonnenburg et al. 2004). The luminal surface of gut epithelium cells prevents the entry of large particles and bacteria, preventing excessive immunological response, which affects gut health. IgA, IgM antibodies regulate the entry of antigen penetration across the epithelium layer. During an immune response, bacteria generate short-chain fatty acids, and these enhance the production of IgA (Li et al. 2019). IgA is crucial to maintain homeostasis of gut microbiota; its deficiencies cause the growth of anaerobic organisms in the gut cells (Suzuki et al. 2004). Short-chain fatty acids like acetate inhibit the growth of other pathogens and viruses, SCFA serves as an energy source for gut microbes (Mezouar et al. 2018).

12.10 Microbiome as a Marker

Most of the research studies explain that a specific microbe colony seems to be either dominant or causal of cancer development and progression. Alteration in the microbiome colonization due to antibiotics, vaccines, host genetics leads to cancers (Shahanavaj et al. 2015). Microbiome, highly populated microorganisms reside within the proximity of epithelium, soon be a way for the personalized medicine development targeting the pathobionts for the cancer progression (Fulbright et al. 2017). The altered microbiome can be a useful marker for diagnosing neoplasm primarily colorectal cancer (CRC), gastric cancer, cervical cancers. *F. nucleatum* found to be highly associated with the CRC in tumoral tissue and feces of the patients in comparison to the control individuals. SCFA, like butyrate, fructose, linoleic acid, acts as a robust diagnostic marker for CRC with low levels found in patients in comparison to control individuals (Saus et al. 2019).

Research studies explain that *B. longum*, *B. adolescentis*, *Parabacteroides merdae, Collinsella aerofaciens* are more populated in the feces of responder patients of melanoma and non-responders have dominated with *Ruminococcus obeum* and *Roseburia intestinalis* colonization (Elkrief et al. 2019).

Higher bacteroidales and low *B. fragilis* composition masking the effect of anti-CTLA-4 in melanoma patients was observed. Butyrate-producing bacterium and Firmicutes like *Faecalibacterium* genus involved with producing a higher response rate with more prolonged progression-free survival (Li et al. 2019).

Pancreatic cancer has been the most prominent and fourth leading cause of death, has some difficulty in early detection due to a lack of specific biomarkers. Recent research studies have been promising to overcome this situation in pancreatic cancer. *Porphyromonas gingivalis*, an oral bacterium is found to be increased in pancreatic cancer. Patients tend to have antibodies against the bacteria *P. gingivalis* ATTC 53978. Also, saliva bacterial biomarkers are specific for the detection of pancreatic cancer (Shahanavaj et al. 2015).

12.11 Microbiome Affecting Cancer

Ongoing cancer research is focused on the human microbiota due to promising results shown in their interaction. The unknown proliferation of the cells is due to external and internal factors of the host affecting the development and progression of cancer. These factors are influenced by the microbiome activities within the host, indirectly affecting cancer. Microbiome in the host affects the remodeling of the tissue-like angiogenesis, a part of the tissue remodeling where adequate blood flow is developed, which is prior necessary for a tumor to get initiated. More investigation is required to understand the mechanism involved between microbiome and angiogenesis interaction (Fulbright et al. 2017). Therefore, the microbiota is observed to be important for the development of the vasculature in the intestines of the host.

Coley, in the nineteenth century, cures malignancies in humans using live cultures. There were few initial failures in the treatment but resulted in a mixture known as Coley's toxin composed of attenuated *Streptococcal* and *Serratia marcescens*. The success rate is 80% with 5 years survival rate treated around 1000 sarcoma patients in the period where the knowledge on cancer is still in its infancy. The mechanism behind the cure was toxins secreted by the composed mixture-induced immune response to fight against the malignant cells.

In 1863, Virchow explained the interrelation between inflammation and cancer onset, based on the studies detailed that carcinogenesis is initiated at the site of chronic inflammation. This concludes the direct microbiome effects on host cell physiology and changes in the equilibrium of the tissues. Modifications in the microbiome may result in undefined local and systemic inflammation and conditions within the host (Shahanavaj et al. 2015).

When cells stop to divide a condition known as cellular senescence, cells in the senescence state secrete growth factors, enable tumor growth, and the intestinal bacteria to induce malignancy. *E. coli* regulates senescence-associated phenotype (SASP) by secreting growth factors inducing tumor development and epithelial proliferation. Therefore, this bystander proliferation and microbial induced cellular senescence mechanisms caused due to microbial and host interactions develop malignancies (Fulbright et al. 2017). Some bacteria within the microbiome can induce chronic inflammation with or without an increase in the ROS, indicating their carcinogenic potential in the host. When the epithelial barriers are damaged by alterations, bacteria that get in direct contact with the host cell secrete toxins, leading to host DNA damage. Bacterial genotoxins like cytolethal distending toxin (CDT) and colibactin cause direct dsDNA damage and instability of the host genome, including phosphorylation of histone proteins and activation ataxia-telangiectasia mutated (ATM)–CHK2 signaling pathway. These genetic changes lead to cell swelling and cell cycle arrest at G2/M phase. Other toxins like *B. fragilis* toxin and cytotoxic necrotizing factors affect the cellular responses and thus indirectly play a role in tumorigenesis (Schwabe and Jobin 2013).

Studies explain there is an increase in the interferon a/b signals in lung stromal cells, which aid in resist Influenza virus infection due to the gut microbiome. Researchers observed FMT enhances the immune system by altering the tumor microbiota. These promising observations allow defining cancer treatment by modifying the tumor immune micro-environment using the gut microbiota (Shui et al. 2020).

The gut microbiome has a significant impact on treating cancer and related toxicities in cancer-related therapies (Helmink et al. 2019). This explains that the gut microbiome has potential in overall cancer therapy. Gut microbiome alters the gut-associated lymphoid tissue and mucosa immune function through the interaction with PAMPs and antigen-presenting cells and TLRs, triggering an innate response in the host. These immune activities result in accelerated antitumor immune function with the low number of myeloid-derived suppressor cells (MDSCs) and high levels of tumor infiltration lymphocytes (TILs) (Helmink et al. 2019). Scientists revealed that patients have a higher diversity of the bacteria in their gut, who responded to the anti-PD-1 ICIs therapy compared to the non-responders. The diversity of the microbiota mainly includes an abundance of *Ruminococcus, Faecalibacterium, Clostridiales* (Gopalakrishnan et al. 2018). Further studies on the microbiota diversity generate more customized and increase the efficiency of cancer immunotherapy. Immunotherapy as a cancer treatment is an efficient way of utilizing the patient immune system to generate an antitumor effect with less adverse effects. Different approaches like sensitizing tumor cells as non-self to the immune system, immune checkpoint inhibitors (ICIs), a novel therapeutic agent with promising clinical results in malignancies. Monoclonal antibodies blocking PD-1/ PD-L-1, CTLA-4 blockade sensitize cancer cells to the patient immune system. Recent research studies explain that the gut microbiome affects the therapeutic efficacy of ICIs against cancer (Li et al. 2019). *B. fragilis* colonized in the mouse gut flora increased TH1 responses in the lymph nodes near to the tumor to enhance the efficacy of the CTLA-4 immune checkpoint inhibitor blockade (Elkrief et al. 2019).

The microbiome has proven to be a double-sided sword in cancer studies; wild type mice can combat carcinogenesis compared to germ-free mice, on the other end it can promote carcinogenesis by inducing inflammation to intestinal cells (Li et al. 2019) when there are alterations in the microbiome due to environmental or intrinsic factors affecting microbial structure (Zechner 2017). The chronic inflammation caused due to microbial dysbiosis has been known to promote cancer in the site of inflammation and also enhance the accumulation of *E. Coli.*

H. pylori is carcinogenic bacteria interacting with cell growth signaling pathways. Certain bacteria and viruses are known to cause fatal disease or chronic inflammation, as primary and secondary effects would be carcinogenic nature (Li et al. 2019). *H. pylori* have cytotoxin associated gene A (CagA) which produce virulence protein VacA, ureas, NapA2; the Vac A modulates β-catenin, resulting in inflammation and carcinogenesis (Rajagopala et al. 2017).

Fusobacterium nucleatum, an enterotoxigenic bacteria, when fed to ApcMin/+ mice, showed characteristics like human colorectal cancer with an abundance of the same bacteria in the tissues; however, few other mouse models did not exhibit any tumorigenesis (Kostic et al. 2013b). Other studies showed *F. nucleatum* virulence protein FadA activates the β-catenin pathway; alteration with NF-kB leads to inflammation and promotes a favorable tumor environment. Moreover, other virulence proteins like RadD induce the formation of biofilm from different bacteria, FaP2 binds to Gal-GalNAc, which promotes colonization of *F. nucleatum;* besides, it inhibits NK cells (Rajagopala et al. 2017). These studies suggest more details are required in signaling pathways between bacterial cells and host immune cells.

Certain bacterial species induce proinflammatory toxins, alteration in signaling pathways, also the production of genotoxic substances (Helmink et al. 2019). Some microbes are known to cause cancer other than inducing inflammation. Microbes produce toxic substances and some microbes themselves, when mixed with blood, get carried to distant locations in the body and can cause cancer (Rajagopala et al. 2017). Human papillomavirus, hepatitis B and C viruses, human cell leukemia virus, Epstein–Barr virus (EBV), Kaposi sarcoma-associated virus (KSHV) human, T lymphotropic virus one and all known to cause cancer in humans. EBV is associated with gastric cancer (Rajagopala et al. 2017). Virus composition in the human virome has been unexplored. Fungi and protozoa research studies are to be focused on knowing different microbiome genome interaction with host (Elinav et al. 2019).

12.12 Microbiome in Cancer Immunotherapy

12.12.1 CD47 Blockade with *Bifidobacteria*

The effect of immunotherapy in patients is influenced by the host gut's ability to resist invading pathogens and response to treatment. In the malignant mouse model, scientists found that anaerobic bacteria travel to tumor sites and boost effectiveness against immunotherapy. In tumor-bearing mice, the absence of gut bacteria did not respond to anti-CD47 antibiotics (Shi et al. 2020). Bifidobacteria present in the human gut travels and accumulates at the tumor site and blocks CD47 to increase the response against immunotherapy via stimulators of interferon genes (STING). In a similar study, mice with inactive STING pathways showed no benefit from bacteria-immunotherapy combined approach. STING is a transmembrane protein present on macrophages, T cells, dendritic cells. STING stimulates innate immune genes with respect to invading viruses, bacteria into the host. STING is activated by certain cyclic dinucleotides (CDNs) produced by certain bacteria, followed by subsequent reaction process type I interferons (IFNs) are secreted outside the cytoplasm of the cell (Barber 2015). Type I interferons are antiviral cytokines and regulate adaptive immune systems (Haller et al. 2006).

CD47 (cluster of domains) is a transmembrane protein; it is present in different cell types (Zhang et al. 2019) (Zhang et al. 2020). CD47 is an immunoglobulin known as integrin associated protein (IAP). It is overexpressed in cancerous cells to avoid immune responses by acting as self-cells. High levels of CD47 in cancerous cells mask immunotherapy and its prognosis. One of its ligand, known as signal regulatory protein α (SIRPα) is a transmembrane protein present on myeloid cells such as monocytes, macrophages, granulocytes, and myeloid dendritic cells. Formation of CD47 and SIRPα signaling complex inhibits the build-up of myosin IIA in phagocytic synapses, which acts as a "do not eat me" signal. Blocking this CD47 has potential in cancer treatment and has been used in various immunotherapies. Monoclonal antibodies against this complex have proven to be an effective therapy for solid tumor and hematologic malignancies (Folkes et al. 2018).

12.12.2 PD-L1 Blockade Assisted with *Bifidobacterium*

Bifidobacterium (gram +ve) found in the healthy gastrointestinal tract, which helps in digestion and produces vitamin K and B, codes for carbohydrate digestive enzymes, also used in probiotics (O'Callaghan and van Sinderen 2016). This organism has shown antitumor activity when subjected to mice with melanoma. *Bifidobacterium*, along with programmed cell death protein1 ligand (PD-L1), abrogated cancer with enhanced CD8 + T cells (Sivan et al., 2015).

12.12.3 CTLA-4 Blockade Assisted with *Bacteroidales*

CTLA-4 binding achieved with monoclonal antibody studied in patients with III/IV stage melanoma faced effects on gastrointestinal immunity (Berman et al. 2010). In fecal microbial transplantation (FMT) study conducted in mice proved the microbial influence of blocking of CTLA-4. *B. fragilis*, *B. thetaiotaomicron,* and *Burkholderiales* played a significant role in antitumor activity with the help of interleukin 12(IL-12) dependent T cells (Vétizou et al. 2015). Contradicting the above statement, *Bacteroides fragilis* is an enterotoxigenic bacteria; its abundance was co-related to colorectal cancer by a study conducted in 150 humans (Purcell et al. 2017).

12.12.4 Short-Chain Fatty Acids in Treatment for Cancer

The most prominently studied SCFAs are acetate, butyrate, and propionate compared to valerate and caproate. The abundant SCFA like acetate, butyrate, and propionate is produced in the ratio of 60:20:20 (Chambers et al. 2018). Acetate, butyrate, propionate, valerate, and caproate were used in a study to understand the effects of SCFA in apoptosis and cancer. This study concluded that the butyrate was more potent to compare to propionate and valerate to induce cell growth arrest and differentiation in colon cancer cell lines. A related study showed that this ability of SCFA depends on histone hyperacetylation effects, alteration in cell cycle regulators p21 and CB1. Butyrate enhanced histone acetylation compared to other SCFA and increased the rate of programmed cell death. The exact mechanisms of action is not well known; it has been proposed that butyrate modifies chromatin structure by inhibiting histone deacetylase resulting in hyperacetylation of core proteins. During histone acetylation, the DNA becomes loosely packed to histone protein and is available for transcription of specific genes like cell regulators, chemokines (Hinnebusch et al. 2002).

The chemokines expressed by the epithelial cells like IL-8 and MCP-1 are found to attract neutrophils and monocytes (Fusunyan et al. 1999). Butyrate, with the help of p21 protein downregulated Cyclin B1(CB1), is found in a study conducted on HT-29 cells. CB1 is a crucial component for health development; its increased levels are found in colon cancer (Hinnebusch et al. 2002). It can control p53 mitotic cell division through regulating CB1 levels and preventing neoplastic transformation (Innocente et al. 1999). Cyclin B1 plays a critical role in cell cycle progression from G2 to M phase, with the involvement of NF-κB. Studies are explaining that CB1 has induced tumor malignancy in esophageal cancer (Zhan et al. 2012).

It is studied that chronic intestinal inflammation causes cancer in the intestine; also, it leads to pattern alteration in epithelial differentiation leading to an undifferentiated state. Interleukin-8, a proinflammatory cytokine induces differentiation in epithelial cells, butyrate has shown inhibition of IL-8 also can induce differentiation of cells in vivo (Huang et al. 1997). SCFA are known to directly activate G-coupled protein receptors like GPR43, GPR109A, and GPR41, which activate anti-inflammatory cascades (Venegas et al. 2019) (Lazar et al. 2018). The detailed illustration is in Fig. 12.2.

12.13 Future Perspectives

We are in the era of the microbiome, which has more positive preclinical and clinical research in treating cancer. Furthermore, few challenges are upfront to know how to regulate gut microbiota and the interaction of other genomes in the microbiome to improve the efficacy of cancer immunotherapy. Targeting cancer immunotherapy through the microbiome can be more successful and improve immune surveillance when the favorable components of the microbiome are completely defined. FMT in anticancer therapy acts as a promising way to treat cancers if the donor composition is well known. The favorable bacteria composed of *Akkermansia muciniphila, Bifidobacteria* spp., *E. hirae*, and *Bacteroides* spp. are found to impact malignant cells effectively. Finalizing the set of microbes for treating cancer can be done by filling the research gap, knowing the interaction of other microbial genomes with hosts like viruses, archaea, protists, and fungi would be more promising in treating cancers and building personalized medicine.

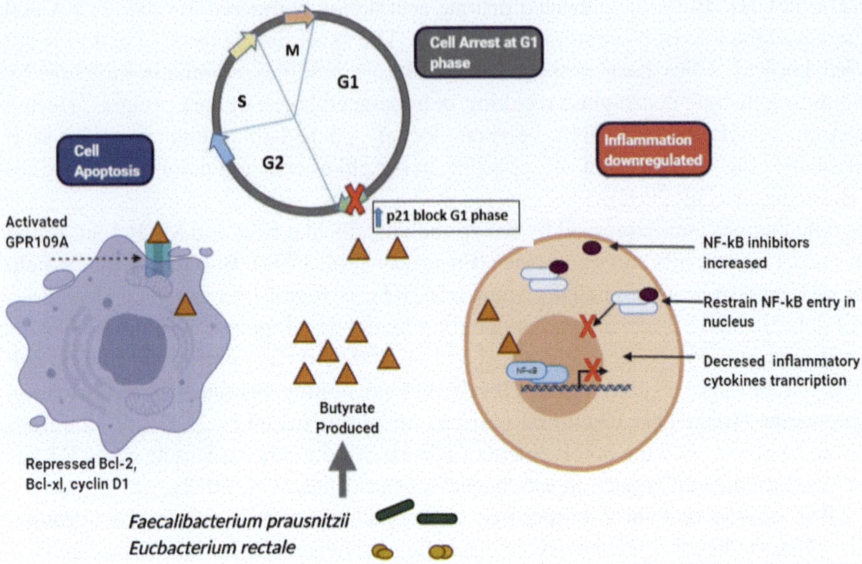

Fig. 12.2 SCFA production and its impact in different ways on the immune system of the host. When the host consumes a high fiber diet, interbacterial fermentation by *Faecalibacterium prausnitzii, Eubacterium rectale* produces high SCFA like butyrate and activates some immune proteins to kill cancer cells by apoptosis, ell cycle arrest at G1 phase, by downregulation of inflammation (Canani et al. 2011) (Segain et al. 2000)

12.14 Conclusion

Microbiomes within the body can be a good source in treating cancer growth without any adverse effect on the host body. The host–microbiome plays a crucial role in maintaining homeostasis of the immune system and its study can be an efficient and economical way of developing a treatment for cancer and other microbial diseases. The interaction between the host immunity, microbiome, and cancer progression is explained to an extent, but more studies are to be performed. Microbiomes have been affected by many factors, these alterations modifying the favorable microbiome to unhealthy ones. Different microbes act as a marker specific to cancer and have been used as an early diagnostic route to detect them. Immunotherapy is an existing way of treating cancers, microbiome playing a considerable role in the effect of immunotherapy enhances the antitumor effects in the patients. To conclude, we are in a state of a holistic vision of using the microbiome as a strategy in cancer immunotherapy. In the coming years, more studies on other genomes of the microbiomes and their interaction would strengthen the knowledge on the microbiome and make it a promising way to treat cancer.

Acknowledgments The corresponding author Dr. Pola Sudhakar would like to express his deepest thanks to authorities of Andhra University, Visakhapatnam, India for providing necessary facilities to carry out the research work and for extending constant support.

Conflict of Interest The authors declare that there is no potential conflict of interest.

References

Allaband C, McDonald D, Vázquez-Baeza Y, Minich JJ, Tripathi A, Brenner DA, Loomba R, Smarr L, Sandborn WJ, Schnabl B, Dorrestein P, Zarrinpar A, Knight R (2019) Microbiome 101: studying, analyzing, and interpreting gut microbiome data for clinicians. Clinical Gastroenterology and Hepatology 17(2):218–230. https://doi.org/10.1016/j.cgh.2018.09.017

Atarashi K, Nishimura J, Shima T, Umesaki Y, Yamamoto M, Onoue M, Yagita H, Ishii N, Evans R, Honda K, Takeda K (2008) ATP drives lamina propria TH17 cell differentiation. Nature 455:808–812. https://doi.org/10.1038/nature07240

Barber GN (2015) STING: infection, inflammation and cancer. Nature Reviews Immunology 15 (12):760–770. https://doi.org/10.1038/nri3921

Berman D, Parker SM, Siegel J, Chasalow SD, Weber J, Galbraith S, Targan SR, Wang HL (2010) Blockade of cytotoxic T-lymphocyte antigen-4 by ipilimumab results in dysregulation of gastrointestinal immunity in patients with advanced melanoma. Cancer Immun 10:11

Canani RB, Di Costanzo M, Leone L, Pedata M, Meli R, Calignano A (2011) Potential beneficial effects of butyrate in intestinal and extraintestinal diseases. World J Gastroenterol 17 (12):1519–1528. https://doi.org/10.3748/wjg.v17.i12.1519

Cebra JJ (1999) Influences of microbiota on intestinal immune system development. Am J Clin Nutr 69:1046–1051. https://doi.org/10.1093/ajcn/69.5.1046s

Cerdó T, García-Santos JA, Bermúdez MG, Campoy C (2019) The role of probiotics and prebiotics in the prevention and treatment of obesity. Nutrients 11(3):635. https://doi.org/10.3390/nu11030635

Chambers ES, Preston T, Frost G, Morrison DJ (2018) Role of gut microbiota-generated short-chain fatty acids in metabolic and cardiovascular health. Current Nutrition Reports 7(4):198–206. https://doi.org/10.1007/s13668-018-0248-8

Chen W, Wang S, Wu Y, Shen X, Guo Z, Li Q, Xing D (2020) Immunogenic cell death: a link between gut microbiota and anticancer effects. Microb Pathog 141:103983. https://doi.org/10.1016/j.micpath.2020.103983

Dominguez-Bello MG, De Jesus-Laboy KM, Shen N, Cox LM, Amir A, Gonzalez A, Bokulich NA, Song SJ, Hoashi M, Rivera-Vinas JI, Mendez K, Knight R, Clemente JC (2016) Partial restoration of the microbiota of cesarean-born infants via vaginal microbial transfer. Nat Med 22 (3):250–253

Elinav E, Garrett WS, Trinchieri G, Wargo J (2019) The cancer microbiome. Nat Rev Cancer 19 (7):371–376. https://doi.org/10.1038/s41568-019-0155-3

Elkrief A, Derosa L, Zitvogel L, Kroemer G, Routy B (2019) The intimate relationship between gut microbiota and cancer immunotherapy. Gut Microbes 10(3):424–428. https://doi.org/10.1080/19490976.2018.1527167

Fessler J, Matson V, Gajewski TF (2019a) Exploring the emerging role of the microbiome in cancer immunotherapy. Journal for Immuno Therapy of Cancer 7(1):1–15. https://doi.org/10.1186/s40425-019-0574-4

Fessler J, Matson V, Gajewski TF (2019b) Exploring the emerging role of the microbiome in cancer immunotherapy. J Immunother Cancer 7(1):1–15. https://doi.org/10.1186/s40425-019-0574-4

Folkes AS, Feng M, Zain JM, Abdulla F, Rosen ST, Querfeld C (2018) Targeting CD47 as a cancer therapeutic strategy: the cutaneous T-cell lymphoma experience. Current Opinion in Oncology 30:332. https://doi.org/10.1097/CCO.0000000000000468

Fulbright LE, Ellermann M, Arthur JC (2017) The microbiome and the hallmarks of cancer. PLoS Pathog 13:1–6. https://doi.org/10.1371/journal.ppat.1006480

Fusunyan RD, Quinn JJ, Fujimoto M, MacDermott RP, Sanderson IR (1999) Butyrate switches the pattern of chemokine secretion by intestinal epithelial cells through histone acetylation. Mol Med 5(9):631–640. https://doi.org/10.1007/bf03402075

Garud NR, Pollard KS (2020) Population genetics in the human microbiome. Trends Genet 36 (1):53–67. https://doi.org/10.1016/j.tig.2019.10.010

Gopalakrishnan V, Helmink BA, Spencer CN, Reuben A, Wargo JA (2018) The influence of the gut microbiome on cancer, immunity, and cancer immunotherapy. Cancer Cell 33(4):570–580. https://doi.org/10.1016/j.ccell.2018.03.015

Gorjifard S, Goldszmid RS (2016) Microbiota—myeloid cell crosstalk beyond the gut. J Leukoc Biol 100(5):865–879. https://doi.org/10.1189/jlb.3ri0516-222r

Haller O, Kochs G, Weber F (2006) The interferon response circuit: induction and suppression by pathogenic viruses. Virology 344(1):119–130. https://doi.org/10.1016/j.virol.2005.09.024

Helmink BA, Khan MAW, Hermann A, Gopalakrishnan V, Wargo JA (2019) The microbiome, cancer, and cancer therapy. Nat Med 25:377. https://doi.org/10.1038/s41591-019-0377-7

Hinnebusch BF, Meng S, Wu JT, Archer SY, Hodin RA (2002) The effects of short-chain fatty acids on human colon cancer cell phenotype are associated with histone hyperacetylation. J Nutr 132(5):1012–1017. https://doi.org/10.1093/jn/132.5.1012

Huang N, Katz JP, Martin DR, Wu GD (1997) Inhibition of IL-8 gene expression in Caco-2 cells by compounds which induce histone hyperacetylation. Cytokine 9:27. https://doi.org/10.1006/cyto.1996.0132

Innocente SA, Abrahamson JLA, Cogswell JP, Lee JM (1999) p53 regulates a G2 checkpoint through cyclin B1. Proc Natl Acad Sci USA 96:2147. https://doi.org/10.1073/pnas.96.5.2147

Kho ZY, Lal SK (2018) The human gut microbiome—a potential controller of wellness and disease. Frontiers in Microbiology 9:1835. https://doi.org/10.3389/fmicb.2018.01835

Kim D, Zeng MY, Núñez G (2017) The interplay between host immune cells and gut microbiota in chronic inflammatory diseases. Experimental and Molecular Medicine 30:492–506. https://doi.org/10.1038/emm.2017.24

Kostic AD, Chun E, Robertson L, Glickman JN, Gallini CA, Michaud M, Clancy TE, Chung DC, Lochhead P, Hold GL, El-Omar EM, Brenner D, Fuchs CS, Meyerson M, Garrett WS (2013a) Fusobacterium nucleatum potentiates intestinal tumorigenesis and modulates the tumor-immune microenvironment. Cell Host and Microbe 14(2):207–215. https://doi.org/10.1016/j.chom.2013.07.007

Kostic AD, Howitt MR, Garrett WS (2013b) Exploring host-microbiota interactions in animal models and humans. Genes Dev 27(7):701–718. https://doi.org/10.1101/gad.212522.112

Lazar V, Ditu LM, Pircalabioru GG, Gheorghe I, Curutiu C, Holban AM, Picu A, Petcu L, Chifiriuc MC (2018) Aspects of gut microbiota and immune system interactions in infectious diseases, immunopathology, and cancer. Frontiers in Immunology 9:1830. https://doi.org/10.3389/fimmu.2018.01830

Li W, Deng Y, Chu Q, Zhang P (2019) Gut microbiome and cancer immunotherapy. Cancer Lett 447:41–47. https://doi.org/10.1016/j.canlet.2019.01.015

Lv J, Guo L, Liu JJ, Zhao HP, Zhang J, Wang JH (2019) Alteration of the esophageal microbiota in Barrett's esophagus and esophageal adenocarcinoma. World J Gastroenterol 25(18):2149–2161. https://doi.org/10.3748/wjg.v25.i18.2149

Methé BA, Nelson KE, Pop M, Creasy HH, Giglio MG, Huttenhower C, Gevers D, Petrosino JF, Abubucker S, Badger JH, Chinwalla AT, Earl AM, Fitzgerald MG, Fulton RS, Hallsworth-Pepin K, Lobos EA, Madupu R, Magrini V, Martin JC, White O (2012) A framework for human microbiome research. Nature 486(7402):215–221. https://doi.org/10.1038/nature11209

Mezouar S, Chantran Y, Michel J, Fabre A, Dubus JC, Leone M, Sereme Y, Mège JL, Ranque S, Desnues B, Chanez P, Vitte J (2018) Microbiome and the immune system: from a healthy steady-state to allergy associated disruption. Human Microbiome Journal 10:11–20. https://doi.org/10.1016/j.humic.2018.10.001

O'Callaghan A, van Sinderen D (2016) Bifidobacteria and their role as members of the human gut microbiota. Frontiers in Microbiology 2015:01030. https://doi.org/10.3389/fmicb.2016.00925

Prosperi D (2020) Preface: Frontiers in cancer immunotherapy: understanding the role of gut microbiota. Curr Pharm Biotechnol 21(1):2–2. https://doi.org/10.2174/138920102101191209163306

Purcell RV, Pearson J, Aitchison A, Dixon L, Frizelle FA, Keenan JI (2017) Colonization with enterotoxigenic Bacteroides fragilis is associated with early-stage colorectal neoplasia. PLoS One 12(2):e0171602. https://doi.org/10.1371/journal.pone.0171602

Rajagopala SV, Vashee S, Oldfield LM, Suzuki Y, Venter JC, Telenti A, Nelson KE (2017) The human microbiome and cancer. Cancer Prev Res 10(4):226–234. https://doi.org/10.1158/1940-6207.CAPR-16-0249

Rhee K-J, Sethupathi P, Driks A, Lanning DK, Knight KL (2004) Role of commensal bacteria in development of gut-associated lymphoid tissues and preimmune antibody repertoire. J Immunol 172(2):1118–1124. https://doi.org/10.4049/jimmunol.172.2.1118

Roy S, Trinchieri G (2017) Microbiota: a key orchestrator of cancer therapy. Nat Rev Cancer 17 (5):271–285. https://doi.org/10.1038/nrc.2017.13

Saus E, Iraola-Guzmán S, Willis JR, Brunet-Vega A, Gabaldón T (2019) Microbiome and colorectal cancer: roles in carcinogenesis and clinical potential. Mol Asp Med 69:93–106. https://doi.org/10.1016/j.mam.2019.05.001

Schwabe RF, Jobin C (2013) The microbiome and cancer. Nat Rev Cancer 13(11):800–812. https://doi.org/10.1038/nrc3610

Seekatz AM, Aas J, Gessert CE, Rubin TA, Saman DM, Bakken JS, Young VB (2014) Recovery of the gut microbiome following fecal microbiota transplantation. MBio 5(3):e00893. https://doi.org/10.1128/mBio.00893-14

Segain JP, Raingeard de la Blétière D, Bourreille A, Leray V, Gervois N, Rosales C, Ferrier L, Bonnet C, Blottière HM, Galmiche JP (2000) Butyrate inhibits inflammatory responses through NFkappaB inhibition: implications for Crohn's disease. Gut 47(3):397–403

Shahanavaj K, Gil-Bazo I, Castiglia M, Bronte G, Passiglia F, Carreca AP, Del Pozo JL, Russo A, Peeters M, Rolfo C (2015) Cancer and the microbiome: potential applications as new tumor biomarker. Expert Review of Anticancer Therapy 15(3):317–330. https://doi.org/10.1586/14737140.2015.992785

Shi Y, Zheng W, Yang K, Harris KG, Ni K, Xue L, Lin W, Chang EB, Weichselbaum RR, Fu YX (2020) Intratumoral accumulation of gut microbiota facilitates CD47-based immunotherapy via STING signaling. J Exp Med 217(5):e20192282. https://doi.org/10.1084/jem.201922

Shui L, Yang X, Li J, Yi C, Sun Q, Zhu H (2020) Gut microbiome as a potential factor for modulating resistance to cancer immunotherapy. Front Immunol 10:1–16. https://doi.org/10.3389/fimmu.2019.02989

Sivan A, Corrales L, Hubert N, Williams JB, Aquino-Michaels K, Earley ZM, Benyamin FW, Lei YM, Jabri B, Alegre ML, Chang EB, Gajewski TF (2015) Commensal Bifidobacterium promotes antitumor immunity and facilitates anti-PD-L1 efficacy. Science 350 (6264):1084–1089. https://doi.org/10.1126/science.aac4255

Sonnenburg JL, Angenent LT, Gordon JI (2004) Getting a grip on things: how do communities of bacterial symbionts become established in our intestine? Nature Immunology 5:569–573. https://doi.org/10.1038/ni1079

Suzuki K, Meek B, Doi Y, Muramatsu M, Chiba T, Honjo T, Fagarasan S (2004) Aberrant expansion of segmented filamentous bacteria in IgA-deficient gut. Proc Natl Acad Sci USA 101(7):1981–1986. https://doi.org/10.1073/pnas.0307317101

Temraz S, Nassar F, Nasr R, Charafeddine M, Mukherji D, Shamseddine A (2019) Gut microbiome: a promising biomarker for immunotherapy in colorectal cancer. International Journal of Molecular Sciences 20:4155. https://doi.org/10.3390/ijms20174155

Ursell LK, Metcalf JL, Parfrey LW, Knight R (2012) Defining the human microbiome. Nutr Rev 70:38–44. https://doi.org/10.1111/j.1753-4887.2012.00493.x

Venegas DP, De La Fuente MK, Landskron G, González MJ, Quera R, Dijkstra G, Harmsen HJM, Faber KN, Hermoso MA (2019) Short chain fatty acids (SCFAs) mediated gut epithelial and immune regulation and its relevance for inflammatory bowel diseases. Frontiers in Immunology 10:277. https://doi.org/10.3389/fimmu.2019.00277

Vétizou M, Pitt JM, Daillère R, Lepage P, Waldschmitt N, Flament C, Rusakiewicz S, Routy B, Roberti MP, Duong CPM, Poirier-Colame V, Roux A, Becharef S, Formenti S, Golden E, Cording S, Eberl G, Schlitzer A, Ginhoux F et al (2015) Anticancer immunotherapy by CTLA-4 blockade relies on the gut microbiota. Science 350(6264):1079–1084. https://doi.org/10.1126/science.aad1329

Vieira AT, Teixeira MM, Martins FS (2013) The role of probiotics and prebiotics in inducing gut immunity. Frontiers in Immunology 4:445. https://doi.org/10.3389/fimmu.2013.00445

Wu HJ, Wu E (2012) The role of gut microbiota in immune homeostasis and autoimmunity. Gut Microbes 3(1):4–14. https://doi.org/10.4161/gmic.19320

Zechner EL (2017) Inflammatory disease caused by intestinal pathobionts. Current Opinion in Microbiology 35:64–69. https://doi.org/10.1016/j.mib.2017.01.011

Zhan QM, Wang LH, Song YM, Ou YQ, Jiang J, Fan J, Wang JB, Shen J (2012) Esophageal carcinoma. In: Recent Advances in Cancer Research and Therapy. Elsevier, Burlington. https://doi.org/10.1016/B978-0-12-397833-2.00018-2

Zhang W, Huang Q, Xiao W, Zhao Y, Pi J, Xu H, Zhao H, Xu J, Evans CE, Jin H (2020) Advances in anti-tumor treatments targeting the CD47/SIRPα Axis. Front Immunol 11:1–15. https://doi.org/10.3389/fimmu.2020.00018

Zhang X, Wang Y, Fan J, Chen W, Luan J, Mei X, Wang S, Li Y, Ye L, Li S, Tian W, Yin K, Ju D (2019) Blocking CD47 efficiently potentiated therapeutic effects of anti-angiogenic therapy in non-small cell lung cancer. J Immunother Cancer 7(1):346. https://doi.org/10.1186/s40425-019-0812-9

Zhou A, Tang L, Zeng S, Lei Y, Yang S, Tang B (2020) Gut microbiota: a new piece in understanding hepatocarcinogenesis. Cancer Lett 474:15–22. https://doi.org/10.1016/j.canlet.2020.01.002

Zitvogel L, Ma Y, Raoult D, Kroemer G, Gajewski TF (2018) The microbiome in cancer immunotherapy: diagnostic tools and therapeutic strategies. Science 359(6382):1366–1370. https://doi.org/10.1126/science.aar6918

Insight into the Animal Models for Microbiome Studies

13

Chanda Vikrant Berde, P. Salvi Sagar, V. Kajarekar Kunal, A. Joshi Suyoj, and B. Berde Vikrant

Abstract

The microbiomes, including bacteria, fungi, and viruses, exist within and on all the organisms, which is the current field of research. Particularly of interest are microbiome of human and its direct impact on human health. The health and fitness of animals, including humans, are influenced by the existence and composition of microbial communities of the host. To date, maximum microbiome research has been focused on the mouse as a model organism for studying the mechanisms of different processes occurring in the microbial communities. Mouse microbiome models have also been the primary choice for performing preclinical tests for studying relationships between the microbiomes and host physiological, metabolic, immune, and neurologic phenotypes. These were also used for developing methodologies to correct functional abnormalities in these communities that lead to disease. The mouse, however, is not a perfect model for studying different aspects of the microbiome and for studying the host stimuli and environmental responses. Hence, researchers have been conducting microbiome studies using other animals as well, for example, zebrafish, pigs, and *Drosophila*. This chapter summarizes the microbiome studies conducted using different models and an insight into its advantages.

Keywords

Microbiome · Animals · Models · Health · Microflora · Host–microbe interaction

C. Vikrant Berde (✉)
Marine Microbiology, School of Earth, Ocean and Atmospheric Sciences (SEOAS), Goa University, Taleigao Plateau, Goa, India

P. Salvi Sagar · V. Kajarekar Kunal · A. Joshi Suyoj
Gogate Jogalekar College, Ratnagiri, Maharashtra, India

B. Berde Vikrant
Department of Zoology, Arts, Commerce and Science College, Lanja, Maharashtra, India

P. V. Bramhachari (ed.), *Microbiome in Human Health and Disease*,
https://doi.org/10.1007/978-981-16-3156-6_13

13.1 Introduction

Microbiomes including bacteria, fungi, and viruses live inside and on all organisms and are a growing area of research. Particularly of interest is the human microbiome because of its direct impact on the human health. Humans carry trillions of microbes inhabiting our bodies, resulting in creating complex, body-habitat specific, and adaptive ecosystems. These systems are finely adapted to continuously changing host physiology. The presence or absence of residential microbial communities has a direct impact on the health and fitness of the animal. Dysbiosis or dysbacteriosis is the microbial imbalance or maladaptation of the body, for example, the microbiota composition during a number of diseases such as inflammatory bowel disease, multiple sclerosis, types 1 and 2 diabetes, etc. (Berg 1996; Qin et al. 2010; Turnbaugh and Gordon 2009; Delzenne et al. 2011; Kau et al. 2011). The microbiota of humans plays an important role in possibly causing, spreading, and prevention of human illness (Lai et al. 2014; Norman et al. 2014; Palm et al. 2015). Gut microflora is now known as an important factor in etiology of a number of human diseases such as obesity (Turnbaugh and Gordon 2009; Delzenne et al. 2011), inflammation (Kau et al. 2011; Garrett et al. 2010), metabolic syndrome (Kau et al. 2011; Cani et al. 2012), and colorectal cancer (Arthur and Jobin 2013; Macdonald and Wagner 2012). Use of humanized mouse has brought great advancement in the field of gut microbiology and associated health outcomes (Turnbaugh et al. 2009; Goodman et al. 2011; Gootenberg and Turnbaugh 2011). Normally, these models are made by seeding germ-free mice with the bacteria derived from human. Therefore, they provide a solid system for studying different interactions between human microbiome and chronic diseases where use of humans as subject is not possible.

A number of potential features of healthy microbiome have been proposed. These potential features include prevalent organisms or molecular pathways (Cani et al. 2012) and usual ecological properties, such as diversity or stability (Garrett et al. 2010; Arthur and Jobin 2013). Normally, microbiomes show a great degree of diversity irrespective of the presence or absence of the diseased condition (Garrett et al. 2010; Macdonald and Wagner 2012). This characteristic of the microbiomes creates complication in identification of simple microbial agents causing disease or present in diseased state.

Most of the studies carried out regarding microbiome to date have focused on mouse as a model for studying how different mechanisms occur in the microbial communities. Mouse microbiome models have been first choice for carrying out not only preclinical tests for studying relationships between community and hosts metabolic, physiological, immune, and neurologic phenotypes but also for developing different methodologies to correct functional abnormalities in microbial communities, which results in causation of disease. Many reports have stated that mouse is not a perfect model for studying various aspects of microbiome and also for studying host stimuli and environmental response. Researchers are carrying out microbiome studies in other animals as well as in zebrafish, pigs, and *Drosophila*.

Development of different animal models for microbiota studies allows studying of microbiota subsets, i.e., causative vs. correlative factors in diseased states, and also offers a system to reveal putative therapeutics.

13.2 Mouse as the Animal Model for Human Microbiota Studies

The advantages of mouse models are numerous, and also, the expanse of research on the gastroenterology, genetics, and immunology of mice is much more than any other model. The genotypes and phenotypes range offered by mouse models out-do all other model organisms. Thus, mice models have played a very vital role in the research concerning human gut microbiota.

The physiology and anatomical structures of humans and mouse are quite similar, and thus, this is one of the reasons behind more use of mouse as a model organism in biomedical research. In the case of mice and humans, gastrointestinal tract is made of anatomically similar organs. The prominent differences in anatomy of both human and mice intestinal tract are because of their diets, patterns of feeding, metabolic requirements, and size of the body.

Mice and humans share the average ratio of intestinal surface area: body surface area (Casteleyn et al. 2010), but this ratio varies between different sections of gut between these two organisms. The average small intestine:colon length ratio is 2.5:7 in mice and humans, respectively (Treuting and Dintzis 2012), and the surface ratio of small intestine:colon is only 18 in mice as compared to 400 in humans (Casteleyn et al. 2010).

The cecum in mouse is relatively large in comparison to its total gastrointestinal tract. Cecum is recognized as the chief site for the fermentative decomposition of plant material and the synthesis of vitamin K and B, which is reabsorbed via coprophagy. The human cecum is small, having anatomically similar structure to colon and doesn't hold any clear function (Treuting and Dintzis 2012).

At microscopic level, there are numerous differences in the structure of intestinal tract of humans and mice (Treuting and Dintzis 2012). The colon of mouse consists of thin muscularis mucosae, whereas the human colon is covered with thicker mucosal wall. There are a number of transverse folds present along colonic mucosa in humans. In the case of mice, transverse folds are found only in cecum and proximal colon. These differences in the compartmentalization and structure of colon might contribute to creating different ecological microniches holding a variety of microbial communities.

The surface of mice intestinal crypts in mucin-producing goblet cells of proximal colon are abundantly present, whereas their number decreases at the base of crypt, in distal colon and rectum. In the case of humans goblet cells, they are profusely present from cecum to rectum.

There is a difference in the presence of another type of intestinal epithelial cell, i.e., the Paneth cell in mouse and human. The role of Paneth cells is to secrete antimicrobial components in the lumen of small intestine. In the case of humans,

Paneth cells are present in the cecum and proximal colon. Paneth cells are uniquely present in the cecum in mouse but are not present in entire colonic mucosa.

There is a difference in location and amount of defensins produced by Paneth cells, and their secretion and storage have been found to be different in human and mice (Cunliffe et al. 2001; Ghosh et al. 2002; Ouellette and Selsted 1996). The dissimilarities in location and amount of Paneth cells and goblet cells indicate difference in local immune responses, which might contribute to composition of intestinal microbiota.

Laboratory mice have been instrumental for understanding role of normal flora in many aspects of human physiology, which includes studies like angiogenesis by Stappenbeck et al. (2002) and Reinhardt et al. (2012), bone mineral density studies reported by Cho et al. (2012) and Sjogren et al. (2012a, b), and studies related to innate and adaptive immune function (Garrett et al. 2010; Littman and Pamer 2011; Hooper et al. 2012).

Mice are an important model system for studying host-microbiota interactions that are applicable to human biology (Spor et al. 2011) because of

1. Mice share around 99% genes with humans.
2. These genes have key similarities with human gut microbiome at phylum through family level.

There are a number of characters of mice genetics that mark the mice as a powerful model system for studying genetics of humans in the interactions between host and microbiota. The availability of both inbred and outbred strains, numerous collections of knockout, knock-in, and transgenic mutants (International Knockout Mouse Consortium [IKMC, http:// www.knockoutmouse.org]), available data and work done by Knockout Mouse Project [KOMP, https://www.komp.org], and data available at Mutant Mouse Regions Resource Center [MMRRC, http://www.mmrrc.org] also make mice a preferred study model.

13.3 Germ-Free and Antibiotics Treatment Models

Germ-free (GF) animals are devoid of any microorganism in its lifetime (Wostmann 1996; Yi and Li 2012). For studying the interactions between a host and its microbiota, germ-free animals are important experimental aids. They are colonized with specific microorganisms and are then referred to as gnotobiotic (Fritz et al. 2013; Smith et al. 2007). Germ-free mice are bred in isolators that prevent entry of any microorganisms. Hence, together with these special facilities, monitoring for contamination using different methods, along with the cost, labor, and skills required to maintain them, makes GF mice quite expensive (Fontaine et al. 2015; Nicklas et al. 2015).

To remove the microbiota from the model animals, specific antibiotics having different mechanisms of action are utilized, such as antibiotic polymyxin B to target the Gram-negative bacteria and Vancomycin for Gram-positive bacteria (Atarashi et al. 2011; Schubert et al. 2015). Thus, the composition of the animal gut may be manipulated by using the combination of antibiotics as per the requirements (Schubert et al. 2015; Zackular et al. 2016).

13.3.1 GF Mice as Experimental Models

GF animal models have been used to study and understand host-microbiota interactions in various fields of study, which include lipid metabolism (Nicholson et al. 2012), cardiology (Stepankova et al. 2010), neurogastroenterology (Diaz Heijtz et al. 2011; Al-Asmakh et al. 2012; Neufeld et al. 2011; McVey Neufeld et al. 2015), reproductive biology (Al-Asmakh et al. 2014; Shimizu et al. 1998), bone homeostasis (Sjogren et al. 2012a, b), and so on. Another interesting observation in the humanized rats was the maintenance of some metabolic activities in the gut microflora transferred from humans to rats. The activities such as production of equol (Bowey et al. 2003) and reduction of cholesterol (Gérard et al. 2004) have been reported to be maintained in the microflora.

Mice models are among the best preferred tools for studying microbiota-associated human diseases, by understanding the host-microbe interactions as monocolonization of single bacteria is possible. The mice are made germ-free and then inoculated with human gut microflora. These are termed the humanized gnotobiotic models (Goodman et al. 2011). This model thus helps in recapitalization of microbiota composition of the human gut.

The research findings on GF, however, cannot be directly utilized for treatment purposes. The reasons being that the bacterial species in mice gut are not found in humans, and this microflora is influenced by numerous factors involving anatomy, behavior, etc. (Gordon and Pesti 1971; Sommer and Backhed 2013; Kostic et al. 2013; Gootenberg and Turnbaugh 2011). In spite of these drawbacks, the GF mice is the most preferred model system for studying host-microbe interactions.

13.4 Other Models

There are numerous invertebrate model species, which are often used in the studies related to certain interactions between the host and its microbiota. The selection of these invertebrate species is dependent on two factors, namely, the innate immune system (Chu and Mazmanian 2013) and a highly restricted gut microbiota (Chaston and Goodrich-Blair 2010). 'Humanized' animals, i.e., models with human microflora, have been established to understand the human microbiome under controlled conditions, utilizing highly researched and genetically manipulable mice and rats, in addition to pigs, dogs, etc. (Hazenberg et al. 1981; Hirayama 1999; Bowey et al. 2003; Gérard et al. 2004; Kibe et al. 2005; Pang et al. 2007).

Some of the invertebrate models used are as follows:

Numerous systems have been studied, which include *Heterorhabditis bacteriophora* and *Steinernema carpocapsae* and their respective symbionts,

Photorhabdus luminescens and *Xenorhabdus nematophila* (Clarke 2008; Wollenberg et al. 2016; Singh et al. 2015; Sicard et al. 2004), *Hirudo verbana* and *Aeromonas veronii* (Rio et al. 2009; Graf et al. 2006), and *Euprymna scolopes* and *Vibrio fischeri* (McFall-Ngai 2014; Schleicher and Nyholm 2011). The most commonly employed model organisms are, however, *Drosophila melanogaster* and *Caenorhabditis elegans*. The ease of rendering these models germ-free (Kietz et al. 2018; Berg et al. 2016), their small size, and freedom from regulatory concerns as in vertebrate models make these models advantageous over the other. The drawbacks are the gastrointestinal anatomical differences with the host, the differences in the microbiota of the gut, and inability to carry out certain studies in these systems such as adaptive immunity in humans.

The gut microflora of invertebrates has fewer microbial species, and the composition is dependent on the environment (McFall-Ngai 2007). In the case of vertebrates, adaptive immune response plays a role in the establishment and development of microbiota of the gut (McFall-Ngai 2007; Maynard et al. 2012). This leads to zebrafish *Danio rerio*, the simplest vertebrate system and having a diverse microbiota, being preferred as model system for host-symbiont relationship studies.

13.4.1 Zebrafish (*Danio rerio*)

Over the last few decades, Zebrafish use in research has increased progressively. Zebrafish is advantageous owing to the limited requirements of less space, availability in large numbers, cost effectiveness, and high prolificacy. The preference for zebrafish is also due to many similarities to its mammalian hosts (Trede et al. 2004; Norton et al. 2008; Alsop and Vijayan 2009; Wong et al. 2013). Homology is seen in the adaptive immune system and in the digestive system. Organs similar to mammals Zebrafish are pancreas, gall bladder, liver, and intestine. Also, the intestinal epithelial cells consist of absorptive enterocytes, goblet cells, and enteroendocrine cells and share similarity to mammals.

Rawls et al. (2007) focused on the transparent nature of the zebrafish, which allows real-time visualization of fluorescently labeled microbes lining the gut, throughout. The external fertilization of Zebrafish is followed by development of the embryo. Transparency of the embryo and larvae permits for the visualization of developing cells and the successive development of microflora using time-lapse microscopy. Studies concerning the host genes or signaling pathways that are regulated by the gut microbiota can be performed (Patton and Zon 2001). Genetic screening methods include mutagenesis, retrovirus-based insertional mutagenesis, zinc finger nucleases, morpholino-based gene knockdown, role of RNAi in function loss (Amacher 2008; Nasevicius and Ekker 2000; De Rienzo et al. 2012), and genome editing using TALEN system (Bedell et al. 2012). Thus, the zebrafish has a numerous features that make it an attractive experimental system.

Pham et al. (2008) have reported studies on early postembryonic development of Zebrafish using relatively simple methods for the development of green fluorescent (GF) and gnotobiotic zebrafish. GF zebrafish larvae are obtained by surface sterilization of embryos with various antibiotics (Bates et al. 2006; Davis et al. 2016a, b). These GF larvae may be labeled, or the microflora of the larvae may be fluorescently labeled in the larva and visualized, through the transparent body (Bates et al. 2007; Russo et al. 2015; Singer et al. 2010). By using these techniques, the role of microbiota can be determined in various disease conditions such as inflammatory bowel disease (IBD) (Yang et al. 2014; Geiger et al. 2013; Brugman et al. 2009), effect of probiotic bacteria on stress- and anxiety-related behavior (Davis et al. 2016a, b), metabolism and reproduction (Qin et al. 2014; Giorgini et al. 2010), and immunity and pathogen resistance (Wang et al. 2016; Qin et al. 2017). Limitations in the use of zebrafish as study model for microbiota-related research are the differences in environmental conditions and exposures.

13.4.2 Fruit Fly (Drosophila)

Numerous studies have been carried out on the gut microflora of Drosophila in order to understand the host-microbe interaction. The composition of gut microflora has been studied, and many studies have shown that the microbial community is less complex as compared to those found in mammalian gut (Corby-Harris et al. 2007; Cox and Gilmore 2007; Ren et al. 2007; Ryu et al. 2008; Chandler et al. 2011; Wong et al. 2011). Drosophila is therefore being looked upon as a model for host microbial interaction studies.

It was observed in a couple of studies that the gut microflora composition was highly subjective to diet and bacteria belonging to families Acetobacteraceae, Lactobacillales, and Enterobacteriaceae were most dominant (Corby-Harris et al. 2007; Chandler et al. 2011). The microflora of the gut is aerobic in nature and is easily cultured in the laboratory, which has made possible to have microbial stocks of the microbiota available for studies on host-microbial interaction (Chandler et al. 2011; Charroux and Royet 2012; Shin et al. 2011). Thus, these factors have made Drosophila, a model for studies on symbiosis, with huge potential to disclose new insights into host–symbiont interactions.

13.4.3 Dogs (*Canis familiaris*)

There are a number of factors that make dogs the preferred models over other models. Gastrointestinal tract of the dogs with respect to size and structure is fairly identical to humans, in that dogs are monogastric like humans, cecum more developed than human cecum (Song et al. 2013; Misic et al. 2015), and both suffer from diet-induced periodontal disease (Gorrel 1998; Harvey 1998). Common oral flora includes *Streptococcus, Staphylococcus, Pseudomonas, Actinomyces, Pasteurella, Neisseria*, and *Porphyromonas* spp. (Dewhirst et al. 2010; Sakamoto et al. 2005), while common lung microflora includes *Pseudomonas, Streptococcus, Prevotella,* and *Fusobacterium* (Ericsson et al. 2016; Erb-Downward et al. 2011). Thus, dogs are an ideal model species for finding the microbiota present in other internal organ systems such as the respiratory tract, intestinal tract, skin, etc. Also, studies on beneficial effects of the gut microflora and probiotic benefits are being carried out using canine models. There are numerous limitations to the usage of dogs as study models. Like rats, dogs are costly and require housing facilities.

13.4.4 Rabbits (*Oryctolagus cuniculus*)

Rabbits are used not so commonly as animal models, more so due to the cost factor. Rats or mice are preferred to rabbits. The same is true for research concerning GF models. Rabbits have been used for research on infectious diseases caused by GI tract pathogens (De and Chatterje 1953). The rabbits have well-developed system, and hand-rearing cesarean-born rabbits are maintained as GF models to be used in the studies (Lanning et al. 2000; Schousboe et al. 2001). A part of the small intestine is reported being used in experiments on studying the effects of pathogens like *Vibrio cholerae*, etc. (Taylor et al. 1958; Duncan et al. 1968; Arm et al. 1965; Sanyal et al. 1995; Melling et al. 1976). The technique involves ligation of small intestine and inoculation of pathogen under study in the portion of intestine and placing it back in the abdomen (De and Chatterje 1953). There are, of course, limitations to the use of rabbits as model species in host-microbiota interactive research mainly including their cost, relative to rodents.

13.4.5 Pigs (*Sus scrofa domesticus*)

Pigs are omnivorous and have anatomical and physiological similarities with human gut. Their gut microflora has been well-characterized (Zhao et al. 2015), and the composition is similar to human gut microflora (Panasevich et al. 2018; Pedersen et al. 2013; Ji et al. 2018). Microbial community is similar to the human donor developed in germfree piglets. Hence, these GF animals are excellent models for studies dealing with the effect of dietary changes on the establishment of the gut microbiome. Apart from these, other features that make pigs a model of choice for studies are size, physiology, developmental stages relative to humans, and ability to manipulate their genome (Perleberg et al. 2018; Ryu et al. 2018). Hence, they have been used in a number of studies such as gastrointestinal immuno-ontogeny (Sinkora and Butler 2016), diet-induced obesity (Turnbaugh et al. 2006), xenotransplantation (Vodicka et al. 2005), gastrointestinal physiology (Roura et al. 2016), and cardiovascular physiology (Hughes 1986; Gallo et al. 2017). Effects of resistant starch (Haenen et al. 2013), high- and low-fat diets (Heinritz et al. 2016), antibiotics (Looft et al. 2012; Allen et al. 2011), prebiotics (Berding et al. 2016), probiotics (Barszcz et al. 2016; Riboulet-Bisson et al. 2012; Shen et al. 2010; Zhang et al. 2014; Wen et al. 2014), and myriad other compounds (Liu et al. 2012) on the GM of pigs. Pigs have been successfully colonized with human microflora (Wang and Donovan 2015). The housing and feeding costs as well as the size of the pig models make them undesirable as study models.

13.5 Conclusions

The microbiome field has undergone a big change in recent years. The studies have mainly focused our attention on the role played by the host microbiota in the maintenance of host health. The role of the microflora of the host in the initiation and propagation of disease has also been highlighted by this study. These interactions are studied using model systems. Fundamental discoveries in microbiome research can be made using the most controllable animal systems, including nonmammal vertebrates such as the zebrafish and invertebrates such as Drosophila and *Caenorhabditis elegans*. The early studies of humanized animal models have permitted assessment of the human microbiome that would be challenging to achieve using customary human cohort studies or in vitro model systems.

Each one of the animal models described in this chapter displays some resemblance to the physiology of the human digestive system, thus providing valuable knowledge from diverse angles about the gut microbiota in health and disease. The information obtained from these studies has diversified our understanding of the human gut microbiota in general. Although much of the research to date has focused on the human microbiome, similar metagenomic studies can be applied to understand better animals of agricultural importance as well as pets.

It can be concluded that depending on the study, the animal model may be selected. Dogs may be used for studies related to host-associated microbiota and interactions between host and microbes, while for research concerning nutrition, omnivorous animals such as pigs may be used. The fruit fly offers the advantage of microbial manipulability in the perspective of a genetically manipulable host, while the zebrafish is good for experiments requiring a greater degree of genetic tractability than that of complex vertebrates. Thus, depending on the line of research, the most appropriate model for the study may be selected.

Acknowledgement The authors are grateful to their respective institutions for encouragement and support.

Conflict of Interest The authors declare no conflict of interests.

References

Al-Asmakh M, Anuar F, Zadjali F, Rafter J, Pettersson S (2012) Gut microbial communities modulating brain development and function. Gut Microbes 3:366–373

Al-Asmakh M, Stukenborg JB, Reda A, Anuar F, Strand ML, Hedin L (2014) The gut microbiota and developmental programming of the testis in mice. PLoS One 9:e103809

Allen HK, Looft T, Bayles DO (2011) Antibiotics in feed induce prophages in swine fecal microbiomes. MBio 2:e00260

Alsop D, Vijayan MM (2009) Molecular programming of the corticosteroid stress axis during zebrafish development. Comp Biochem Physiol A Mol Integr Physiol 153:49–54

Amacher SL (2008) Emerging gene knockout technology in zebrafish: zinc-finger nucleases. Brief Funct Genomic Proteomic 7:460–464

Arm HG, Floyd TM, Faber JE (1965) Use of ligated segments of rabbit small intestine in experimental shigellosis. J Bacteriol 89:803–809
Arthur JC, Jobin C (2013) The complex interplay between inflammation, the microbiota and colorectal cancer. Gut Microbes 4:253–258
Atarashi K, Tanoue T, Shima T, Imaoka A, Kuwahara T, Momose Y (2011) Induction of colonic regulatory T cells. Science 331:337–342
Barszcz M, Taciak M, Skomial J (2016) The effects of inulin, dried Jerusalem artichoke tuber and a multispecies probiotic preparation on microbiota ecology and immune status of the large intestine in young pigs. Arch Anim Nutr 70:278–292
Bates JM, Akerlund J, Mittge E (2007) Intestinal alkaline phosphatase detoxifies lipopolysaccharide and prevents inflammation in zebrafish in response to the gut microbiota. Cell Host Microbe 2:371–382
Bates JM, Mittge E, Kuhlman J (2006) Distinct signals from the microbiota promote different aspects of zebrafish gut differentiation. Dev Biol 297:374–386
Bedell VM, Wang Y, Campbell JM, Poshusta TL, Starker CG, Krug RG II, Tan W, Penheiter SG, Ma AC, Leung AYH (2012) *In vivo* genome editing using a high-efficiency TALEN system. Nature 491:114–118
Berding K, Wang M, Monaco MH (2016) Prebiotics and bioactive milk fractions affect gut development, microbiota, and neurotransmitter expression in piglets. J Pediatr Gastroenterol Nutr 63:688–697
Berg M, Stenuit B, Ho J (2016) Assembly of the *Caenorhabditis elegans* gut microbiota from diverse soil microbial environments. ISME J 10:1998–2009
Berg RD (1996) The indigenous gastrointestinal microflora. Trends Microbiol 4:430–435
Bowey E, Adlercreutz H, Rowland I (2003) Metabolism of isoflavones and lignans by the gut microflora: a study in germ-free and human flora associated rats. Food Chem Toxicol 41:631–636
Brugman S, Liu KY, Lindenbergh-Kortleve D (2009) Oxazolone-induced enterocolitis in zebrafish depends on the composition of the intestinal microbiota. Gastroenterologia 137:1757–1767
Cani PD, Osto M, Geurts L, Everard A (2012) Involvement of gut microbiota in the development of low grade inflammation and type 2 diabetes associated with obesity. Gut Microbes 3:279–288
Casteleyn C, Rekecki A, Van der Aa A, Simoens P, Van den Broeck W (2010) Surface area assessment of the murine intestinal tract as a prerequisite for oral dose translation from mouse to man. Lab Anim 44:176–183
Chandler JA, Lang JM, Bhatnagar S, Eisen JA, Kopp A (2011) Bacterial communities of diverse *Drosophila* species: ecological context of a host-microbe model system. PLoS Genet 7: e1002272
Charroux B, Royet J (2012) Gut-microbiota interactions in nonmammals: what can we learn from drosophila? Semin Immunol 24:17–24
Chaston J, Goodrich-Blair H (2010) Common trends in mutualism revealed by model associations between invertebrates and bacteria. FEMS Microbiol Reviews 34:41–58
Cho I, Yamanishi S, Cox L, Methe BA, Zavadil J, Li K, Gao Z, Mahana D, Raju K, Teitler I (2012) Antibiotics in early life alter the murine colonic microbiome and adiposity. Nature 488:621–626
Chu H, Mazmanian SK (2013) Innate immune recognition of the microbiota promotes host-microbial symbiosis. Nat Immunol 14:668–675
Clarke DJ (2008) Photorhabdus: a model for the analysis of pathogenicity and mutualism. Cell Microbiol 10:2159–2167
Corby-Harris V, Pontaroli AC, Shimkets LJ, Bennetzen JL, Habel KE, Promislow DEL (2007) Geographical distribution and diversity of bacteria associated with natural populations of *Drosophila melanogaster*. Appl Environ Microbiol 73:3470–3479
Cox CR, Gilmore MS (2007) Native microbial colonization of *Drosophila melanogaster* and its use as a model of *Enterococcus faecalis* pathogenesis. Infect Immun 75:1565–1576

Cunliffe RN, Rose FR, Keyte J, Abberley L, Chan WC, Mahida YR (2001) Human defensin 5 is stored in precursor form in normal Paneth cells and is expressed by some villous epithelial cells and by metaplastic Paneth cells in the colon in inflammatory bowel disease. Gut 48:176–185

Davis DJ, Bryda EC, Gillespie CH (2016a) Microbial modulation of behavior and stress responses in zebrafish larvae. Behav Brain Res 311:219–227

Davis DJ, Doerr HM, Grzelak AK (2016b) Lactobacillus plantarum attenuates anxiety-related behavior and protects against stress-induced dysbiosis in adult zebrafish. Sci Rep 6:33726

De Rienzo G, Gutzman JH, Sive H (2012) Efficient shRNA mediated inhibition of gene expression in zebrafish. Zebrafish 9:97–107

De SN, Chatterje DN (1953) An experimental study of the mechanism of action of *Vibrio cholerae* on the intestinal mucous membrane. J Pathol Bacteriol 66:559–562

Delzenne NM, Neyrinck AM, Bäckhed F, Cani PD (2011) Targeting gut microbiota in obesity: effects of prebiotics and probiotics. Nat Rev Endocrinol 7:639–646

Dewhirst FE, Chen T, Izard J (2010) The human oral microbiome. J Bacteriol 192:5002–5017

Diaz Heijtz R, Wang S, Anuar F, Qian Y, Bjorkholm B, Samuelsson A, Hibberd ML, Forssberg H, Pettersson S (2011) Normal gut microbiota modulates brain development and behavior. Proc Natl Acad Sci 108:3047–3052

Duncan CL, Sugiyama H, Strong DH (1968) Rabbit ileal loop response to strains of *Clostridium perfringens*. J Bacteriol 95:1560–1566

Erb-Downward JR, Thompson DL, Han MK (2011) Analysis of the lung microbiome in the "healthy"smoker and in COPD. PLoS One 6:e16384

Ericsson AC, Personett AR, Grobman ME (2016) Composition and predicted metabolic capacity of upper and lower airway microbiota of healthy dogs in relation to the fecal microbiota. PLoS One 11:e0154646

Fontaine CA, Skorupski AM, Vowles CJ, Anderson NE, Poe SA, Eaton KA (2015) How free of germs is germ-free? Detection of bacterial contamination in a germ free mouse unit. Gut Microbes 6:225–233

Fritz JV, Desai MS, Shah P, Schneider JG, Wilmes P (2013) From meta-omics to causality: experimental models for human microbiome research. Microbiome 1:14

Gallo M, Poser H, Bottio T (2017) The Vietnamese pig as a translational animal model to evaluate tissue engineered heart valves: promising early experience. Int J Artif Organs 40:142–149

Garrett WS, Gordon JI, Glimcher LH (2010) Homeostasis and inflammation in the intestine. Cell 140:859–870

Geiger BM, Gras-Miralles B, Ziogas DC (2013) Intestinal upregulation of melanin-concentrating hormone in TNBS-induced enterocolitis in adult zebrafish. PLoS One 8:e83194

Gérard P, Béguet F, Lepercq P, Rigottier-Gois L, Rochet V, Andrieux C, Juste C (2004) Gnotobiotic rats harboring human intestinal microbiota as a model for studying cholesterol-to-coprostanol conversion. FEMS Microbiol Ecol 47:337–343

Ghosh D, Porter E, Shen B, Lee SK, Wilk D, Drazba J, Yadav SP, Crabb JW, Ganz T, Bevins CL (2002) Paneth cell trypsin is the processing enzyme for human defensin-5. Nat Immunol 3:583–590

Giorgini E, Conti C, Ferraris P (2010) Effects of *Lactobacillus rhamnosus* on zebrafish oocyte maturation: in FTIR imaging and biochemical analysis. Anal Bioanal Chem 398:3063–3072

Goodman AL, Kallstrom G, Faith JJ, Reyes A, Moore A, Dantas G, Gordon JI (2011) Extensive personal human gut microbiota culture collections characterized and manipulated in gnotobiotic mice. Proc Natl Acad Sci U S A 108:6252–6257

Gootenberg DB, Turnbaugh PJ (2011) Companion animals symposium: humanized animal models of the microbiome. J Anim Sci 89:1531–1537

Gordon HA, Pesti L (1971) The gnotobiotic animal as a tool in the study of host microbial relationships. Bacteriol Rev 35:390–429

Gorrel C (1998) Periodontal disease and diet in domestic pets. J Nutr 128:2712S–2714S

Graf J, Kikuchi Y, Rio RV (2006) Leeches and their microbiota: naturally simple symbiosis models. Trends Microbiol 14:365–371

Haenen D, Zhang J, Souza da Silva C (2013) A diet high in resistant starch modulates microbiota composition, SCFA concentrations, and gene expression in pig intestine. J Nutr 143:274–283

Harvey CE (1998) Periodontal disease in dogs. Etiopathogenesis, prevalence, and significance. Vet Clin North Am Small Anim Pract 28:1111–1128

Hazenberg MP, Bakker M, Verschoor-Burggraaf A (1981) Effects of the human intestinal flora on germ-free mice. J Appl Bacteriol 50:95–106

Heinritz SN, Weiss E, Eklund M (2016) Intestinal microbiota and microbial metabolites are changed in a pig model fed a high-fat/low-fiber or a low-fat/high-fiber diet. PLoS One 11: e0154329

Hirayama K (1999) Ex-germfree mice harboring intestinal microbiota derived from other animal species as an experimental model for ecology and metabolism of intestinal bacteria. Exp Anim 48:219–227

Hooper LV, Littman DR, Macpherson AJ (2012) Interactions between the microbiota and the immune system. Science 336:1268–1273

Hughes HC (1986) Swine in cardiovascular research. Lab Anim Sci 36:348–350

Ji Y, Guo Q, Yin Y (2018) Dietary proline supplementation alters colonic luminal microbiota and bacterial metabolite composition between days 45 and 70 of pregnancy in Huanjiang mini-pigs. J Anim Sci Biotechnol 9:18

Kau AL, Ahern PP, Griffin NW, Goodman AL, Gordon JI (2011) Human nutrition, the gut microbiome and the immune system. Nature 474:327–336

Kibe RM, Sakamoto H, Yokota H, Ishikawa Y, Aiba Y, Koga Y, Benno (2005) Movement and fixation of intestinal microbiota after administration of human feces to germ free mice. Appl Environ Microbiol 71:3171–3178

Kietz C, Pollari V, Meinander A (2018) Generating germfree drosophila to study gut-microbe interactions: protocol to rear Drosophila under axenic conditions. Curr Protoc Toxicol 77:e52

Kostic AD, Howitt MR, Garrett WS (2013) Exploring host microbiota interactions in animal models and humans. Genes Dev 27:701–718

Lai HC, Young J, Lin CS, Chang CJ, Lu CC, Martel J (2014) Impact of the gut microbiota, prebiotics, and probiotics on human health and disease. Biom J 37:259–268

Lanning D, Sethupathi P, Rhee KJ (2000) Intestinal microflora and diversification of the rabbit antibody repertoire. J Immunol 165:2012–2019

Littman DR, Pamer EG (2011) Role of the commensal microbiota in normal and pathogenic host immune responses. Cell Host Microbe 10:311–323

Liu H, Ivarsson E, Dicksved J (2012) Inclusion of chicory (*Cichorium intybus* L.) in pigs' diets affects the intestinal microenvironment and the gut microbiota. Appl Environ Microbiol 78:4102–4109

Looft T, Johnson TA, Allen HK (2012) In-feed antibiotic effects on the swine intestinal microbiome. Proc Natl Acad Sci U S A 109:1691–1696

Macdonald RS, Wagner K (2012) Influence of dietary phytochemicals and microbiota on colon cancer risk. J Agric Food Chem 60:6728–6735

Maynard CL, Elson CO, Hatton RD, Weaver CT (2012) Reciprocal interactions of the intestinal microbiota and immune system. Nature 489:231–241

McFall-Ngai M (2007) Adaptive immunity: care for the community. Nature 445:153

McFall-Ngai M (2014) Divining the essence of symbiosis: insights from the squid-vibrio model. PLoS Biol 12:e1001783

McVey Neufeld KA, Perez-Burgos A, Mao YK, Bienenstock J, Kunze WA (2015) The gut microbiome restores intrinsic and extrinsic nerve function in germ-free mice accompanied by changes in calbindin. Neurogastroenterol Motil 27:627–636

Melling J, Capel BJ, Turnbull PC (1976) Identification of a novel enterotoxigenic activity associated with *Bacillus cereus*. J Clin Pathol 29:938–940

Misic AM, Davis MF, Tyldsley AS (2015) The shared microbiota of humans and companion animals as evaluated from *Staphylococcus* carriage sites. Microbiome 3:2

Nasevicius A, Ekker SC (2000) Effective targeted gene 'knockdown' in zebrafish. Nat Genet 26:216–220

Neufeld KM, Kang N, Bienenstock J, Foster JA (2011) Reduced anxiety-like behavior and central neurochemical change in germ-free mice. Neurogastroenterol Motil 23:255–264

Nicholson JK, Holmes E, Kinross J, Burcelin R, Gibson G, Jia W, Pattersson S (2012) Host-gut microbiota metabolic interactions. Science 336:1262–1267

Nicklas W, Keubler L, Bleich A (2015) Maintaining and monitoring the defined microbiota status of gnotobiotic rodents. ILAR J 56:241–249

Norman JM, Handley SA, Virgin HW (2014) Kingdom agnostic metagenomics and the importance of complete characterization of enteric microbial communities. Gastroenterology 146:1459–1469

Norton WH, Folchert A, Bally-Cuif L (2008) Comparative analysis of serotonin receptor (HTR1A/HTR1B families) and transporter (slc6a4a/b) gene expression in the zebrafish brain. J Comp Neurol 511:521–542

Ouellette AJ, Selsted ME (1996) Paneth cell defensins: endogenous peptide components of intestinal host defense. FASEB J 10:1280–1289

Palm NW, de Zoete MR, Flavell RA (2015) Immune microbiota interactions in health and disease. Clin Immunol 159:122–127

Panasevich MR, Wankhade UD, Chintapalli SV (2018) Cecal versus fecal microbiota in Ossabaw swine and implications for obesity. Physiol Genomics 50:355–368

Pang X, Hua X, Yang Q, Ding D, Che C, Cui L, Jia W, Bucheli P, Zhao L (2007) Inter-species transplantation of gut microbiota from human to pigs. ISME J 1:156–162

Patton EE, Zon LI (2001) The art and design of genetic screens: zebrafish. Nat Rev Genet 2:956–966

Pedersen R, Ingerslev HC, Sturek M (2013) Characterisation of gut microbiota in Ossabaw and Gottingen minipigs as models of obesity and metabolic syndrome. PLoS One 8:e56612

Perleberg C, Kind A, Schnieke A (2018) Genetically engineered pigs as models for human disease. Disease Mod Mech 11:dmm030783

Pham LN, Kanther M, Semova I, Rawls JF (2008) Methods for generating and colonizing gnotobiotic zebrafish. Nat Protoc 3:1862–1875

Qin C, Xu L, Yang Y (2014) Comparison of fecundity and offspring immunity in zebrafish fed *Lactobacillus rhamnosus* CICC 6141 and Lactobacillus casei BL23. Reproduction 147:53–64

Qin C, Zhang Z, Wang Y (2017) EPSP of L. casei BL23 protected against the infection caused by *Aeromonas veronii* via enhancement of immune response in zebrafish. Front Microbiol 8:2406

Qin J, Li R, Raes J, Arumugam M, Burgdorf KS, Manichanh C, Nielsen T, Pons N, Levenez F, Yamada T, MetaHIT Consortium (2010) A human gut microbial gene catalogue established by metagenomic sequencing. Nature 464:59–65

Rawls JF, Mahowald MA, Goodman AL, Trent CM, Gordon JI (2007) In vivo imaging and genetic analysis link bacterial motility and symbiosis in the zebrafish gut. Proc Natl Acad Sci U S A 104:7622–7627

Reinhardt C, Bergentall M, Greiner TU, Schaffner F, Ostergren-Lunden G, Petersen LC, Ruf W, Backhed F (2012) Tissue factor and PAR1 promote microbiota-induced intestinal vascular remodelling. Nature 483:627–631

Ren C, Webster P, Finkel SE, Tower J (2007) Increased internal and external bacterial load during *Drosophila* aging without life-span trade-off. Cell Metab 6:144–152

Riboulet-Bisson E, Sturme MH, Jeffery IB (2012) Effect of *Lactobacillus salivarius* bacteriocin Abp118 on the mouse and pig intestinal microbiota. PLoS One 7:e31113

Rio RV, Maltz M, McCormick B (2009) Symbiont succession during embryonic development of the European medicinal leech, *Hirudo verbana*. Appl Environ Microbiol 75:6890–6895

Roura E, Koopmans SJ, Lalles JP (2016) Critical review evaluating the pig as a model for human nutritional physiology. Nutr Res Rev 29:60–90

Russo P, Iturria I, Mohedano ML (2015) Zebrafish gut colonization by mCherry-labelled lactic acid bacteria. Appl Microbiol Biotechnol 99:3479–3490

Ryu J, Prather RS, Lee K (2018) Use of gene-editing technology to introduce targeted modifications in pigs. J Anim Sci Biotechnol 9:5

Ryu JH, Kim SH, Lee HY, Bai JY, Nam YD, Bae JW, Lee DG, Shin SC, Ha EM, Lee WJ (2008) Innate immune homeostasis by the homeobox gene caudal and commensal-gut mutualism in *Drosophila*. Science 319:777–782

Sakamoto M, Umeda M, Benno Y (2005) Molecular analysis of human oral microbiota. J Periodontal Res 40:277–285

Sanyal SC, Singh SJ, Sen PC (1995) Enteropathogenicity of *Aeromonas hydrophila* and *Plesiomonas shigelloides*. J Med Microbiol 8:195–198

Schleicher TR, Nyholm SV (2011) Characterizing the host and symbiont proteomes in the association between the bobtail squid, *Euprymna scolopes*, and the bacterium, *Vibrio fischeri*. PLoS One 6:e25649

Schousboe LP, Rasmussen LM, Ovesen T (2001) Induction of mucin and adhesion molecules in middle ear mucosa. Acta Otolaryngol 121:596–601

Schubert AM, Sinani H, Schloss PD (2015) Antibiotic-induced alterations of the murine gut microbiota and subsequent effects on colonization resistance against *Clostridium difficile*. MBio 6:e00974-15

Shen J, Zhang B, Wei H (2010) Assessment of the modulating effects of fructo-oligosaccharides on fecal microbiota using human flora-associated piglets. Arch Microbiol 192:959–968

Shimizu K, Muranaka Y, Fujimura R, Ishida H, Tazume S, Shimamura T (1998) Normalization of reproductive function in germfree mice following bacterial contamination. Exp Anim 47:151–158

Shin SC, Kim SH, You H, Kim B, Kim AC, Lee KA, Yoon JH, Ryu JH, Lee WJ (2011) *Drosophila* microbiome modulates host developmental and metabolic homeostasis via insulin signaling. Science 334:670–674

Sicard M, Ferdy JB, Pages S (2004) When mutualists are pathogens: an experimental study of the symbioses between *Steinernema* (entomopathogenic nematodes) and *Xenorhabdus* (bacteria). J Evol Biol 17:985–993

Singer JT, Phennicie RT, Sullivan MJ (2010) Broad-host range plasmids for red fluorescent protein labeling of gram-negative bacteria for use in the zebrafish model system. Appl Environ Microbiol 76:3467–3474

Singh S, Orr D, Divinagracia E (2015) Role of secondary metabolites in establishment of the mutualistic partnership between *Xenorhabdus nematophila* and the entomopathogenic nematode *Steinernema carpocapsae*. Appl Environ Microbiol 81:754–764

Sinkora M, Butler JE (2016) Progress in the use of swine in developmental immunology of B and T lymphocytes. Dev Comp Immunol 58:1–17

Sjogren K, Engdahl C, Henning P, Lerner UH, Tremaroli V, Lagerquist MK (2012a) The gut microbiota regulates bone mass in mice. J Bone Miner Res 27:1357–1367

Sjogren K, Engdahl C, Henning P, Lerner UH, Tremaroli V, Lagerquist MK, Backhed F, Ohlsson C (2012b) The gut microbiota regulates bone mass in mice. J Bone Miner Res 27:1357–1367

Smith K, McCoy KD, Macpherson AJ (2007) Use of axenic animals in studying the adaptation of mammals to their commensal intestinal microbiota. Semin Immunol 19:59–69

Sommer F, Backhed F (2013) The gut microbiota – masters of host development and physiology. Nat Rev Microbiol 11:227–238

Song SJ, Lauber C, Costello EK (2013) Cohabiting family members share microbiota with one another and wit their dogs. Elife 2:e00458

Spor A, Koren O, Ley R (2011) Unravelling the effects of the environment and host genotype on the gut microbiome. Nat Rev Microbiol 9:279–290

Stappenbeck TS, Hooper LV, Gordon JI (2002) Developmental regulation of intestinal angiogenesis by indigenous microbes via Paneth cells. Proc Natl Acad Sci 99:15451–15455

Stepankova R, Tonar Z, Bartova J, Nedorost L, Rossman P, Poledne R (2010) Absence of microbiota (germ-free conditions) accelerates the atherosclerosis in ApoE-deficient mice fed standard low cholesterol diet. J Atheroscler Thromb 17:796–804

Taylor J, Maltby MP, Payne JM (1958) Factors influencing the response of ligated rabbit-gut segments to injected *Escherichia coli*. J Pathol Bacteriol 76:491–499

Trede NS, Langenau DM, Traver D (2004) The use of zebrafish to understand immunity. Immunity 20:367–379

Treuting PM, Dintzis SM (2012) Lower gastrointestinal tract, in comparative anatomy and histology – a mouse and human atlas. In: Dintzis SM, Frevert CW, Liggitt HD, Montine KS, Treuting PM (eds) , 1st edn. Elsevier, Amsterdam

Turnbaugh PJ, Gordon JI (2009) The core gut microbiome, energy balance and obesity. J Physiol 587:4153–4158

Turnbaugh PJ, Ley RE, Mahowald MA (2006) An obesity- associated gut microbiome with increased capacity for energy harvest. Nature 444:1027–1031

Turnbaugh PJ, Ridaura VK, Faith JJ, Rey FE, Knight R, Gordon JI (2009) The effect of diet on the human gut microbiome: a metagenomic analysis in humanized gnotobiotic mice. Sci Transl Med 1:ra14

Vodicka P, Smetana K Jr, Dvorankova B (2005) The miniature pig as an animal model in biomedical research. Ann N Y Acad Sci 1049:161–171

Wang M, Donovan SM (2015) Human microbiota-associated swine: current progress and future opportunities. ILAR J/National Res Council Inst Lab Animal Res 56:63–73

Wang Y, Ren Z, Fu L (2016) Two highly adhesive lactic acid bacteria strains are protective in zebrafish infected with *Aeromonas hydrophila* by evocation of gut mucosal immunity. J Appl Microbiol 120:441–451

Wen K, Tin C, Wang H (2014) Probiotic *Lactobacillus rhamnosus* GG enhanced Th1 cellular immunity but did not affect antibody responses in a human gut microbiota transplanted neonatal gnotobiotic pig model. PLoS One 9:e94504

Wollenberg AC, Jagdish T, Slough G (2016) Death becomes them: bacterial community dynamics and stilbene antibiotic production in cadavers of *Galleria mellonella* killed by *Heterorhabditis* and *Photorhabdus* spp. Appl Environ Microbiol 82:5824–5837

Wong CNA, Ng P, Douglas AE (2011) Low-diversity bacterial community in the gut of the fruitfly *Drosophila melanogaster*. Environ Microbiol 13:1889–1900

Wong RY, Oxendine SE, Godwin J (2013) Behavioral and neurogenomic transcriptome changes in wild-derived zebrafish with fluoxetine treatment. BMC Genomics 14:348

Wostmann BS (1996) Germfree and gnotobiotic animal models: background and applications. CRC Press, Boca Raton

Yang Y, Tomkovich S, Jobin C (2014) Could a swimming creature inform us on intestinal diseases? Lessons from zebrafish. Inflamm Bowel Dis 20:956–966

Yi P, Li L (2012) The germfree murine animal: an important animal model for research on the relationship between gut microbiota and the host. Vet Microbiol 157:1–7

Zackular JP, Baxter NT, Chen GY, Schloss PD (2016) Manipulation of the gut microbiota reveals role in colon tumorigenesis. mSphere 1:e00001

Zhang H, Wang H, Shepherd M (2014) Probiotics and virulent human rotavirus modulate the transplanted human gut microbiota in gnotobiotic pigs. Gut Pathog 6:39

Zhao W, Wang Y, Liu S (2015) The dynamic distribution of porcine microbiota across different ages and gastrointestinal tract segments. PLoS One 10:e0117441

Bioinformatics Algorithms and Software for Predicting Microbiomes

14

Shrikant Pawar, Yong Chiang Tan, and Chandrajit Lahiri

Abstract

Over the last decade, bioinformatics approaches have been used extensively for analyzing microbiomes from various sources. Microbiomes can vary from external skin to internal gut, with each of the tissue demanding different analytical techniques. Of these, the next-generation sequencing strategy is the current trend instead of amplicon sequencing. It is a principal method for phenotypic trait characterization of strains, as exemplified by the shotgun metagenomics of probiotics. Moreover, quantifying abundance is crucial and can be portrayed using different computational steps enabling gene predictions and annotations. This profiling or metagenome assembly is achievable by utilizing varying algorithms. In fact, bioinformatics implications are tremendous in microbiome predictions and this review focuses on some of the salient strategies. Essentially, we present comprehensive coverage of the current next-generation sequencing platforms. To this end, we add up the open-source databases and bioinformatics algorithms and software tools for microbiome prediction and analysis. Finally, we delineate the bioinformatics challenges for discovering microbiomes.

Shrikant Pawar and Yong Chiang Tan contributed equally with all other contributors.

S. Pawar
Yale Center for Genome Analysis (YCGA), Yale University, New Haven, CT, USA
e-mail: shrikant.pawar@yale.edu

Y. C. Tan · C. Lahiri (✉)
Department of Biological Sciences, Sunway University, Petaling Jaya, Selangor, Malaysia
e-mail: chandrajitl@sunway.edu.my

P. V. Bramhachari (ed.), *Microbiome in Human Health and Disease*,
https://doi.org/10.1007/978-981-16-3156-6_14

Keywords

Bioinformatics · Microbiomes · Next-generation sequencing · Metagenomics · Metaproteomics

14.1 Introduction

Microbes are ubiquitous on the Earth, residing even in extreme environments and within a multicellular organism. The presence of normal human microbiota is crucial in maintaining human health, and the alteration in microflora composition, or dysbiosis, can cause several health complications. These can range from mental issues such as depression to physiological changes such as hypertension, carcinogenesis, and obesity (Yang et al. 2015; Sobhani et al. 2011; Aron-Wisnewsky et al. 2019). Recent researches have been hugely focusing on the impact of microbiota composition on disease progression due to the discovery of various host-microbiota links, such as the gut-lung axis and brain-gut axis (Budden et al. 2017; Mayer and Tillisch 2011; Quigley 2017). For instance, a study by Riley et al. (2013) has unearthed excessive exposure to low-dose antibiotics, which leads to gut microflora dysbiosis, as a probable cause of the prevalence of obesity in the USA. Moreover, disruptions of microflora composition can stimulate microbial pathogen introduction, thus leading to severe infections like *Pseudomonas aeruginosa*-associated pneumonia (Rosa et al. 2020). In the light of the close relation between microbiota composition and health complications, microbiome detection has become a crucial field of medical research, especially in disease diagnosis and discovery of new intervention strategies (Malla et al. 2019; Shi et al. 2015).

In microbiome detection, cultivation approaches have been traditionally hugely utilized (Boase et al. 2013; Sibley et al. 2011). For example, one of the important research areas of utilizing such approaches is determining the efficacy of probiotics. In this process characterization of different strains is carried out for determining overall composition (de Dios Miranda et al. 2019). Research has shown that probiotics usually benefit humans. The identification of novel probiotics, however, from different bacterial strains, is a necessity but a laborious task (LaPierre et al. 2019) with testing different cultures and libraries. From the prepared culture libraries, bioinformatics analysis, if performed on different strains for characterization, can lead to several potential candidates which then have to pass through different regulations and clinical trials to be approved in the market (FEEDAP et al. 2018). Therein, whole-metagenome sequencing can provide strain characterization as well as a screening of the traits based on gene expression levels (Berendsen et al. 2016). This itself results as an extension of bioinformatics applications to analyze and predict microbiome utilizing data mining and machine learning (Zeevi et al. 2015).

The presence of unculturable or hard to culture microbes, however, has impeded the utilization of cultivation approaches in in-depth microbiome analyses hoping to detect every microbe involved in the sample. Therefore, culture-independent approaches, especially with the help of bioinformatics, have gained huge attention and became the main trend of microbiome study due to indiscriminative coverage of detection. Therein, metagenomic and metaproteomic studies have been developed, of which microbiome detection can be done at the level of nucleic acid (DNA or RNA) and proteins, respectively, via next-generation sequencing (NGS) and mass spectrometry (MS) (Lepage et al. 2013; Zhang and Figeys 2019; Zhang et al. 2016). For downstream analysis of the metagenomic and metaproteomic data, bioinformatics approaches have been utilized via many tools and algorithms which have been developed in the hope of increasing efficacy and accuracy. Therefore, herein, we mainly review the various bioinformatics tools and algorithms for microbiome detection and analysis. Meanwhile, we also briefly mention the recently used NGS and MS platforms in microbiome detections.

14.2 Microbiomes

Microbiome composition has been an important area of study not only in life science researches, but also in geology, archaeology, environmental studies, and other fields of research (Ramsøe et al. 2020; Vick et al. 2018). This is due to the huge coverage of microbial population and their indispensable roles in maintaining the normal functioning of their residing environment while making sure of their survival. The ubiquitous property of microorganisms indicates that they possess specific functionalities for survival in extreme environments which can be further manipulated for the benefits of humanity.

Environmental microbiomes. The symbiotic presence of microbial communities has been crucial for the natural environment in carrying out recycling functions. Examples of these are the mediation of global carbon and nitrogen cycles by the land and ocean microbes (Canfield et al. 2010; Gougoulias et al. 2014; Sulman et al. 2014; Worden et al. 2015). Symbiotic microbiota residing in the soil has been aiding the growth of plants, especially in terms of nutrient uptake and immune system, thereby contributing to the smooth recycling of chemicals in the global food cycle (Finkel et al. 2019; Hacquard et al. 2017; Müller et al. 2016). Therefore, understanding crucial environmental microbiomes is important for the manipulation of natural resources in researches. To cater to the need, plant microbial fuel cell (PMFC) was being proposed and developed for a potent new and clean source of electricity (Lu et al. 2015; Timmers et al. 2012; Wetser et al. 2015) through the utilization of plant microbiota comprising the bacterial and archaeal community.

Human or animal microbiomes. The interactions between animal and symbiotic microflora are crucial in maintaining the health and normal function of the host. Different organs are comprised of different microflora compositions, thereby supporting the normal activity of the respective organs in different ways. For instance, gut microbiome has been crucial in helping food digestion, conferring immunity against various types of diseases, and altering normal behavior of the host (Cryan and Dinan 2012; El Kaoutari et al. 2013; Kosiewicz et al. 2011; Million et al. 2018). Moreover, vaginal microbiota, especially comprising lactobacilli, has been crucial in preventing infections such as vulvovaginal candidiasis (VVC) (Oerlemans et al. 2020; Tortelli et al. 2020). The involvement of gut microbiome in drug metabolism has led to the establishment of a framework by Javdan et al. (2020) to consider gut microbiome composition upon drug discovery. Therefore, it is crucial to explore the microbiome for better diagnosis and cope with health complications due to dysbiosis.

14.3 Microbiome Analysis by Sequencing

New techniques for microbiome studies have evolved with long-read sequencing including Pacific Biosciences (PacBio) and Oxford Nanopore Technology (ONT) platforms which are at the forefront of sequencing technology. PacBio has its unique advantages wherein high-quality full-length sequences can be sequenced with multiple reads which are currently unavailable with short-read sequencing (Saulnier et al. 2011). Sequencing costs have reduced significantly in the last couple of years allowing researchers to generate replicates and improve the depth of sequencing for microbiome samples thereby leading to improvised downstream analyses. Machine learning algorithms like support vector machines, random forests, neural networks on next-generation sequencing data have improvised disease state classifications and predictions of different pathologies (van Dijk et al. 2014).

Next-generation sequencing (NGS) has made microbiome research more comprehensive. Illumina and Ion Torrent technologies, which came before Pacific Biosciences and Oxford Nanopore Technologies, were only capable of generating millions of short reads while newer technologies generate hundreds of long reads per run. Thus, time and cost efficiency have been greatly increased thereafter. For instance, in Illumina sequencers, DNA fragments attached to the glass slides are amplified with fluorescently labelled nucleotides binding to the complementary DNA sequence producing short reads (300 bp) (van Dijk et al. 2014). Pacific Biosciences on the other hand works through engineered DNA polymerase working on concentrated genomic DNA producing long reads (10–15 kb) but, unfortunately, with a high error rate (Koren et al. 2012). Oxford Nanopore technologies work on single molecule sequencing, where a single DNA strand goes through a protein nanopore, wherein an enzyme usually helps in unwinding the double strand (Jain et al. 2015).

Due to the presence of non-culturable microbial contaminants, NGS stood out to be a better approach in microbiome analysis due to its indiscriminative coverage of detection. A study by Lewis et al. (2020) has proposed the utilization of Illumina ScriptSeq in detecting pathogens in fresh produce. The proposed NGS detection has manifested detection not only in bacterial pathogens (*Salmonella*) but also viral pathogens (phage MS2), of which such simultaneous detection of the interkingdom microbiome could not be achieved in a single culture environment through cultivation approach. Moreover, body fluid microbiome detection has been arising as detection for infectious diseases. Utilizing Illumina MiSeq 16S rRNA sequencing, NGS is a promising tool in the detection of the pathogenesis of continuous ambulatory peritoneal dialysis (CAPD) (Kim et al. 2020). Besides, a new pathogenesis detection and classification through a search-based approach, involving NGS, were established by Su et al. (2020). In this case NGS was firstly conducted, and a microbiome novelty score (MNS) was utilized in differentiating diseased from healthy genomic samples. Then, the diseased genomic samples were further classified based on references.

Furthermore, NGS technology has been utilized in probiotics and gut microbiome detection. As different composition of gut microbiota affects the health of host differently, and probiotics have been proven to improve human health, tracking of gut microbiota composition and probiotic efficacy became an important task in monitoring a person's health (Kumar et al. 2020; Rawi et al. 2020). Utilizing 16S rRNA metagenomic sequencing, Suez et al. (2018) monitored the effects of probiotics introduction and autologous fecal-microbiome transplantation on gut microbiome recovery after antibiotic introduction. Moreover, another research using a similar method has proven the importance of personalized probiotic administration due to person-specific colonization resistance (Zmora et al. 2018). Besides, through gut metagenomic analyses, bacterial species, involved in pregnancy-related iron-deficiency anemia, were discovered (Celis and Relman 2020).

14.4 Microbiome Analysis by Mass Spectrometry

Besides NGS, which processes the sample DNA and RNA molecules for metagenomic analysis, mass spectrometry (MS) plays a huge role in microbiome prediction by processing protein molecules on samples for metaproteomic analyses. Proteins are the main functional component of the central dogma of molecular biology, and different DNA sequence codes for different amino acid sequences. Therefore, by analyzing the protein or peptide composition in a sample, information on taxonomy until functions and metabolic activities of the sampled microbiome can be predicted, through amino acid sequences, and protein identification and quantification, respectively. The increased importance of metaproteomic analysis can be seen by the recent development of microbiome prediction tools (detailed in section "Metaproteomic Analysis") with MS spectra as direct input.

In metaproteomic analysis, liquid chromatography-mass spectrometry (LCMS) is usually carried out. For MS, a triple-quadrupole mass spectrometer like SCIEX Triple Quad™ 7500 is usually utilized for quantifications. The shortcoming of triple-quadrupole MS is that it utilizes selected reaction monitoring (SRM) which specifically detects only the selected compounds (Picó 2020). Fortunately, high-resolution mass spectrometry (HRMS) is recently being introduced as an alternative to the traditional MS due to its untargeted screening capabilities. LC-HRMS has been prevalently utilized in the metaproteomic analysis of biological samples as described by Pezzatti et al. (2020). Currently marketed HRMS instruments that are frequently utilized in metaproteomic studies are the Orbitrap and Q-ToF mass analyzers. Although Orbitrap confers higher resolution analysis compared to Q-ToF mass analyzers, both instruments have high accuracies in mass quantification.

14.5 Metagenomic Analyses in Microbiome Prediction

The metagenomic analysis is a crucial procedure in microbiome prediction mainly for taxonomical identification and diversity quantification. By analyzing DNA or RNA information extracted from a sampled microbiome, sequential information can lead to the identification of important microbial taxa present in the sample. Figure 14.1 summarizes the main workflow of metagenomic analyses in microbiome prediction. It comprises the usage of QUAST, a software package that handles the metagenomic analysis of detected genomic data, and consists of QUAST (genomic data quality assessment), MetaQUAST (match metagenomic data to references), QUAST-LG (QUAST but specifically for large data), and Icarus (visualizing tool for QUAST results) (Gurevich et al. 2013). Alignment of the original reads to the assembled data for the detection of structural variants in microbiomes is performed by MetaQUAST, where assembly refers to the alignment of genomes to the reference. This software utilizes de *Bruijn* graph data structure for assembly, which can

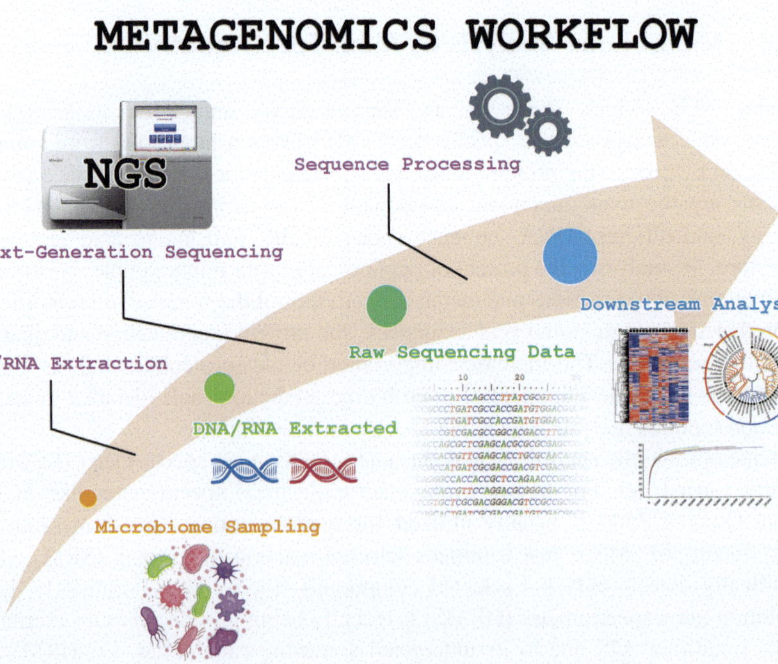

Fig. 14.1 Schematic diagram showing an overview of metagenomic analyses workflow

generate multiple genome tables and plots for understanding diversity in species, can auto-download reference sequences for unknown species and can detect chimeric regions reporting interspecies misassemblies (Zerbino and Birney 2008). MetaVelvet is another tool that extends MetaQUAST application utilizing single de novo genome assembler Velvet (Zerbino and Birney 2008). MetaVelvet decomposes de *Bruijn* graph into sub-graphs thereby isolating every species genome from short reads. The main advantages of MetaVelvet are that it utilizes multispecies metagenomic assemblies and assembles read data with longer N50 sizes compared to the single-genome assemblers and can avoid chimeric scaffolds forming longer scaffolds with an increase in the predicted number of genes. Moreover, it can also be widely applicable to metagenomic analysis (taxonomic content, functional composition, etc.).

Another application utilizing the *de Bruijn* graph is Ray Meta, it utilizes parallel computing for handling large datasets, the *k*-mers are utilized for calculating coverage distributions, followed by aligning reads against reference genomes. It can also utilize multiple cores to decrease memory requirements and computing time in a high-performance computing environment (Boisvert et al. 2012). Ray Meta has wide applications ranging from de novo genome, metagenome, transcriptome assembly, quantification of the contig, microbiome consortia members, transcript expression abundances, taxonomy, and gene ontology profiling of samples. Ray Meta is a scalable distributed technique with faster fetching of data yielding taxonomic profiles by graph coloring with unique colors for *k*-mers to identify taxons at low taxonomic ranks. It is available as a distributed software having a message-passing interface (MPI) implemented Open-MPI library with a division of tasks in several workers with message aggregation strategy. Storage of *k*-mers is achieved with double hashing with the utilization of pointers for compacting memory on high-performance computing applications.

Another important tool developed on some of the existing tools is MetAMOS. It can perform microbiome genome assembly using sequencing technology used by the investigator (Treangen et al. 2013). MetAMOS is an automated, reproducible, and traceable assembly method generating assembly, scaffolds, variant motifs, annotations, and other analysis reports. A workflow in MetAMOS is a text file with input sequences for filtering, assemblers, read mapping, classifier, annotation, validation, and scaffolders. MetAMOS essentially has two main components, initPipeline (initializing sequence libraries) and a runPipeline (actual run). Although running MetAMOS is relatively easier, installation requires specific packages and versions like Java (6+), perl (5.8.8+), python (2.7.3+), R (2.11.1+ with PNG support), gcc (4.7+ for full functionality), curl, and wget. The most common cause of a failed run for MetAMOS is a missing package or dependency.

There are several algorithms for microbiome assembly, the one which stands out is binning. This clusters sequences into groups, after which assembly can be performed on genomes. The importance of binning is that it prevents the mixing of different genomes, so ideally each bin consists of a single genome overcoming the problem of wrong assemblies connecting contigs. PHYSCIMM is another program that essentially does binning and clustering to characterize microbial composition and species classification (Kelley and Salzberg 2010). PHYSCIMM utilizes supervised learning to improve clustering from well-characterized genera. While it is important to not mix different genomes, some investigators may be interested in multiple strains of a species with identical 16S rRNA sequences, to analyze these couple of software, StrainPhlAn and PanPhlAn can be applied to shotgun microbiome data (Scholz et al. 2016; Truong et al. 2017). Another binning program worth mentioning is CONCOCT which uses sequence coverages throughout multiple samples for clustering microbiomes (Alneberg et al. 2014). The CONCOCT program uses Gaussian mixture models to cluster contigs into genomes with clusters assigned with a variational Bayesian approach. CONCOCT's precision of the clusters is 0.988, with a recall of the same cluster at 0.998 and an adjusted Rand index (precision and recall combined) of 0.983 (Alneberg et al. 2014). Thus, CONCOCT has been effectively utilized for automatically clustering contigs into genomes, linking cluster abundances to environmental variables, evaluating clusters with single-copy core genes, and evaluating clusterings by comparing them to known genome assignments. Once the assembly is completed, genes and regulatory elements need to be annotated.

As mentioned above, short reads are relatively difficult to assemble as they are usually fragmented. To overcome this problem, MetaGene Annotator, a microbiome gene finding algorithm from short sequences could be utilized (Noguchi et al. 2008). It utilizes a self-training model from inputs for predictions with a capacity to detect prophage and horizontally transferred genes. It can also analyze ribosomal binding sites with a prediction of the translation start sites with 96% and 93% sensitivity and specificity, respectively. MetaGene Annotator works well on longer genomic sequences for precise annotations with statistical models of prophage genes in lateral gene transfers. The ribosomal binding sequences of the 3′ tail of 16S RNA predict translation starts of genes improving prediction accuracies of genes. Glimmer also does a similar analysis with bacterial gene prediction via clustering data belonging to the same organism (Kelley et al. 2012). Glimmer gene prediction identifies >99% of the genes in prokaryotic genomes but falls short of accuracy with fragmented and error-prone sequences. It does classification and clustering of the sequences before gene prediction, utilizing a probabilistic model for predicting gene length and start/stop presence with insertion, deletion, and stop codon substitution errors. Glimmer also utilizes Markov models for understanding gene composition from available training data and utilizes a flexible open reading framework to capture ribosomal binding sites followed by scoring its log-likelihood ratio. Comparing other known programs, other researchers have found that Glimmer predicts several more genes from 454 reads of the human gut microbiome (Kelley et al. 2012).

Predicting genes can also be performed using FragGeneScan without assembly, wherein, unlike binning, it utilizes hidden Markov models (HMMs) in overcoming any frameshift sequencing errors (Rho et al. 2010). It specifically targets genes from short reads without compromising increasing sequencing error rates. It combines a sequencing error model with a hidden Markov model for prediction in short reads. Authors state that it can outperform MetaGene with 62% accuracy on 1% sequencing errors and can predict significantly more genes than MetaGene with many genes with no possible homologs.

A command-line-based pipeline, SHOGUN, was developed by Hillmann et al. (2020) for metagenomic analysis of WMS data. After filtering of contamination in raw sequencing data input, SHOGUN carries out sequence alignment with three different algorithms which are Bowtie2, BURST, and UTree. Eventually, taxonomic annotation and gene abundance prediction are generated using the last-common ancestor (LCA) algorithm and Bayesian redistribution (BD), respectively. Similarly, iMAP is another command-line-based pipeline that handles taxonomic assignments of metagenomic data (Buza et al. 2019). With demultiplexed sequencing data as input, iMAP utilizes QIIME2 pipeline and mothur-based taxonomic annotation regarding SILVA and Greengenes classifiers, respectively (DeSantis et al. 2006; Yilmaz et al. 2014). Besides OTU abundance generation through sequential binning, iMAP also estimates microbiome diversity through principal component analysis (PCA), principal-coordinate analysis (PCoA), and non-metric multidimensional scaling (NMDS) analysis of Bray–Curtis dissimilarity coefficients. Thereafter, phylogenetic annotation is carried out through iTOL tree viewer.

BiomMiner is a command-line-based tool mainly developed for downstream analysis of metagenomic data although it provides an optional built-in upstream analysis besides the processing protocol recommended by the sequencer (Shamsaddini et al. 2020). It has five analysis modules, which are Overview, Alpha Diversity, Beta Diversity, Differential Abundance Analysis, and Machine Learning. The Overview module summarizes the taxonomic abundance and computes a rarefaction curve through Mothur v1.34. Alpha Diversity module computes the diversity estimate value and statistically (Kruskal–Wallis test) compares between two datasets for differential analysis. Differential Abundance Analysis module utilizes Metastats, Linear discriminate analysis Effect Size (LEfSe), and Kruskal–Wallis test algorithms to compute differences in abundance between two datasets. Beta Diversity module computes beta diversity indices through PCoA and NMDS and then quantifies the similarities in OTU features between two datasets. Machine Learning module integrates the random forest (RF) and supports vector machine (SVM) in assigning "importance" value to each OTU, which indicates the accuracy of the search.

While most metagenomic analysis tools are either command-line based or with a user-unfriendly interface which is challenging for beginner researchers, recently introduced tools are aiming at improving the graphical user interface (GUI) while utilizing or pipelining present algorithms. One such tool is the MicrobiomeAnalyst (Chong et al. 2020; Dhariwal et al. 2017). Microbiome Analyst online server has four modules, which are Marker Data Profiling (MDP), Projection with Public Data (PPD), Shotgun Data Profiling (SDP), and Taxon Set Enrichment Analysis (TSEA). All modules aim to confer functional annotation to metagenomic data. The MDP module analyses the marker-gene abundance data, upon which species diversity (alpha and beta diversity) and functional predictions are carried out. PPD prints the summary of the submitted 16S rRNA data and compares it with the public database to probably derive novel insights. SDP conducts differential functional analysis with the aid of statistics on gene list or gene abundance data annotated through KEGG Ortholog (KO), COG, or EC. TSEA module enriches taxa list with their associated implications on a host like diseases. As a result, MicrobiomeAnalyst can infer host–microbiome interactions from functional assignments of metagenomic data.

14.6 Metaproteomic Analyses

Besides metagenomics which analyzes and categorizes genomic data at the sequential and taxonomical level, the metaproteomic approach (Fig. 14.2) which utilizes protein information has been developed along the way of microbiome detection and analysis. EggNOG-mapper predicts and annotates functional information of protein sequences based on the eggNOG ("evolutionary genealogy of genes: Non-supervised Orthologous Groups") HMM (Hidden Markov Models) and protein databases (Huerta-Cepas et al. 2019; Huerta-Cepas et al. 2017). Other than functional annotation, eggNOG-mapper can also be utilized in orthology analysis. Instead of the traditional homology-based approaches, eggNOG utilizes either HMMs or Double Index Alignment of NGS Data (DIAMOND) approach in ortholog search, of which DIAMOND is recommended in large data size due to its speed advantage over HMMs but with less sensitivity. In orthology assignments, eggNOG outperformed homology-based approaches, which are Basic Local Alignment Search Tool (BLAST) and InterProScan, in terms of precision and speed. EggNOG accepts not only protein sequences as input but also coding sequences from metagenomic and metatranscriptomic data. MetaGOmics is another traditional metaproteomic analysis tool, which is an online pipeline utilizing the BLAST algorithm to search through UniProtKB for Gene Ontology (GO) assignments (Riffle et al. 2018). Therein, taxonomical and functional annotations can therefore be achieved, and a directed acyclic graph (DAG) will be plotted for visualization and analysis of the GO annotation. However, MetaGOmics only allows FASTA file input.

Proteomic studies usually have mass spectrometry peptide spectra as output data, and therefore, for convenience, MS spectra, as input for analyses, are usually preferred. Trans-Proteomic Pipeline (TPP) is a pipeline that comes along with a set of tools for MS data analysis (Deutsch et al. 2015). TPP can process MS spectral data into a list of identified proteins. From MS spectral data, TPP will initially conduct peptide identification, validation, and quantification, followed by protein assignment such that finally an identified protein list is generated. After the protein list is generated, further downstream analysis is usually carried out through external tools for taxonomical and functional annotation. For instance, Rabe et al. (2019) utilize PROPHANE for downstream analysis of TPP resulting in oral microbiome analysis. PROPHANE confers downstream metaproteomic analysis with FASTA input along with report files from external upstream analysis like MetaProteomeAnalyzer (MPA), Scaffold, Proteome Discoverer, and tabular reports. PROPHANE integrates the lowest common ancestor (LCA) algorithm into protein identification and reports protein quantitation through normalized spectral abundance factors (NSAF). In PROPHANE taxonomical and functional assignments, database search will be conducted through three algorithms (DIAMOND,

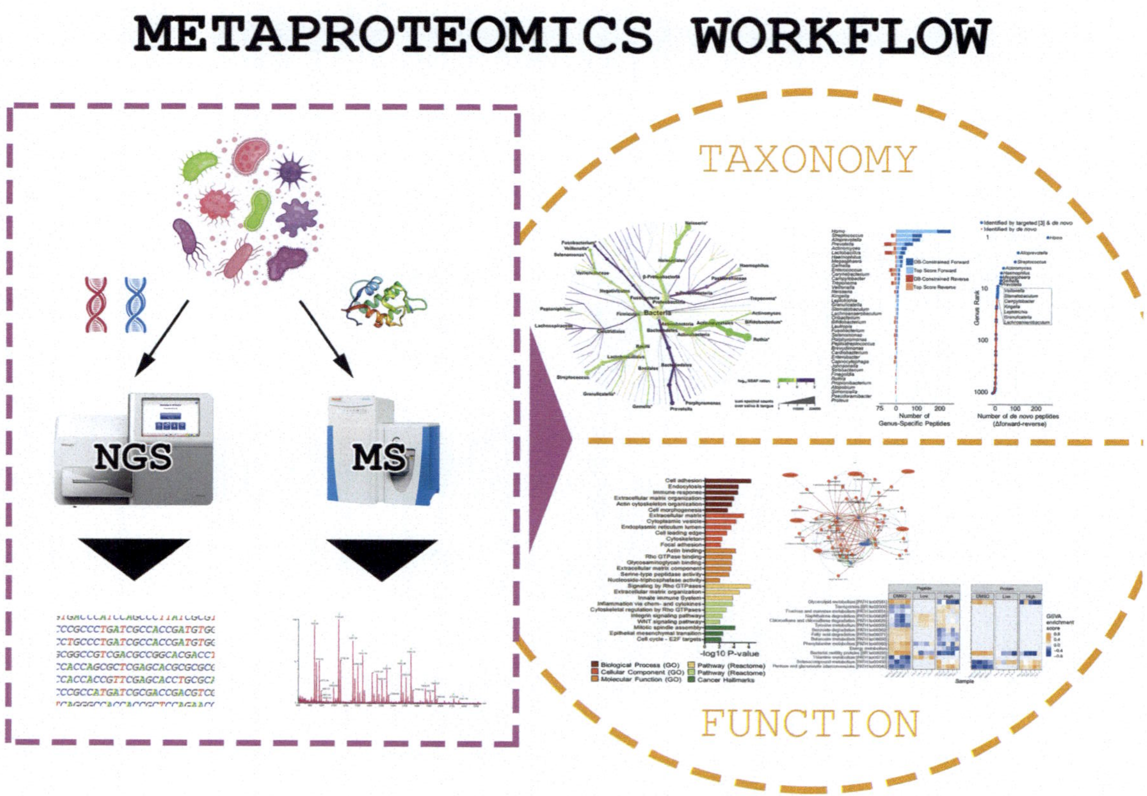

Fig. 14.2 Schematic diagram showing an overview of metaproteomic analyses workflow

eggNOG-mapper, and HMMER3) along with their associated databases such as non-redundant (nr) proteins from NCBI, UniProtKB, SwissProt, and eggNOG.

Besides that, there are some tools capable of carrying out data processing from MS spectral input to functional and taxonomical annotations. MetaProteomeAnalyzer (MPA) is a traditional pipeline-based tool which utilizes MS spectra as the initial input, followed by database search through UniProt, KEGG Pathways, Enzyme Commission (EC) Classification, and NCBI Taxonomy databases (Muth et al. 2015). As a result, taxonomic and pathway classification along with protein identification can be retrieved. Moreover, PeptideShaker is a tool for general proteomics data analysis (Vaudel et al. 2015). PeptideShaker utilizes several search engines, then estimates error probabilities through target-decoy search approach to finalize the peptide-spectrum match (PSM) list, resulting in more accurate final search results. PeptideShaker results output can be exported in Cytoscape (Proteomic network analysis and visualization) and Nonlinear (LC-MS data analysis and visualization) formats for further analyses, besides the commonly used text and graphic formats. Besides, UniPept is another pipeline-based tool in metaproteomic analyses (Singh et al. 2019). In UniPept, the MS input data will be searched for references through GO terms and EC classifications, therefore resulting in functional assignment of the analyzed metaproteomic sample. Thereafter, taxonomy and protein identification will be generated. Van Den Bossche et al. (2020) have introduced a pipeline combining MPA and PeptideShaker for MS input data processing, with UniPept for downstream processing and visualizations due to its user-friendly graphical output.

MetaLab also allows automated pipeline-based metaproteomic data analysis with MS peptide spectra as input (Cheng et al. 2017). Upon raw data input, MetaLab processes the data through two separate approaches, which are MetaPro-IQ iterative search and spectral clustering, to narrow down the input data prior to database search, which increases the temporal efficacy of the search (Zhang et al. 2016). Thereafter, peptide identification and quantification are carried out through maxLFQ algorithm (label-free XIC intensity-based quantification) or MaxQuant (isotope labelling quantification), followed by taxonomy analysis by searching through the built-in pep2tax database (Cox et al. 2014; Tyanova et al. 2016). An updated version, MetaLab 2.0, allows the detection of post-translational modified (PTM) proteomes, which is a feature that is rarely seen in other metaproteomic tools and it allows more accurate and higher coverage of proteomic identification (Cheng et al. 2019). MetaLab results can be exported in Biological Observation Matrix (BIOM) format, and therefore can be subjected to downstream analysis such as the MEGAN tool as recommended by the author (Cheng et al. 2017; Huson et al. 2016).

As enzymatic digestion of proteomic samples, prior to MS, is usually carried out in metaproteomic analyses wherein traditional tools like MPA are unable to handle such cases with trimmed peptides. Therefore, to cope up with this problem, PepFunk has been created by Simopoulos et al. (2020) for gut microbiome metaproteomic analyses. It utilizes mass spectrometry (MS) data as input and is capable of handling trypsin-digested protein samples. By automatically searching through their modified KEGG database, the Integrated human gut microbial Gene Catalog (IGC) database, functional assignments can be achieved with KEGG pathways as pepFunk output. The IGC database was curated by the in silico trypsin digestion of proteins in KEGG database. Sample treatment examples, which starts with microbiome cultivation followed by trypsin digestion and mass spectrometry, were tested and elucidated by Simopoulos et al. (2020). Like pepFunk, ProteoClade enables users to search through in silico enzymatically digested databases using MS data as input (Mooradian et al. 2020). Instead of the KEGG database, ProteoClade Database (PCDB) originated from UniProt and NCBI databases and can be utilized as a reference for the metaproteomic search. Therefore, the taxonomic classification of the detected peptides is applicable herein. Upon benchmarking, ProteoClade has manifested less RAM consumption and less duration of a process compared to UniPept and MetaProteomeAnalyzer (MPA).

14.7 Combined Analysis

M^2IA developed by Ni et al. (2020) is a pipeline-based tool that integrates metagenomic and metabolomic analysis in microbiome detection. M^2IA is mainly used for differential analysis of metagenomic and metabolomic information between two datasets. M^2IA requires 16S rRNA sequencing data (OTU table in .txt and respective sequences in .fasta or .fna), metabolome data (mass spectrometry data in . csv), and a tabulated sample information (in .csv) as described by Ni et al. (2020). The overall similarity between two datasets in M^2IA is quantified through Coinertia analysis (CIA) and Procrustes analysis (PA) approaches, of which a correlation coefficient (R-value) ranging from 0 to 1, and 1 being exactly similar, is computed. Besides that, M^2IA also carries out several pairwise statistical tests to predict the correlation between two datasets, as well as between microbiome and metabolome. Moreover, a network plot based on a relationship between microbes and metabolites will be plotted for visualization.

14.8 Algorithms and Tools

Several softwares utilizing varied algorithms are presently available for analyzing and detecting microbiomes. Table 14.1 summarizes the important tools explained in this paper.

14.9 Open-Source Databases

With information on enzymatic pathways or classes, users can utilize some specific online repositories for performing microbiome analysis. Some of the important tools include antiSMASH, dbCAN, and Resfams (Gibson et al. 2015; Blin et al. 2017) which focus on metabolite synthesis and resistance pathways in different microbiomes. Another tool with the incorporation of 12 such databases for proteins

Table 14.1 A summary of microbiome analysis tools

Types	Tools	Usage
Metagenomics		
Tool within software package	MetaQUAST	Utilizes *de Bruijn* graph data structure for assembly
Standalone tool	MetaVelvet	Utilizes single de novo genome assembler velvet
Software package	Ray	Parallel computing for handling large datasets
Toolkit	MetAMOS	Perform microbiome genome assembly using sequencing technology used by the investigator
Standalone tool	PHYSCIMM	Does binning and clustering
Standalone tools	StrainPhlAn and PanPhlAn	Shotgun microbiome data
Standalone tool	CONCOCT	Uses sequence coverages throughout multiple samples for clustering microbiomes
Online tool	MetaGeneAnnotator	Microbiome gene finding algorithm from short sequences
Software	Glimmer	Bacterial gene prediction via clustering data
Standalone tool	FragGeneScan	Utilizes hidden Markov models (HMMs) overcoming any frameshift sequencing errors
Pipeline	SHOGUN	Taxonomic annotation by LCA and gene abundance prediction by Bayesian redistribution
Pipeline	iMAP	Taxonomic assignment, microbiome diversity, and phylogenetic annotation
Standalone tool	BiomMiner	16S rRNA taxonomic abundance and diversity analysis
Online tool	MicrobiomeAnalyst	Infer host–microbiome interaction through metagenomic data
Metaproteomics		
Standalone tool	EggNOG-mapper	Utilizes HMMs and DIAMOND approach in orthology and functional predictions
Online tool	MetaGOmics	Utilizes BLAST to search through UniProtKB database
Pipeline with tool package	Trans-proteomic pipeline	Protein identification from MS spectral input
Online tool	PROPHANE	Downstream analysis tool for taxonomical and functional annotation
Standalone tool	MetaProteomeAnalyzer	Search through UniProt, KEGG pathways, EC classification, and NCBI taxonomy databases
Standalone tool	PeptideShaker	Multiple search engine with target-decoy search method of error estimation
Online tool	UniPept	Search through GO terms and EC classifications

(continued)

Table 14.1 (continued)

Types	Tools	Usage
Standalone tool	MetaLab	Iterative search or spectral clustering for MS spectra input, supports both label-free and isotope labelling quantification MetaLab 2.0 allows PTM proteome identification
Online tool	PepFunk	Search through enzymatically digested KEGG database
Standalone tool	ProteoClade	Search through enzymatically digested UniProt and NCBI databases
Combined		
Online tool	M^2IA	Combined metagenomic and metabolomic analysis

is InterPro, which integrates information with protein function, families, publication information, protein domain information, etc. (Mitchell et al. 2015). Querying with a name or sequence information from its online portal produces possible matches. Besides that Reactome is an open-source curated repository for interpreting interactions between biomolecules (Jassal et al. 2020). KEGG or Kyoto Encyclopedia of Genes and Genomes combines multiple databases with genomics and proteomics data and also incorporates information about genes, proteins, and their associated pathways, diseases, and drug databases (Kanehisa et al. 2012).

MediaWiki software is another attempt in improving deficiencies from the above portals where a user can curate or edit the pathway structure and the child product of this software is an open-source pathway analysis platform, WikiPathways (Kelder et al. 2012). With an improvised strategy, MetaCyc is another open-source tool with pathways curated from all domains of life with prioritization on metabolic pathways from published experimental datasets (Caspi et al. 2016). The general idea behind all these tools is an online annotation that is user-friendly and fast, as prediction and annotation files for assembly are usually huge such that querying them on personal computers is impossible. Considering this aspect, the Meta4 web application was developed (Richardson et al. 2013). It can search for thousands of genes and proteins and retrieve their original raw files. Moreover, it can also be installed on a personal server with a user-friendly interface. It can perform several features like BLAST against proteins or nucleotides and can also set up a private account for unpublished user data.

Furthermore, researchers have been aggregating microbiome analysis data in databases such as MGnify to ease future taxonomic annotations (Mitchell et al. 2020). MGnify users can search microbiome datasets through their location of sample collection. Besides, the expanded Human Oral Microbiome Database (eHOMD) consists of 16S rRNA reference sequences of the aerodigestive tract microbiome (Chen et al. 2010; Escapa et al. 2018). Similarly, CORE is an oral microbiome 16S rRNA database (Griffen et al. 2011). The BLAST search algorithm is usually employed in microbiome databases.

14.10 Challenges

Microbiomes can vary from external skin to internal gut, where each of the tissue calls for different microbiome analysis techniques. For example, the diversity of human skin microbiomes for exploring skin products may call for one type of longitudinal study, while the gut microbiomes may require another, thereby discovering microbiomes challenging (Zeeuwen et al. 2012). In general, the environment introduces uncharacterized biodiversity which makes metagenomic enzyme characterization challenging. Obtaining complete assemblies of microbiomes is difficult with a different abundant distribution of genomes in the environment (Ayling et al. 2020). Thus, quantifying abundance is important and can be profiled using different computational steps enabling gene predictions and annotations (Fig. 14.3). This profiling or metagenome assembly can be achieved utilizing conventional algorithms like *de Bruijn* graph-based assembly. *Bruijn* graph-based assembly is

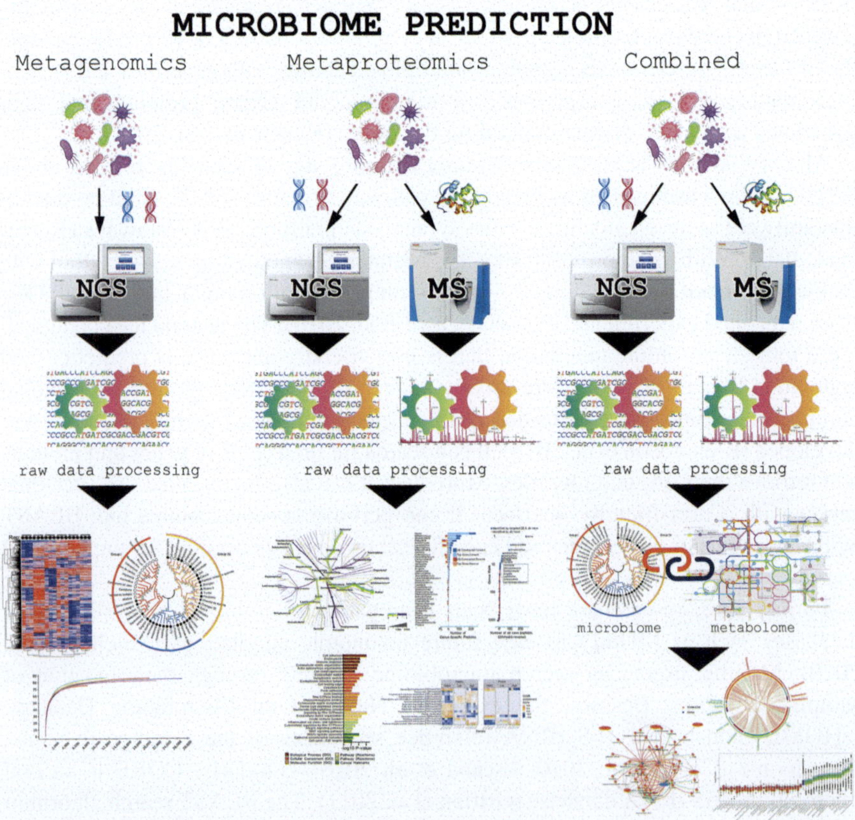

Fig. 14.3 Flowchart of microbiome raw data processing using metagenomics, metaproteomics and combined

the most popular technique owing to its simplicity, scalability, and computational time efficiency (Van der Walt et al. 2017).

14.11 Conclusion

Microbiome prediction is indeed an indispensable and rapidly growing part of medical research. With the inability of cultivation approaches to detect unculturable microbes, culture-independent approaches have emerged as a popular alternative to microbiome predictions and analyses. Herein, we have summarized the important bioinformatics algorithms, software, and portals in extracting putative genes and proteins for functional and taxonomical characterization in microbiomes. The advancement of technology has led to the evolution from short-read to long-read sequencing via the NGS approach, and from MS to HRMS in mass spectrometry, in extracting microbiome information from nucleic acid and protein samples. Meanwhile, new bioinformatics tools and novel algorithms have been frequently introduced for downstream analyses in metagenomics and metaproteomics, with improvements in efficacy and accuracy.

Acknowledgments The author acknowledges the support of Sunway University, Selangor, Malaysia for providing the necessary computational facilities.

References

Alneberg J, Bjarnason BS, De Bruijn I, Schirmer M, Quick J, Ijaz UZ et al (2014) Binning metagenomic contigs by coverage and composition. Nat Methods 11(11):1144–1146

Aron-Wisnewsky J, Prifti E, Belda E, Ichou F, Kayser BD, Dao MC et al (2019) Major microbiota dysbiosis in severe obesity: fate after bariatric surgery. Gut 68(1):70–82

Ayling M, Clark MD, Leggett RM (2020) New approaches for metagenome assembly with short reads. Brief Bioinform 21(2):584–594

Berendsen EM, Boekhorst J, Kuipers OP, Wells-Bennik MH (2016) A mobile genetic element profoundly increases heat resistance of bacterial spores. ISME J 10(11):2633–2642

Blin K, Wolf T, Chevrette MG, Lu X, Schwalen CJ, Kautsar SA et al (2017) antiSMASH 4.0—improvements in chemistry prediction and gene cluster boundary identification. Nucleic Acids Res 45(W1):W36–W41

Boase S, Foreman A, Cleland E, Tan L, Melton-Kreft R, Pant H et al (2013) The microbiome of chronic rhinosinusitis: culture, molecular diagnostics and biofilm detection. BMC Infect Dis 13 (1):210

Boisvert S, Raymond F, Godzaridis É, Laviolette F, Corbeil J (2012) Ray meta: scalable de novo metagenome assembly and profiling. Genome Biol 13(12):1–13

Budden KF, Gellatly SL, Wood DL, Cooper MA, Morrison M, Hugenholtz P, Hansbro PM (2017) Emerging pathogenic links between microbiota and the gut–lung axis. Nat Rev Microbiol 15 (1):55–63

Buza TM, Tonui T, Stomeo F, Tiambo C, Katani R, Schilling M et al (2019) iMAP: an integrated bioinformatics and visualization pipeline for microbiome data analysis. BMC Bioinform 20 (1):1–18

Canfield DE, Glazer AN, Falkowski PG (2010) The evolution and future of Earth's nitrogen cycle. Science 330(6001):192–196

Caspi R, Billington R, Ferrer L, Foerster H, Fulcher CA, Keseler IM et al (2016) The MetaCyc database of metabolic pathways and enzymes and the BioCyc collection of pathway/genome databases. Nucleic Acids Res 44(D1):D471–D480

Celis A, Relman D (2020) Unraveling the role of the gut microbiome in iron-deficiency anemia during pregnancy. FASEB J 34(S1):1–1

Chen T, Yu WH, Izard J, Baranova OV, Lakshmanan A, Dewhirst FE (2010) The human Oral microbiome database: a web accessible resource for investigating oral microbe taxonomic and genomic information. Database 2010:baq013

Cheng K, Ning Z, Zhang X, Li L, Liao B, Mayne J, Figeys D (2017) MetaLab: an automated pipeline for metaproteomic data analysis. Microbiome 5(1):1–10

Cheng K, Ning Z, Zhang X, Li L, Liao B, Mayne J, Figeys D (2019) MetaLab 2.0 enables accurate post-translational modifications profiling in metaproteomics. Journal of the American Society for Mass Spectrometry 31(7):1473–1482

Chong J, Liu P, Zhou G, Xia J (2020) Using microbiome analyst for comprehensive statistical, functional, and meta-analysis of microbiome data. Nat Protoc 15(3):799–821

Cox J, Hein MY, Luber CA, Paron I, Nagaraj N, Mann M (2014) Accurate proteome-wide label-free quantification by delayed normalization and maximal peptide ratio extraction, termed MaxLFQ. Mol Cell Proteomics 13(9):2513–2526

Cryan JF, Dinan TG (2012) Mind-altering microorganisms: the impact of the gut microbiota on brain and behaviour. Nat Rev Neurosci 13(10):701–712

de Dios Miranda J, Seoane JM, Esteban Á, Espí E (2019) 16 microbial exploration techniques: an offshore case study. In: Oilfield microbiology. CRC Press, New York, p 271

DeSantis TZ, Hugenholtz P, Larsen N, Rojas M, Brodie EL, Keller K, Andersen GL (2006) Greengenes, a chimera-checked 16S rRNA gene database and workbench compatible with ARB. Appl Environ Microbiol 72(7):5069–5072

Deutsch EW, Mendoza L, Shteynberg D, Slagel J, Sun Z, Moritz RL (2015) Trans-proteomic pipeline, a standardized data processing pipeline for large-scale reproducible proteomics informatics. Proteom Clin Appl 9(7–8):745–754

Dhariwal A, Chong J, Habib S, King IL, Agellon LB, Xia J (2017) MicrobiomeAnalyst: a web-based tool for comprehensive statistical, visual and meta-analysis of microbiome data. Nucleic Acids Res 45(W1):W180–W188

EFSA Panel on Additives and Products or Substances Used in Animal Feed (FEEDAP), Rychen G, Aquilina G, Azimonti G, Bampidis V, Bastos MDL, Gropp J (2018) Guidance on the characterisation of microorganisms used as feed additives or as production organisms. EFSA J 16(3):e05206

El Kaoutari A, Armougom F, Gordon JI, Raoult D, Henrissat B (2013) The abundance and variety of carbohydrate-active enzymes in the human gut microbiota. Nat Rev Microbiol 11(7):497–504

Escapa IF, Chen T, Huang Y, Gajare P, Dewhirst FE, Lemon KP (2018) New insights into human nostril microbiome from the expanded human oral microbiome database (eHOMD): a resource for the microbiome of the human aerodigestive tract. Msystems 3(6):e00187–e00118

Finkel OM, Salas-González I, Castrillo G, Spaepen S, Law TF, Teixeira PJPL, Dangl JL (2019) The effects of soil phosphorus content on plant microbiota are driven by the plant phosphate starvation response. PLoS Biol 17(11):e3000534

Gibson MK, Forsberg KJ, Dantas G (2015) Improved annotation of antibiotic resistance determinants reveals microbial resistomes cluster by ecology. ISME J 9(1):207–216

Gougoulias C, Clark JM, Shaw LJ (2014) The role of soil microbes in the global carbon cycle: tracking the below-ground microbial processing of plant-derived carbon for manipulating carbon dynamics in agricultural systems. J Sci Food Agric 94(12):2362–2371

Griffen AL, Beall CJ, Firestone ND, Gross EL, DiFranco JM, Hardman JH, Leys EJ (2011) CORE: a phylogenetically-curated 16S rDNA database of the CORE oral microbiome. PLoS One 6(4): e19051

Gurevich A, Saveliev V, Vyahhi N, Tesler G (2013) QUAST: quality assessment tool for genome assemblies. Bioinformatics 29(8):1072–1075

Hacquard S, Spaepen S, Garrido-Oter R, Schulze-Lefert P (2017) Interplay between innate immunity and the plant microbiota. Annu Rev Phytopathol 55:565–589

Hillmann B, Al-Ghalith GA, Shields-Cutler RR, Zhu Q, Knight R, Knights D (2020) SHOGUN: a modular, accurate, and scalable framework for microbiome quantification. Bioinformatics 36 (13):4088–4090

Huerta-Cepas J, Forslund K, Coelho LP, Szklarczyk D, Jensen LJ, Von Mering C, Bork P (2017) Fast genome-wide functional annotation through orthology assignment by eggNOG-mapper. Mol Biol Evol 34(8):2115–2122

Huerta-Cepas J, Szklarczyk D, Heller D, Hernández-Plaza A, Forslund SK, Cook H et al (2019) eggNOG 5.0: a hierarchical, functionally and phylogenetically annotated orthology resource based on 5090 organisms and 2502 viruses. Nucleic Acids Res 47(D1):D309–D314

Huson DH, Beier S, Flade I, Górska A, El-Hadidi M, Mitra S, Tappu R (2016) MEGAN community edition-interactive exploration and analysis of large-scale microbiome sequencing data. PLoS Comput Biol 12(6):e1004957

Jain M, Fiddes IT, Miga KH, Olsen HE, Paten B, Akeson M (2015) Improved data analysis for the MinION nanopore sequencer. Nat Methods 12(4):351–356

Jassal B, Matthews L, Viteri G, Gong C, Lorente P, Fabregat A, Loney F (2020) The reactome pathway knowledgebase. Nucleic Acids Res 48(D1):D498–D503

Javdan B, Lopez JG, Chankhamjon P, Lee YCJ, Hull R, Wu Q, Donia MS (2020) Personalized mapping of drug metabolism by the human gut microbiome. Cell 181(7):1661–1679

Kanehisa M, Goto S, Sato Y, Furumichi M, Tanabe M (2012) KEGG for integration and interpretation of large-scale molecular data sets. Nucleic Acids Res 40(D1):D109–D114

Kelder T, Van Iersel MP, Hanspers K, Kutmon M, Conklin BR, Evelo CT, Pico AR (2012) WikiPathways: building research communities on biological pathways. Nucleic Acids Res 40 (D1):D1301–D1307

Kelley DR, Liu B, Delcher AL, Pop M, Salzberg SL (2012) Gene prediction with glimmer for metagenomic sequences augmented by classification and clustering. Nucleic Acids Res 40(1): e9–e9

Kelley DR, Salzberg SL (2010) Clustering metagenomic sequences with interpolated Markov models. BMC Bioinformatics 11(1):544

Kim YA, Kang EW, Moon HS, Kim D, Yong D (2020) Application of 16S rRNA gene-targeted next-generation sequencing for bacterial pathogen detection in continuous ambulatory peritoneal dialysis peritonitis. Ann Clin Microbiol 23(1):1–10

Koren S, Schatz MC, Walenz BP, Martin J, Howard JT, Ganapathy G, Phillippy AM (2012) Hybrid error correction and de novo assembly of single-molecule sequencing reads. Nat Biotechnol 30 (7):693–700

Kosiewicz MM, Zirnheld AL, Alard P (2011) Gut microbiota, immunity, and disease: a complex relationship. Front Microbiol 2:180

Kumar R, Sood U, Gupta V, Singh M, Scaria J, Lal R (2020) Recent advancements in the development of modern probiotics for restoring human gut microbiome dysbiosis. Indian J Microbiol 60(1):12–25

LaPierre N, Ju CJT, Zhou G, Wang W (2019) MetaPheno: a critical evaluation of deep learning and machine learning in metagenome-based disease prediction. Methods 166:74–82

Lepage P, Leclerc MC, Joossens M, Mondot S, Blottière HM, Raes J et al (2013) A metagenomic insight into our gut's microbiome. Gut 62(1):146–158

Lewis E, Hudson JA, Cook N, Barnes JD, Haynes E (2020) Next-generation sequencing as a screening tool for foodborne pathogens in fresh produce. J Microbiol Methods 171:105840

Lu L, Xing D, Ren ZJ (2015) Microbial community structure accompanied with electricity production in a constructed wetland plant microbial fuel cell. Bioresour Technol 195:115–121

Malla MA, Dubey A, Kumar A, Yadav S, Hashem A, Abd_Allah EF (2019) Exploring the human microbiome: the potential future role of nextgeneration sequencing in disease diagnosis and treatment. Front Immunol 9:2868

Mayer EA, Tillisch K (2011) The brain-gut axis in abdominal pain syndromes. Annu Rev Med 62:381–396

Million M, Tomas J, Wagner C, Lelouard H, Raoult D, Gorvel JP (2018) New insights in gut microbiota and mucosal immunity of the small intestine. Hum Microbiome J 7:23–32

Mitchell A, Chang HY, Daugherty L, Fraser M, Hunter S, Lopez R et al (2015) The InterPro protein families database: the classification resource after 15 years. Nucleic Acids Res 43(D1):D213–D221

Mitchell AL, Almeida A, Beracochea M, Boland M, Burgin J, Cochrane G, Sakharova E (2020) MGnify: the microbiome analysis resource in 2020. Nucleic Acids Res 48(D1):D570–D578

Mooradian AD, Van Der Post S, Naegle KM, Held JM (2020) ProteoClade: a taxonomic toolkit for multi-species and metaproteomic analysis. PLoS Comput Biol 16(3):e1007741

Müller DB, Vogel C, Bai Y, Vorholt JA (2016) The plant microbiota: systems-level insights and perspectives. Annu Rev Genet 50:211–234

Muth T, Behne A, Heyer R, Kohrs F, Benndorf D, Hoffmann M, Rapp E (2015) The MetaProteomeAnalyzer: a powerful open-source software suite for metaproteomics data analysis and interpretation. J Proteome Res 14(3):1557–1565

Ni Y, Yu G, Chen H, Deng Y, Wells PM, Steves CJ, Fu J (2020) M2IA: a web server for microbiome and metabolome integrative analysis. Bioinformatics 36(11):3493–3498

Noguchi H, Taniguchi T, Itoh T (2008) MetaGeneAnnotator: detecting species-specific patterns of ribosomal binding site for precise gene prediction in anonymous prokaryotic and phage genomes. DNA Res 15(6):387–396

Oerlemans EF, Bellen G, Claes I, Henkens T, Allonsius CN, Wittouck S, Lebeer S (2020) Impact of a lactobacilli-containing gel on vulvovaginal candidosis and the vaginal microbiome. Sci Rep 10(1):1–10

Pezzatti J, Boccard J, Codesido S, Gagnebin Y, Joshi A, Picard D et al (2020) Implementation of liquid chromatography–high resolution mass spectrometry methods for untargeted metabolomic analyses of biological samples: a tutorial. Anal Chim Acta 1105:28–44

Picó Y (2020) Chromatography-mass spectrometry: recent evolution and current trends in environmental science. Curr Opin Environ Sci Health 18:47–53

Quigley EM (2017) Microbiota-brain-gut axis and neurodegenerative diseases. Curr Neurol Neurosci Rep 17(12):94

Rabe A, Gesell Salazar M, Michalik S, Fuchs S, Welk A, Kocher T, Völker U (2019) Metaproteomics analysis of microbial diversity of human saliva and tongue dorsum in young healthy individuals. J Oral Microbiol 11(1):1654786

Ramsøe A, van Heekeren V, Ponce P, Fischer R, Barnes I, Speller C, Collins MJ (2020) DeamiDATE 1.0: site-specific deamidation as a tool to assess authenticity of members of ancient proteomes. J Archaeol Sci 115:105080

Rawi MH, Zaman SA, Pa'ee KF, Leong SS, Sarbini SR (2020) Prebiotics metabolism by gut-isolated probiotics. J Food Sci Technol 57(8):2786–2799

Rho M, Tang H, Ye Y (2010) FragGeneScan: predicting genes in short and error-prone reads. Nucleic Acids Res 38(20):e191–e191

Richardson EJ, Escalettes F, Fotheringham I, Wallace RJ, Watson M (2013) Meta4: a web application for sharing and annotating metagenomic gene predictions using web services. Front Genet 4:168

Riffle M, May DH, Timmins-Schiffman E, Mikan MP, Jaschob D, Noble WS, Nunn BL (2018) MetaGOmics: a web-based tool for peptide-centric functional and taxonomic analysis of metaproteomics data. Proteomes 6(1):2

Riley LW, Raphael E, Faerstein E (2013) Obesity in the United States–dysbiosis from exposure to low-dose antibiotics? Front Public Health 1:69

Rosa CP, Pereira JA, Cristina de Melo Santos N, Brancaglion GA, Silva EN, Tagliati CA et al (2020) Vancomycin-induced gut dysbiosis during Pseudomonas aeruginosa pulmonary infection in a mice model. J Leukoc Biol 107(1):95–104

Saulnier DM, Riehle K, Mistretta TA, Diaz MA, Mandal D, Raza S et al (2011) Gastrointestinal microbiome signatures of pediatric patients with irritable bowel syndrome. Gastroenterology 141(5):1782–1791

Scholz M, Ward DV, Pasolli E, Tolio T, Zolfo M, Asnicar F et al (2016) Strain-level microbial epidemiology and population genomics from shotgun metagenomics. Nat Methods 13 (5):435–438

Shamsaddini A, Dadkhah K, Gillevet PM (2020) BiomMiner: an advanced exploratory microbiome analysis and visualization pipeline. PLoS One 15(6):e0234860

Shi B, Chang M, Martin J, Mitreva M, Lux R, Klokkevold P et al (2015) Dynamic changes in the subgingival microbiome and their potential for diagnosis and prognosis of periodontitis. MBio 6 (1):e01926–e01914

Sibley CD, Grinwis ME, Field TR, Eshaghurshan CS, Faria MM, Dowd SE et al (2011) Culture enriched molecular profiling of the cystic fibrosis airway microbiome. PLoS One 6(7):e22702

Simopoulos CM, Ning Z, Zhang X, Li L, Walker K, Lavallée-Adam M, Figeys D (2020) pepFunk: a tool for peptide-centric functional analysis of metaproteomic human gut microbiome studies. Bioinformatics 36(14):4171–4179

Singh RG, Tanca A, Palomba A, Van der Jeugt F, Verschaffelt P, Uzzau S et al (2019) Unipept 4.0: functional analysis of metaproteome data. J Proteome Res 18(2):606–615

Sobhani I, Tap J, Roudot-Thoraval F, Roperch JP, Letulle S, Langella P et al (2011) Microbial dysbiosis in colorectal cancer (CRC) patients. PLoS One 6(1):e16393

Su X, Jing G, Sun Z, Liu L, Xu Z, McDonald D et al (2020) Multiple-disease detection and classification across cohorts via microbiome search. Msystems 5(2):e00150–e00120

Suez J, Zmora N, Zilberman-Schapira G, Mor U, Dori-Bachash M, Bashiardes S et al (2018) Post-antibiotic gut mucosal microbiome reconstitution is impaired by probiotics and improved by autologous FMT. Cell 174(6):1406–1423

Sulman BN, Phillips RP, Oishi AC, Shevliakova E, Pacala SW (2014) Microbe-driven turnover offsets mineral-mediated storage of soil carbon under elevated CO 2. Nat Clim Chang 4 (12):1099–1102

Timmers RA, Rothballer M, Strik DP, Engel M, Schulz S, Schloter M et al (2012) Microbial community structure elucidates performance of Glyceria maxima plant microbial fuel cell. Appl Microbiol Biotechnol 94(2):537–548

Tortelli BA, Lewis WG, Allsworth JE, Member-Meneh N, Foster LR, Reno HE et al (2020) Associations between the vaginal microbiome and Candida colonization in women of reproductive age. Am J Obstet Gynecol 222(5):471–4e1

Treangen TJ, Koren S, Sommer DD, Liu B, Astrovskaya I, Ondov B et al (2013) MetAMOS: a modular and open source metagenomic assembly and analysis pipeline. Genome Biol 14(1):R2

Truong DT, Tett A, Pasolli E, Huttenhower C, Segata N (2017) Microbial strain-level population structure and genetic diversity from metagenomes. Genome Res 27(4):626–638

Tyanova S, Temu T, Cox J (2016) The MaxQuant computational platform for mass spectrometry-based shotgun proteomics. Nat Protoc 11(12):2301

Van Den Bossche T, Verschaffelt P, Schallert K, Barsnes H, Dawyndt P, Benndorf D et al (2020) Connecting MetaProteomeAnalyzer and PeptideShaker to Unipept for seamless end-to-end metaproteomics data analysis. J Proteome Res 19(8):3562–3566

Van der Walt AJ, Van Goethem MW, Ramond JB, Makhalanyane TP, Reva O, Cowan DA (2017) Assembling metagenomes, one community at a time. BMC Genomics 18(1):1–13

van Dijk EL, Auger H, Jaszczyszyn Y, Thermes C (2014) Ten years of next-generation sequencing technology. Trends Genet 30(9):418–426

Vaudel M, Burkhart JM, Zahedi RP, Oveland E, Berven FS, Sickmann A et al (2015) PeptideShaker enables reanalysis of MS-derived proteomics data sets. Nat Biotechnol 33 (1):22–24

Vick SH, Greenfield P, Tran-Dinh N, Tetu SG, Midgley DJ, Paulsen IT (2018) The coal seam microbiome (CSMB) reference set, a lingua franca for the microbial coal-to-methane community. Int J Coal Geol 186:41–50

Wetser K, Sudirjo E, Buisman CJ, Strik DP (2015) Electricity generation by a plant microbial fuel cell with an integrated oxygen reducing biocathode. Appl Energy 137:151–157

Worden AZ, Follows MJ, Giovannoni SJ, Wilken S, Zimmerman AE, Keeling PJ (2015) Rethinking the marine carbon cycle: factoring in the multifarious lifestyles of microbes. Science 347(6223):1257594

Yang T, Santisteban MM, Rodriguez V, Li E, Ahmari N, Carvajal JM et al (2015) Gut dysbiosis is linked to hypertension. Hypertension 65(6):1331–1340

Yilmaz P, Parfrey LW, Yarza P, Gerken J, Pruesse E, Quast C et al (2014) The SILVA and "all-species living tree project (LTP)" taxonomic frameworks. Nucleic Acids Res 42(D1):D643–D648

Zeeuwen PL, Boekhorst J, van den Bogaard EH, de Koning HD, van de Kerkhof PM, Saulnier DM et al (2012) Microbiome dynamics of human epidermis following skin barrier disruption. Genome Biol 13(11):1–18

Zeevi D, Korem T, Zmora N, Israeli D, Rothschild D, Weinberger A et al (2015) Personalized nutrition by prediction of glycemic responses. Cell 163(5):1079–1094

Zerbino DR, Birney E (2008) Velvet: algorithms for de novo short read assembly using de Bruijn graphs. Genome Res 18(5):821–829

Zhang X, Figeys D (2019) Perspective and guidelines for metaproteomics in microbiome studies. J Proteome Res 18(6):2370–2380

Zhang X, Ning Z, Mayne J, Moore JI, Li J, Butcher J et al (2016) MetaPro-IQ: a universal metaproteomic approach to studying human and mouse gut microbiota. Microbiome 4(1):1–12

Zmora N, Zilberman-Schapira G, Suez J, Mor U, Dori-Bachash M, Bashiardes S et al (2018) Personalized gut mucosal colonization resistance to empiric probiotics is associated with unique host and microbiome features. Cell 174(6):1388–1405

The manufacturer's authorised representative in the EU is Springer Nature Customer Service Centre GmbH, Europaplatz 3, 69115 Heidelberg, Germany. If you have any concerns regarding our products, please contact ProductSafety@springernature.com

Printed and bound by CPI Group (UK) Ltd, Croydon, CR0 4YY
30/09/2026
02225247-0004